PUBLIC ACCEPTANCE OF NEW TECHNOLOGIES:
An International Review

Public Acceptance of
NEW TECHNOLOGIES

AN INTERNATIONAL REVIEW

Edited by
ROGER WILLIAMS
and
STEPHEN MILLS

CROOM HELM
London • Sydney • Wolfeboro, New Hampshire

© The Technical Change Centre 1986
Croom Helm Ltd, Provident House, Burrell Row,
Beckenham, Kent, BR3 1AT
Croom Helm Australia Pty Ltd, Suite 4, 6th Floor,
64-76 Kippax Street, Surry Hills, NSW 2010, Australia

British Library Cataloguing in Publication Data

Public acceptance of new technologies: an
international review.
1. Technological innovations — Social
aspects
I. Mills, Stephen II. Williams, Roger, *1942*
600   T173.8
ISBN 0-7099-4319-9

Croom Helm, 27 South Main Street,
Wolfeboro, New Hampshire 03894-2069, USA

Library of Congress Cataloging in Publication Data

Public acceptance of new technologies.

Papers presented at an international conference held
in Jan 1985, sponsored by the Technical Change Centre.
Includes index.
1. Technological innovations — public opinion —
congresses. I. Mills, S.C. (Stephen C.) II. Williams,
Roger 1942-   . III. Technical Change Centre.
T173.8.P83   1986   338.9   86-8813
ISBN 0-7099-4319-9

Printed and bound in Great Britain by
Biddles Ltd, Guildford and King's Lynn

CONTENTS

TABLES

When The Technical Change Centre (TCC) was commissioned to investigate the attitudinal factors which might help to explain the differences in growth-rates in the countries represented at the conference of Heads of Government at Versailles in 1982 - viz. Canada, France, West Germany, Italy, Japan, the UK and the USA - it was hoped that it would be possible for research organisations in several of those countries to co-operate with TCC in a common programme of research. When that did not prove possible it was decided to invite experts from the countries involved in the Versailles summit, and from five other countries, to prepare national essays on attitudes to technical change. This volume includes those essays, together with an introduction and conclusion by Roger Williams, an Honorary Fellow, and Stephen Mills, a Principle Research Fellow of the TCC.

Another volume written by myself (with special appendices contributed by Miss Jenny-Bryan-Brown, Professor Roderick Martin, Dr Nuala Swords-Isherwood, Mr David Sawers, Mrs Julian Swann, and Professor Roger Williams and Mrs Rebecca Smellie), has a more detailed analysis of the attitudinal factors which have influenced growth-rates in the UK. It includes some comparative material based mainly on attitude surveys prepared by or for the European Commission.

A third volume, by Professor Roger Williams and Mrs Rebecca Smellie, is entitled <u>Boiler Suits or Business Suits? Technology Education in the UK.</u>

Although these three volumes are separate publications and not interconnected volumes 1, 2 and 3, together they constitute the TCC's final report on the project Comparative National Assessments of Public Attitudes to New Technologies.

Chapter One

INTRODUCTION

Stephen Mills and Roger Williams

Men, my brothers, men the workers, ever reaping something new:
That which they have done but earnest of the things that they shall do:

For I dipt into the future, far as human eye could see,
Saw the Vision of the world, and all the wonder that would be;

Tennyson, Locksley Hall

Books require explanations, and the explanation for
this one is that it grew out of the Public
Acceptance of New Technologies (PANTs) project,
which in turn was a result of the 1982 Versailles
'Economic Summit'. The Declaration of the Heads of
Government (1) after that summit argued that
revitalisation of the world economy would largely
depend upon the 'exploitation of scientific and
technological development'. It referred to 'the
immense opportunities presented by the new
technologies, particulary for creating employment',
and it stressed that each country would need 'to
create the economic, social and cultural conditions
which allow these technologies to develop and
flourish'.

The Working Group which was subsequently
established reached eighteen broad conclusions, one
of which was that 'The fate of our scientific and
technological innovations is largely a function of
the willingness of the public to accept them. More
attention to the problem of public acceptance of new
technology is needed'. (Technology, Growth and
Employment, Cmnd 8818, March 1983). The Group
argued that new technologies had often presented
problems of public acceptance, specifically when
they came to be regarded as unacceptably risky or
threatening to the environment or to jobs. It felt
that people were often ambivalent in their

1

attitudes, perhaps resisting new technology in the workplace while accepting it enthusiastically at home, and it suggested that much of the more vocal opposition came from minority groups, with the fears expressed going beyond technical considerations to broad political grievances about social values and democratic processes. Judging that there was little real understanding of the factors which shape public attitudes to new technologies, the Group called for a programme of studies in this area.

In the event, the UK took responsibility for this sector of the Working Group's concern, the Department of Trade and Industry commissioning, via the Economic and Social Research Council, four projects (2), one of which was entrusted to the Technical Change Centre (TCC). In order to ensure an international dimension to its study, the TCC decided to sponsor an international conference on the subject (3). The resulting conference was held in January 1985 and this book is its most obvious fruit, although invaluable links were also formed between those who took part, and it is confidently expected that these will be built upon further in future.

Conference participants were each asked to address in their papers four broad questions, as follows:

In Country X:

1. What are the main conclusions to be drawn from such survey data as exists about public attitudes towards technology? Is it possible to distinguish between the attitudes of the public at large and those of elites or specially interested publics? between the public as consumers, employees, environmentally concerned citizens ect? between attitudes now and, say, ten years ago? between attitudes towards different technologies?

2. Are there technologies the public response towards which is of special interest? e.g. opposition to nuclear power, strong support of personal computers, tacit acceptance of fast-food technology, indifference towards apparent safety problems of particular aircraft or motor vehicles.

3. What discernible parts do the following institutions or structures play in shaping public attitudes towards technical change?: the political system (rhetoric and

reality); the status and reward systems in the society; the educational system (at all levels); the character of the economy; the media (including advertising); the legal system.

4. What specific and significant steps (if any) have been taken either to make new technologies more acceptable or to address latent or manifest public concerns? e.g. distrust of expertise, fear of jobs losses, distress at physical hazards, loss of amenity, general socio-psychological alienation.

Notwithstanding the importance of the overall subject, the expenditure of large sums of money on new research was obviously not appropriate or possible at the beginning, and it was also recognised that data and time limitations would mean that the various authors would not be able to treat the above four questions equally, or indeed some of the questions or parts of them at all. This was thought less important than arriving at as perceptive an audit as possible of where overall each country stood in the mid-eighties. And despite the governmental origins of the project, contributors were categorically not asked to recommend how new technologies be made more acceptable. In effect, each was encouraged to decide for himself what was and what was not currently important in the country about which he was writing. The participants were drawn from right across the social sciences to give as wide a perspective as possible on a problem which, it was clear, cut completely across the usual disciplinary boundaries.

This book contains twelve country reviews, all of the countries dealt with being advanced industrial capitalist states, but with the coverage including the small as well as the large, four of the five continents, and southern as well as northern Europe. Clearly, this still leaves gaps, and it would be attractive to repair this shortcoming in a later volume, but with the countries included here having a population in excess of half a billion, we would claim a sound initial foundation.

During the course of the conference it became apparent, as we had suspected it might, that the case of nuclear power was of exceptional importance. We therefore decided that it would be valuable to include a chapter cutting the same basic problem the

other way, that is, a single technology across
several states. Once again, time constraints meant
that only four of the countries treated in the book
could be analysed in this nuclear power chapter, but
the experiment still seemed useful in its own right
and it is also suggested interesting possibilities
for future work: a cross-national comparison of the
public acceptance of each of the main currently
controversial technologies.

As was only to be expected, because for example
of different cultures and intellectual styles, it
emerged at the conference that significant
definitional problems attached to each of the three
concepts 'public', 'acceptance' and 'new
technologies'. What exactly is a 'new technology'?
Strictly, for instance, a technology like nuclear
power is no longer really new; many new technologies
grow out of existing ones and the exact point of
emergence may not be very clear; and one has to
consider the application of new technologies in old
industries as well as their contribution to the
creation of wholly new industries. In effect, a
generous construction is put upon the term 'new
technology' in this book, newness like beauty
essentially being regarded as lying in the eye of
the beholder.

'Public' again is a rather inexact word. The
reality is that, first, 'the public' is a function
of, at least, context (producers, consumers,
generally interested or directly affected parties)
of the particular technology or technologies at
issue, and of time and media attention; and
secondly, that the public is made up of distinct
individuals who may and often do differ in their
views in line with the various demographic variables
(age, sex, occupation, education, location) as well
as in response to personality variables and actual
life experiences. People, as the writers and their
readers would both insist, are not molecules whose
cumulative actions and reactions neatly fit the
formulae of some statistical mechanics, yet perforce
they - we, since we are all involved - have to be
delineated in these terms, and this can be the
source not only of descriptive error, but also of
subsequent insensitivity and failure in prediction.
After all, even in physics the advent of the
uncertainty principle has curtailed the possibility
of prediction at the level of the individual atom.

But the biggest difficulty of all arises with
the term 'acceptance'. Simply taking the British
interpretation - and as we have said, in other

4

cultures other interpretations apply - the relevant definition in the <u>Concise</u> <u>Oxford</u> <u>Dictionary</u> reads as follows:

> 1. consent to receive (gift, thing delivered, payment, pleasure, duty); favourable reception (of person or thing, by or with person); affirmative answer to invitation. 2. approval, belief; toleration.

It may also be worth quoting the definition of 'acceptable' in the same dictionary: 'worth accepting; pleasing, welcome; tolerable'. The reason for citing these definitions is twofold. First, they bring out that the degree of acceptability is to be seen as effectively intrinsic to the thing (technology) being offered, whereas acceptance essentially inheres in the individual or group to whom that thing is offered or upon whom it impinges. And secondly, both words imply a spectrum of states, ranging as it were from outright opposition by some, via a passive, and quite probably grudging, condition, in which the thing at issue is just, but only just, tolerated, to one in which it is cautiously or even positively welcomed, ultimately to the point where those it will most affect are actively seeking it. The first of these conditions carries at least a suggestion of manipulation by those sponsoring whatever it may be, whereas the last emphasises, indeed <u>actually requires</u>, the full involvement of those who are going to be affected. The gulf between these extremes is wide, and it would be narrow and unsatisfactory to put only the former construction upon acceptance in the case of technology, to the point in fact where doing so would undermine the fundamental requirements of liberal democracy. By acceptance, therefore, it is advisable to mean a condition firmly predicated upon both as full information as possible, and wide public involvement.

The point is that technology is a change-producing factor of unequalled importance and potential. Placed on a firm foundation since the Second World War, technological development has become the key to international competitiveness, to social betterment, and indeed to economic and political power more widely. As a factor capable of being used for good or ill, technology has no rival.

The true democrat therefore rejoices when he finds that public acceptance is a central concern of government: really, he can wish for nothing less for a force of this potency.

In the remainder of this Introduction we briefly summarise the ground covered in each of the national chapters. However, these summaries are no more than trailers which necessarily leave out much of interest from chapters rich in their information and insight, and the reader is strongly urged to consult the chapters themselves before coming to conclusions. In our final chapter we offer our own overview of what makes for acceptance, as well as offering a tentative 'ideal model' of conditions which, the evidence of this book at least suggests to us, would enhance the relationship between the public and new technology in democratic societies. But we would still want this to be regarded as a first word, and certainly we do not pretend that it is a last one.

BRITAIN

Williams is explicitly concerned with attitudinal reasons for Britain's economic decline, his point of departure being the general circumstances of innovation. He outlines the influence which governments, businessmen and workers respectively can have on new technology, and draws out also the role of the educational system and parental attitudes, quoting the Finniston report on the unfavourable position of the engineer in Britain. He next demonstrates the concern of successive British governments with technology, citing both Research and Development (R & D) statistics and institutional initiatives, and mentioning also differences between the Conservative and Labour parties on this question. Noting that Britain's educational system was 'not appropriate to the new conditions of industrial innovation' as these emerged in the late nineteenth century, he reviews government efforts in this direction since the Second World War, dwelling on the continuing low output of engineers and the inadequacies of post-employment training. He refers to the limited formal qualifications and 'anti-intellectual attitudes' of British managers, suggests that British industry should do more R & D, and quotes surveys which reveal that British managers are aware of the fact that their firms are insufficiently innovative. On workers' resistance to change, Williams cites authors who believe this to be a special problem for Britain, and he argues that when there are strong unions doubtful about technical change, then employers will tend to delay the introduction of new technology or else accept overmanning. In his review of the available survey data, Williams concludes that while only a small proportion of workers oppose technical change, most recognising its importance to competitiveness, there remains much pessimism about the associated employment implications. In his treatment of the environmental movement, Williams bases himself on Cotgrove's study, pointing out that there has been a decline in environmentalism since Cotgrove completed his inquiry, and that even before, the actual impact of environmentalism on public policy had been small. The live issues he notes in this connection are nuclear power and acid rain, but even here he sees doubts resting more on economics in the one case and lack of scientific knowledge in the other, than upon public attitudes.

USA

Nelkin reviews from its sixties origins the American public's concern with the social and environmental effects of technology, demonstrating how surveys in the 1970s revealed a widespread ambivalence about technology. This ambivalence seems to her to have had several origins: incidents highlighting technology's dangers; scientific advances which seemed to point to hazards where none had previously been suspected; disappointment with the actual achievements of technology; distrust of increased dominance by experts, especially following the Vietnam War; and an increased tendency for American politics to become issue-centred. Nelkin notes that the project delays caused by technological controversies in the seventies led many to conclude that the US was becoming 'risk aversive', and she identifies three means through which public acceptance was promoted: the analytical and institutional development of technology - and risk-assessment; legislative and other measures to extend public participation; and renewed efforts by sponsors of technology to market their products. In the 1980s, Nelkin sees the Reagan administration as having tried to depoliticise technological choice by emphasising analytical techniques, buttressing this with public relations efforts to regain support for technological growth, and mobilising the media to this end and the related one of furthering scientific and computer literacy. While uncertain about the impact of these initiatives, she observes that there is now much less overt dissent, and although acid rain and toxic wastes have come to exercise public opinion, she finds no equivalent to the anti-nuclear movement of the last decade, nuclear power as a target having been effectively scratched by the absence of plans for new plant. Issue which once provoked protest are now, she says, 'virtually ignored', and the lack of resistance to the 'Star Wars' initiative suggests to her 'a tendency toward technological fatalism'. Yet an underlying ambivalence towards technology, she argues, remains; perhaps her most striking conclusion being that moral and religious values are currently playing a large role in shaping the American public's response to new technology.

JAPAN

Nisihira and Dore find no instances in Japan's last century of 'bone-headed' resistance to science-based innovation, only of protests based on strong personal interest or public spiritedness. Public awareness of science and technology, they feel, has been even greater in Japan than elsewhere - 'Herman Kahn sells more in translation than Rachel Carson' - for the specific reason that 'catching up with the West' has been a Japanese goal for more than a century. Starting as a humble adaptor of things foreign, the Japanese, it seems, have almost come to believe in the Western stereotype of them as still no more than imitators. Nisihira and Dore argue that although the oil crisis and fast growth elsewhere in Asia finally ended a decade during which Japan had been the envy of the world for her growth-rates, in the 1980s it became clear that Japan could compete in the biggest leaague, that of scientific and technological leadership. As a result, they say, Japanese self-confidence, as revealed by surveys, has greatly increased, a development they quote a further survey to show has nevertheless not made the Japanese more nationalistic, but instead rather more disposed towards international co-operation. Nisihira and Dore explain how heightened awareness of science and technology is fully reflected in Japanese books and media, and they underline how this translates into investment and a responsive, though not undiscriminating, market for new technology. With respect to computers too they give data to show that a discriminating appreciation of the dangers involved accompanies a broadly positive outlook.

They also mention surveys which have tried to get at deeper general attitudes to science and technology - here it seems that Japanese sentiments 'fluctuate with the mood of the times', so that, for instance, there was a peak around 1973 in the feeling that, though bringing convenience, science and technology also dehumanise life. Specific fears about job losses caused by technology appear to have increased suddenly in 1983. In the medical field, Nisihira's and Dore's figures make the Japanese look doubtful about test-tube babies, yet more relaxed than Europeans about the effect of new discoveries on the human personality.

WEST GERMANY

Petermann and Thurn, who discuss at length 'acceptance' as a topic of research, argue that although the debate over technology has 'lost a lot of its bitterness', the 'silent consensus' which accompanied technological progress in the 1950s and 60s has not returned and that the public acceptance issue instead retains substantial latent importance. They see a 'socially responsible and politically legitimate approach to new technologies, especially on the part of their promoters', as the 'deciding factor' in public acceptance, and they provide a comprehensive analysis of West German survey data on attitudes to technology. The latter leads them to conclude that although belief in progress is no longer intact and there is now public scepticism with regard to technology, a 'basic faith' in the ability of science to solve problems remains, as well as wide acceptance of technologies on an individual basis. They specifically caution against emphasising the ambivalent and negative views revealed in surveys. They also discuss the role of elites (government, political parties, employers and trade unions) in legitimating the pervasion of society by technology. Noting that it is these elites which 'have the best chances of influencing the public', they find them 'less deeply divided' about technology than is public opinion at large, though they also contrast the way in which microelectronics has 'captivated everyone' with the 'very cool' outlook accorded to nuclear energy. Reviewing the role of the media, Petermann and Thurn note the unsatisfactory nature of research in this area, and tentatively suggest that it is more probable that the media strengthen or weaken existing attitudes than that they actually create opinions. The authors report general agreement in West Germany on the key role of the education system in promoting acceptance of technology, both as regards the inculcation of specific skills and in encouraging 'a sensitivity to the role of technologies in modern life'. They also say the West German education system has 'not yet reacted adequately to the challenge posed by information technology', and they list the measures taken by the state to correct this. They deal at length with nuclear energy and information technologies, and they end their chapter with a brief consideration of technology acceptance against the background of structural problems in society. Here they focus on four aspects: the political system's ability to regulate and justify itself; the extent of public consciousness of the ecological crisis; changing values in modern industrial societies; and the consequences of structural changes in employment. Petermann and Thurn are acutely aware of the need for more social science research in this whole area.

FRANCE
The French authors discuss the public acceptance of new technologies in France in terms of the implications for government decision-making, impacts at the place of employment, and issues for the citizen and consumer. As regards the first of these, they argue that France has done well in those technologies where government planners have dominated, but has been at a disadvantage where international market forces have controlled events. They note that France has experienced its share of public resistance to technical and industrial projects, and they also say that interest groups have never previously been as important in France. In this context too they introduce the results of sociological analysis of local and regional opposition to new installations. They see what they call the 'new middle classes' - engineers, technicians, teachers, researchers etc, as having grown significantly during the Fifth Republic, yet without becoming fully integrated politically, and they conclude that if technology is to prove socially acceptable, then measures will have to be adopted to conserve the environment (though they show that environmental problems have lost some standing with the public) and to enhance political participation. They judge that attitudes to new technologies in France have been changed both by the world economic crisis and by developments in the

state's decision-making structures since 1981, so that whereas in the 1970s outlooks were shaped by anti-industrial tendencies, in the 1980s new technologies are seen as offering a 'vehicle for positive changes in values and lifestyles'. In examining the impact of new technology on employment, the French chapter suggests that the best approach is perhaps via the changes required in skills and qualifications, and they contrast two models: the need for operators to be better trained to handle more complex technology, and the tendency for machines to displace their operators. They are unequivocal about what is happening in the business world: 'the links between qualification, know-how, responsibility, career and wages, are in the process of being dismantled'. Finally, reflecting on the role of citizens and consumers, Moatti et al say that 'social acceptance primarily involves acceptance by teachers', and they add that 'the citizen's problem is edging closer to that of the consumer'. In their opinion, 'Taken as a whole, the commercial aspects of products and services to consumers are now overshadowed by their ideological and cultural ramifications'. They end by seeing a new function for government.

AUSTRALIA
Lamenting the absence of data, Stubbs argues that one would except Australians to be predisposed towards technology, in particular because of the history of their agricultural and mining industries and the country's geographical remoteness. Basing himself on a national survey of 1983, he finds strong support for technological change, though with significant qualifications, his own reservation being that this survey dealt with stated rather than with revealed preferences. Noting that the redistributive effect of technical change makes it likely to create interest groups, Stubbs shows that, with customer pioneering having been done abroad, Australian consumers are enthusiastic accepters of new technology, but that the position facing producers is more complex: they too in public acknowledge the need to apply technology but in private the small size of the domestic market, the high level of multinational investment, and the vested interests created by tariff barriers between them make for a more cautious response, and this divergence tends to be reflected in the public service. Stubbs thinks that, despite occasional major conflicts, senior unionists are generally

realistic in the bargaining process, and he does not believe that labour has inexorably lost out to capital in Australia. His view is that technology has been 'received altogether in a lower key' in Australia than in the USA, environmental groups having been really significant only as regards nuclear power and Tasmanian hydroelectricity. He deals separately with nuclear power, a live issue now only as regards uranium exports, where 'for the moment economics and real politik rule; with medical technology, where he thinks public opinion may have been more receptive than in Britain; with the electronic technologies, where his impression is that Australia will respond as positively as in respect of consumer innovations; and with regulation and road safety, where he notes that although Australia has lagged in enforcement of product safety standards, she has led in mandatory use of seat belts. His overall conclusion is that while more questioning attitudes are emerging, acceptance is 'high and unequivocal', except where the technology may have hidden or belated side effects, or involve asymmetric risks and benefits.

ITALY

Calvi et al begin their treatment of the Italian case by discussing acceptance as a concept, noting that despite an information gap between decision-makers and the public, surveys reveal Italians to be more optimistic and positive than formerly about technology. Calvi acknowledges that traditionally, science and technology were seen as far removed from basic Italian values and culture, but he insists that apart from the odd echo (such as the 'trifling' share of GDP going to research), Italy has changed significantly. Stressing that the country has 'no sign of a strong culture of opposition to scientific development', Calvi explains this by arguing that, having had no 'messianic' expectations of science earlier, Italy has also missed out on the disenchantment produced elsewhere by the failure of these expectations. Attitudes towards technology are, he says, 'more complex and contradictory' than those held towards science. Calvi's conclusion is that, while there are differences between groups, Italians overall are 'convinced that technological development is substantially positive', notwithstanding any adverse social consequences it may have. In the specific instance of nuclear energy, Calvi's data suggests that its image improved between 1980 and 1983 (those

Introduction

in favour up from 32.5 to 41.1 per cent, those
against down from 40.7 to 33.1 per cent); most
remarkably, even when asked whether they would
accept a nuclear plant in their own town, some 60
per cent it seems were not against it. As regards
industry, Calvi finds trade unions reformulating
their position in the face of employers who are
clear that technology must determine the
organisation of labour. He summarises several
surveys of worker attitudes demonstrating, inter
alia, high awareness of the implications of
technical change and a desire for involvement in the
planning of innovation within firms, and he
attributes particular importance to a survey of
industrial researchers because he believes them to
have a key role in influencing the general public.
Tracing the growing attention given by the media in
the last decade to science and technology, Calvi
argues that it is still not enough by comparison
with other countries, and he looks to Italian
universities and schools to diffuse information and
scientific values more widely. He ends by repeating
that his data both debunks the image of a
traditional culture and points to a 'wider scope for
the political sector to act', and it is clear he
suspects that the obstacles to technical change in
Italy lie not with the public but rather with the
political institutions.

SWEDEN
Forslin's starting point is that, engineers and
scientists having been the Swedish hero-figures of
the last century, the environmental concerns of the
seventies abruptly halted public confidence in
technical progress. Noting that the ensueing
'anti-technological era' coincided with increased
access to higher education, he argues that only the
Centre Party successfully channelled the new protest
movement. This movement he sees as having
culminated with Sweden's nuclear power referendum of
1980, the outcome of which was confused. In the
work environment Forslin cites survey evidence to
demonstrate how extensive computer experience has
now become in Sweden, and how optimistic are
people's responses to this. Despite this, another
survey leads him to the view that computers are less
appreciated in Sweden than they are elsewhere in
Europe. He also suggests that since there are 1959
and 1981 surveys which show a similar 60:40 split in
optimistic:pessimistic attitudes towards technology,
the real difference which revealed itself over this

period may have lain in 'the degree of
radicalisation and belligerence' of the 40 per cent.
On Forslin's interpretation of the data, worries
about modern technology correlate more with personal
value-systems than they do with whether or not an
individual has had direct personal experience of
technology. Describing Sweden as the biggest user
of industrial robots per capita in the world, he
refers to a small survey in the car industry to
support his view that there was a negative shift in
opinions about automation between 1974 and 1981.
Forslin also observes that in a society where union
membership approaches 100 per cent, trade union
policy is an important expression of public opinion,
and here it seems that union attitudes have evolved
from acceptance of new technologies conditional upon
it being possible to influence the distribution of
wealth, to a wish, held with increasing intensity,
to 'influence development as such'. Forslin's last
word is that with the totality of the changes being
caused by technology amounting to 'quite a
revolutionary picture', the public may be more aware
of what is happening than is government.

CANADA
Noting that Canada has mostly not been a major
initiator of new technology, Gurstein and Cordell
comment on its switch of dependence from the UK to
the USA, stressing Canada's size, population
dispersal, emphasis on resource extraction, degree
of foreign ownership, two cultures and languages,
and especially the communications impact of the
proximity to the USA, 'the world's most vital
culture'. Gurstein and Cordell find Canadians to be
aware that their firms are less innovative than
foreign ones, and they suggest the country may be
becoming postindustrial without ever having been
fully industrial. They see geography and climate as
having created 'an atmosphere supportive of
technology' - hence the 'relatively quiet acceptance
of nuclear power' - and they describe Canada's
freedom to absorb American technology as a mixed
blessing. Dependence having made Canada 'primarily
a consumer of technology developed elsewhere',
resistance, they feel, has often been due more to
technology's being foreign than to its being new.
Latterly, they regard Canada as having especially
encouraged 'social' technologies, like direct
broadcast satellites, to link the country together.
Having two cultures and languages, they show, has
meant both a time lag and a financial drag in the

dissemination of technology, while also posing special problems for Quebec. Gurstein and Cordell underline the technological role of the Canadian state - in transport, communications, agriculture and energy, and also in assessing the impact of new technologies and funding those advocacy groups which have arisen in response to new technology, and they draw attention to the public's confidence in the technological role of the universities, to the importance of 'old' as well as 'new' technologies, and to the 'almost mystical' concern in Canada with anything, including technology, which impinges on the land. They argue that although TV makes Canada a 'component of the US market' as regards both new technology and opposition to it, attitudes in Canada tend to be more pragmatic and less ideological than in the USA. Dealing specifically with automation, they conclude that Canadians strongly support it while fearing its employment implications. And they contrast Canada's success in communications technology with its failure as regards the content carried by that technology. Gurstein and Cordell clearly attach great importance to Canada's social innovations, notably regional income distribution and the crown corporation, and indeed they end with the thought that 'it is perhaps in the area of the public acceptance of new social technologies that Canada has most to teach the world'.

BELGIUM
According to Eraly, Belgian interest in social attitudes towards new technologies is only now beginning, most Belgians tending to feel both remote from science and technology and powerless before technological change. The desire to participate in the orientation of technological development, most marked amongst the less educationally and economically advantaged and those on the political left, correlates, he says, with apprehension, and he finds a 'profound ambivalence' in public attitudes towards technology. Eraly insists that in Belgium the threat of unemployment dominates public reactions, with technology 'chronically seen as meaning the replacement of workers by machines'. Anxieties about pollution and other technology related effects are, he reports, correspondingly more moderately held, with medical developments (atypically) being strongly supported: however, it seems that Belgium is one of the countries in which nuclear power constitutes a special case, increased polarisation having occurred between 1978 and 1983.

Introduction

Eraly notes that the belief that computers create unemployment is held more firmly in Wallonia than in Flanders, in part at least, he argues, because the recession has been worse in Wallonia and fear correlates with economic vulnerability. The average Belgian is, in his judgement, less concerned about the computer's social impact (including centralised data bases) but entertains hopes of their efficacy in public administration: Belgians also appear to be sceptical about computers in the home, and surprisingly ill-informed about their use in schools. Eraly stresses that attitudes towards computers are not one dimensional, and he quotes surveys which indicate that, while attitudes to computers depend upon the use an individual expects to be able to make of them in furthering his own position, and upon actual experience with them, experience also often eventually leads to people attaching less importance to computers within their overall concerns. Of technology in general Eraly lists several cases to illustrate that 'it is not technology in itself which attracts the interest of public opinion, it is rather the sensational character of a specific issue'. He adds that although government has many possibilities for influencing public opinion, the Belgian government has never been able to formulate a general policy towards technology and its acceptance. Eraly's final conclusions are that pessimism about technology has not in Belgium translated into resistance, that technological decision making is an elite concern, and that, for most Belgians, technology remains something of a 'threatening inevitability'.

NETHERLANDS
Berting states that whereas between 1945 and 1963 the Netherlands' public was optimistic about technology-driven growth, with only mild resistance, from 1963 to 1972 confidence in technology's role was 'clearly waning'. According to him, between 1972 and 1979 the hope was that the social sciences would help political decision-makers cope with the problems of growth and change, but he thinks that since 1979 'faith in the malleability of societal developments has weakened', public acceptance of new technologies at last becoming a prominent political issue. Berting makes the point that concern with the introduction of new technology does not as such show up in polls: in the sixties unemployment and housing were seen as the major issues, with

16

environmental questions coming to the fore in the seventies, and employment again becoming a pressing concern by the eighties. As regards energy, Berting's assessment is that, the 1973 crisis having underlined the Netherlands' vulnerability, by the eighties public apprehension had again fallen. As elsewhere, nuclear power has evidently given most trouble, and a national energy debate in 1983 did not, it seems, significantly change opinions. However, Berting reports that whereas in 1981 69 per cent of one survey expected nuclear plants to be shut down, by 1983 84 per cent of another did not expect this to happen, though half of those questioned were apparently still in favour of closing them at that time. On military technology, while there seems to have been a few cases where Dutch public opinion has opposed arms exports, Berting reiterates that the issue which has generated most opposition has been American Cruise missiles. He records that a majority of the Dutch have favoured a reduced emphasis on nuclear weapons since 1975 but, with some 25 per cent pro-and 35-40 per cent anti-nuclear weapons, the actual effect of the major demonstrations in 1982/3 on public opinion, it seems, remains unclear. Turning finally to information technology, on Berting's analysis, the Dutch debate in this case started late and, despite a number of reports, has 'not aroused much public attention', partly because introduction is seen as inevitable. In his conclusion, Berting notes a 'growing distrust of experts by the general public', and he contrasts the cases of agriculture and sea protection, where public acceptance of technological initiatives is traditional in the Netherlands, with industrial questions involving multinationals, where the public tend to feel that developments cannot be controlled at the national level.

SPAIN

Lopez-Pintor and Ramallo draw attention to the fact that Spanish industrialisation came late, major growth occurring only in the last two decades of the Franco dictatorship, and with political democratisation taking place subsequently against a background of international economic recession. Spain continues, they point out, to have very low investment in science and technology and to be a leading buyer of modern technology. As regards energy, they show that despite a planned increase in nuclear energy, a 1983 moratorium has affected the

17

opening of seven plants. Spanish public opinion, it seems, is becoming more knowledgeable about nuclear power, with mistrust greatest in the middle classes. In respect of science and technology generally there is apparently, as elsewhere, ambivalence, a majority being in favour of increased public expenditure, yet with doubts as to the likely consequences. Support is strongest among elites, ambivalence possibly greatest amongst youth, and according to Lopez-Pintor and Ramallo, an anti-technological factor may have been growing over the last decade. The public as a whole they say remains in favour of industrial modernisation even if this brings more unemployment. They quote survey results to demonstrate that whereas in the early seventies the urban middle class was the main advocate of environmental improvement, by 1982 environmental concerns had 'spread gradually throughout the entire society'. They also suggest that public opinion is alienated in respect of food processing and its regulation. Overall, Lopez-Pintor and Ramallo are clear that, apart from nuclear power, both public policy towards and public opinion about new technology is 'more supportive than negative', the media in effect functioning to reinforce public attitudes.

Notes

1. The countries represented were Canada, France, the Federal Republic of Germany, Italy, Japan, the United Kingdom and the United States of America. Representatives of the European Communities also attended.
2. The other three were studies by the Policy Studies Institute of <u>Information Technology and the Organisation</u> and by PREST (Manchester University) on <u>New Communications Technology and the Consumer</u>, and an analysis by Bill Luckin and Russell Mosely of <u>Attitudes Towards Technical Change Between the Wars.</u>
3. This was one of three initiatives taken by the TCC under the PANTs project, the others both being UK-centred - the one a 'national assessment', the other a national survey of attitudes and technology.

Chapter Two

THE PUBLIC ACCEPTANCE OF NEW TECHNOLOGIES IN THE
UNITED KINGDOM
Bruce Williams

The industrial revolution started in Britain and
between 1770 and 1870 Britain was the leading
industrial nation. But since 1870 the rate of
growth in output per worker has been less in Britain
than in many other industrial nations. The decline
in the dominant position of the British economy
after 1870 was connected with the increasing
importance in technical change of academic science
and chemical and electrical engineering. Economic
historians do not agree on the reasons for the
relative decline of the British economy. Some
explanations concentrate on institutional factors,
others on attitudes, and it is often difficult to
disentangle them. This essay is concerned quite
explicitly with attitudinal factors.
     Opportunities to introduce new or improved
technologies come from the creation of new knowledge
or devices - most of which are now derived from
organised research and development programmes -
which provide the basis of product or process
innovations, or from innovations already made which
might be adopted or adapted. But there would be no
incentive to introduce new consumer products, or to
introduce new methods to reduce the prices of
consumer products, if consumers were not interested
in buying the new products or in responding to lower
prices by buying more of the established products.
There would be no incentive to introduce new
products to be sold to other producers to improve
their processes of production, if the other
producers were not interested in changing their
methods of production.
     The capacity to innovate, or to adapt the
innovations of others, depends on the supplies of
managerial, scientific and technical skills, all of
which depend in increasing measure on the nature and

extent of formal education systems, and on the supply of funds for investment in innovation.

The incentives to innovate may come from the desire to pioneer - what Sombart referred to as the Faust spirit - from an expectation of profit, or from the pressures of competition to adopt new technologies in order to survive.

How Attitudes Influence Innovation

Governments have a powerful influence on invention and innovation through their expenditures on education, scientific research and engineering development. Their tax systems affect the supplies of risk capital and the incentives for firms to spend on R & D and to invest in new technologies. The laws and the administration of laws on monopolies and restraints on trade, and measures to reduce otherwise lasting excess capacity in an industry, whether caused by major changes in technology or in international trade, have a significant influence on the extent and effectiveness of competitive pressures to innovate.

The attitudes of businessmen to the recruitment of manpower at various levels of education, to training schemes within industry, to expenditure on applied R & D, to the investigation of innovations made by other firms which might be adopted or adapted, and to the investment of resources in new technologies which have not yet been proven by experience and which may involve disturbing changes in work organisation, are all important factors in the role of technical change.

The attitude of workers to new technologies can also have a significant influence on the extent and timing of technical change. If workers are organised in (or by) trade unions which are strong enough to insist, as the price of accepting new technologies, on manning levels, or increases in wages, or reductions in hours which make the financial advantages of investment in new technology marginal, then the diffusion of the new technology will be impeded. If managements expect strong worker resistance to new technology they may become reluctant to introduce new technology until it is clear to the workers that 'pressure of competition' makes it necessary to innovate to survive.

There are also attitudinal factors which through the choice of educational programmes can influence rates of innovation. If, as is often alleged to be the case in Britain, the attitudes of parents and teachers are such that the most energetic and mentally able are drawn to the humanities or pure sciences rather than to the applied sciences and technologies, then the capacity to innovate may be weak even though the capacity to make new scientific discoveries and to invent may be very strong. A common explanation of the (not very well justified) view that 'Britain invents, others innovate' is that because of parental attitudes, early specialisation in schools, the tendency in universities to regard engineering as a subordinate branch of science, and the attenuated nature of links between education and industry, too few alpha students enrol in the technologies in universities and polytechnics, and too many employers are mistrustful of graduates and treat them as fit only for specialised production or boffin-type roles. According to the Finniston Committee Report,

compared with continental Europe and the large part of the world which has followed its lead, there have been neither the cultural nor the pecuniary rewards in this country to attract sufficiently the brightest national talents into engineering in industry. Great prestige is attached to science, medicine and the creative arts, so that to be associated with their activities is to share in that esteem, but there is no cultural equivalent in Britain, and hence no basis for similar esteem, to the European concepts conveyed in German by 'Technik' - the synthesis of knowledge from many disciplines to devise technical and economic solutions to practical problems. (1)

There are many different, and at times conflicting, interests in the outcome of innovation. Governments may be expected to welcome innovations which increase growth, tax yields, and strength in international trade and in intergovernment conferences; consumers to welcome innovations which increase incomes and the range and quality of consumer goods, or which improve the environment; workers to welcome innovations which add to their incomes, reduce working hours and sustain employment; and employers to welcome innovations which add to their competitive strength and profits.

There are, however, certain to be other innovations which governments, consumers, workers and employers do not welcome, though their views on which innovations are unwelcome may differ substantially. Governments are not likely to welcome innovations which create regional or cyclical unemployment, or environmental degradation, or additions to the costs of defence, or even innovations which call for major changes in the extent and direction of vocational education in the formal education system. Workers are unlikely to welcome innovations which are deskilling, or reduce their status or bargaining strength, or their prospects of continued employment; or employers to welcome innovations by competitors which reduce their competitive strength and profits, and particularly not if the innovations are made in other countries. A major test of the strength of a government's acceptance of new technology is therefore how it responds to innovations and the diffusion of innovations which create problems for itself or for sections of the community which then demand offsetting action by the government.

Governments and new technology in the UK

One indication of the positive interest of successive British governments in new technologies is the extent of government financial support for research and development activities. In 1981 government expenditure on R & D was 1.34 per cent of gross domestic product (GDP), which was a higher percentage than in the USA, Japan, Italy, Germany or France. However, as in the USA, over one-half of that expenditure was directed towards the creation of new or improved war technologies. The percentage of GDP devoted to civil R & D was considerably less than in France or Germany, though more than in Japan or the USA.

United Kingdom

Table 1: Government Expenditure on R & D in 1981 as Percentage of GDP

	UK	USA	Japan (1980)	Italy	Germany	France
Military	0.70	0.63	0.01	0.04	0.10	0.48
Civil	0.64	0.59	0.52	0.61	1.05	0.81
Total	1.34	1.22	0.53	0.65	1.15	1.29
Total expenditure ($m)	6,256	35,547	5,337	2,957	7,345	6,962

Source: OECD, Science and Technology Indicators, May 1983.

One half of government finance for civil R & D is administered by five Research Councils and the University Grants Committee, and almost all of their funds are used for basic and applied research. (2) As a consequence of Britain's relatively low rate of growth in output in the 1950s and 60s, the proportion spent on basic research has fallen while the proportion spent on applied research and development projects judged by the Research Councils to be relevant to growth in the British economy has increased. That, and measures to reduce real grants to universities as part of policy measures designed to reverse the growth of public expenditure relative to GNP, have raised serious doubts about the capacity of universities to maintain high quality basic research.

Successive British governments have also encouraged the introduction and diffusion of new or improved technologies. In agriculture there is the Agricultural Development and Advisory Service; in agriculture, industry and the services there are tax incentives to invest in new plant and equipment; in industry there has been a variety of schemes to encourage engineering firms to purchase advanced machine tools and, recently, to adopt microprocessors and 'advanced manufacturing systems'. Other measures have included the establishment of the National Research and Development Corporation (NRDC) in 1948 to exploit inventions made in government research centres and universities, the Industrial Reorganisation Corporation (IRC) in 1966 to facilitate rationalisation and faster technical progress, the Economic Development Committees of the National Economic Development Council to consider ways of increasing economic growth in particular sectors of industry, and the National Enterprise Board (NEB) in 1975 to provide finance for new firms in, for example, micro- and bio-technology and for ailing established firms capable of being restored by reorganisation and investment in new technologies. Governments have also encouraged the establishment of American, German and Japanese subsidiary firms to increase investment in new technologies.

There have been differences between Labour and Conservative governments on methods of promoting technical change. The IRC and the NEB were established by Labour governments. A Conservative government abolished the IRC, and at first reduced the assets of the NEB, and then decided to absorb it, and the NRDC, into a new British Technology Group (1984) which is to conform to the present government's policy on privatisation.

Education
Government measures to encourage invention and innovation could hardly be successful in the absence of scientists, engineers, technicians and craftsmen. In Schools and Universities on the Continent (London, 1868) Matthew Arnold wrote of the connection between education and technical change and of the need to improve scientific and technical education. The Report of the Select Committee on Scientific Instruction (1868) and the reports of the Royal Commissions on Scientific Instruction and the Advance of Science (1875) and on Technical Instruction (1884) provided further evidence of the need for major changes in and extensions of education, but during the following 40 years there was little response from parliaments or industrialists, and Britain's capacity for industrial innovation fell away.

From 1870 Britain's growth-rate declined relative to growth-rates in France, Germany and the USA. That relative decline in growth-rates - and actual decline between 1870 and 1913 - took place because innovation came to depend less on skills in mechanical engineering, and more on skills in chemical and electrical engineering which were more 'science-based'. Britain's education system was not appropriate to the new conditions of industrial innovation.

Due to the lessons of the previous 50 years of peace and the success of 'science at war' there was a much more positive response from governments after the Second World War, though the decisions on the expansion and the change in the nature of secondary education were in the event based more on a desire to provide general education for all in a more egalitarian context than they were on considerations of the manpower skills and attitudes conducive to growth and innovation. Evidence of continuing shortages of skilled manual workers led to the creation of Industrial Training Boards to improve apprenticeship and less formal methods of training within industry. The failure of this scheme due to the inadequate response of employers then led to the creation of the Manpower Services Commission and to more positive efforts to improve and relate off-the-job and on-the-job training schemes.

After the Second World War, governments provided funds for large increases in tertiary education, and they also provided guidelines on the proportion of students to be admitted to study the sciences and technologies. Steps were also taken to raise the standards and status of technical education, and some Colleges of Advanced Technology (CATs) were created. Following the report of the Robbins Committee on Higher Education, the CATs became universities, and shortly afterwards a binary system of universities and public-sector institutions was created to promote vocational education in the polytechnics and colleges of higher education which, it was said, would be 'more responsive to industrial needs than the autonomous universities'. However, 'academic drift' in the polytechnics and colleges towards the pattern of education in universities raised doubts about whether the assumption had been well based that local authorities and their technical college traditions would ensure responsiveness to local needs, and there were also suspicions that the constitution and administration of the Council for National Academic Awards was contributing to that drift. The creation of the National Advisory Body for Local Authority Higher Education is intended to bring a stronger central government influence on the polytechnics and colleges of higher education, and the present government intends to use that influence to achieve what it judges to be a need for more directly vocational education and training.

As judged from public expenditure on education, the British government does more than many governments to promote both a capacity to generate opportunities for innovation and a capacity to use them. In 1980 that expenditure (including financial grants to students) was 6.4 per cent of GDP compared to 6.5 in the USA, 5.3 in Japan, 5.1 in Italy, 5.0 in Germany and 3.6 in France. Only in Norway, Sweden, the Netherlands and Canada was public expenditure on education a significantly higher proportion of GDP. (3)

The distribution of graduates between the humanities, sciences and technologies was not, however, ideal from the standpoint of a rapid rate of technical change. In 1976 the number of new British graduates in science was three times higher, but in engineering 50 per cent lower, than in West Germany. The ratio of graduating scientists to engineers is about 3:2 in the UK compared to 1:3 in West Germany. (4)

From the standpoint of technical change, and the effective use of new technologies once introduced, another defect of the British system of education is the low level of provision for post-employment training. Thus, for example, in the British engineering industry 62 per cent of workers, and in metal manufacturing 70 per cent, lack any formal qualifications, compared to 32 per cent and 30 per cent respectively in West Germany. (5)

This defect is due in considerable measure to the attitudes of employers. The 1964 Industrial Training Act placed a statutory responsibility on employers to train workers, and in view of the past record of employers, the decision of the government, as reflected in the 1981 Employment and Training Act, to put the primary responsibility for the content and co-ordination of training upon the voluntary efforts of employers, was somewhat surprising. In 1980 British employers were estimated to be spending about £2.5 billions a year on training as compared with £5.5 billions in Germany (6), although the labour force in Germany was only 2-3 per cent larger than in the UK.

Industry
This lack of commitment to education may be due to the low level of formal qualifications of managers in Britain. The Labour Force Survey 1981 revealed that less than 20 per cent of British managers held university degrees or professional qualifications, and according to an earlier Department of Industry

survey, the proportion of chief executives with university degrees was considerably less than in Belgium, France, West Germany or Sweden. (7)

Many scholars who have attempted to explain Britain's relative industrial decline have concluded that there is a strong anti-intellectual tradition in British industry. The subject of a recent lecture at the Royal Society of Arts was 'British Industry and Anti-Intellectual Tradition'. In this lecture Philip Nind, a former executive of Shell, referred to British businessmen as practical, unideological people, more concerned with the practice than with the theory of ideas, and inclined to regard anything associated with the intellectual as abstract, highbrow and of secondary importance. (8)

That was not meant to be an accurate description of all businessmen, and it is not an accurate description of typical managers in every sector of British industry. But in many sectors of British industry there are predominant attitudes which contribute to the continued weakness of British manufacturers. A plausible explanation of this continued weakness is that, except in a few industries based on R & D activities, the clash of the predominant attitudes in the different worlds of industry and education has been not stimulating but repelling. The Finniston Committee emphasised two factors - the content of courses and the quality of students in engineering - in this negative interaction between higher education and industry. The tendency for universities to treat engineering as a subordinate branch of science, and not as a field providing a synthesis of knowledge from many disciplines to devise technical and economic solutions to practical problems, has done little to persuade top management not to regard engineers as mere purveyors of technical services. This attitude of businessmen is reflected in the status and pay of engineers in industry, which then helps to explain the failure of universities to attract many able and enterprising young people into engineering courses.

Perhaps if the universities and polytechnics had attracted large numbers of able and enterprising young people into engineering courses, a proportion of them would have risen to top management positions, and changed the predominantly 'anti-intellectual attitudes' of British businessmen. The government's decision, following consideration of the Finniston Committee's report, to create an Engineering Council with wide powers to

bring about changes in degree courses and in further
education, and the decisions of the Science and
Engineering Research Council and the Department of
Trade and Industry to operate a Teaching Company
Scheme, are each designed to bring about major
changes of attitudes in both industry and higher
education.

Government expenditure on R & D as a
percentage of GDP in 1981 was given in Table 2.1.
The British percentage was higher than that in the
other countries, though on civil R & D lower than in
France or West Germany. In all countries,
governments finance research both in their own
laboratories and in private industry. In the UK
the percentage of R & D financed by government in
the last 20 years has averaged just under 50 per
cent. R & D performed by British industry from its
own funds changed very little between 1966 and 1968,
fell by 13 per cent between 1968 and 1975, rose by
33 per cent between 1975 and 1981, and then fell by
about 5 per cent between 1981 and 1983. The fall in
expenditure between 1968 and 1975 - when it is
believed that many British firms became disenchanted
with the financial yield from expenditure on R & D -
took place at a time when GDP grew by more than
one-quarter and fixed investment by nearly
one-third, and when real expenditure on R & D in
France, West Germany and Japan was increasing
significantly. The subsequent increase in
expenditure seems to have been due to a realisation
that the financial consequences of a reduction in
research relative to that in competitor countries
would soon lead to further reduction in competitive
strength.

It can be seen from Table 2.2 that
industry-financed R & D is a higher percentage of
GDP than it is in Italy or France (though total
business-financed R & D in France is higher than in
Britain) but it is a substantially lower percentage
of GDP than it is in Japan, the USA or Germany.
Perhaps the right conclusion to be drawn is that
industrial expenditure on R & D is higher than might
be expected from the literature on the
anti-intellectual attitudes of British businessmen,
but lower (at least in absolute terms) than it would
be if British firms were more efficient in making
good use of new technologies, including the new
technologies created, improved or adapted by their
own R & D staff.

United Kingdom

Table 2.2: Industry-financed R & D in 1981 as
Percentage of GDP

	UK	USA	Japan	Italy	Germany	France
Total R & D	2.3	2.7	2.5	1.2	2.6	2.1
Industry-financed	1.0	1.5	1.9	0.6	1.4	0.8

Source: OECD, Science and Technology Indicators, May 1983.

Another popular explanation of Britain's relatively poor record of growth from the introduction of new and improved technologies is that British industrialists are too reluctant to take the risks involved in investment in innovation. A major factor in this aversion to risk is said to be the domination of decision-making by people lacking in technological knowledge. Thus, it is suggested that investment policies informed by understanding of the nature and significance of new technologies

would be less likely to incorporate an undue subjective bias against technological innovations, since they will take greater account of the implications for the company of arriving at a future date without having invested in updated current products and methods in response to market changes. (9)

There is no strong supporting evidence for the view that British managers have an attitude to risk which leads them to reject proposals to invest in innovation unless the prospective rates of return are unusually high, or pay-back periods unusually short by French, German or American standards. There is more evidence to support the view that Japanese firms are prepared to act on the basis of lower prospective yields or longer pay-back periods, but the main reason for that is the different role of the capital market in Japan. The explanation is therefore more institutional than attitudinal. (10)

Two interview surveys by Market and Opinion Research International (MORI) indicate that British managers are aware of the deficiencies of innovational activities in British industry. In 1983 MORI conducted for PA Technology a survey in major British companies. Of the 83 senior managers interviewed, 44 per cent thought that overseas competitors made better use of advanced technology in their industries, and 25 per cent stated that their companies had no defined strategy for innovation and the application of new technology. Forty-six per cent were dissatisfied with the speed of their companies' product development. Whereas 47 per cent believed new technology had had a great impact on their companies' products in the previous 4 or 5 years, only 34 per cent believed that new technology had led to significant improvements in the production processes of their companies. Thirty-nine per cent thought that their own companies had the most technologically advanced products but, in line with the comment on process and product innovation, 44 per cent of the managers interviewed thought that overseas competitors had the most technologically advanced production processes.

Early in 1984 MORI carried out for PA Technology a more extensive study, and interviewed 100 senior managers in Britain, the USA, West Germany, Belgium and Australia. The companies sampled came from the manufacturing, electronics, telecommunications, and process computer industries. One in three respondents stated that new technology had led to significant product innovations in their company in the last 4-5 years. But whereas only 18 per cent of British managers reported such changes, 45 per cent did so in West Germany. Just over one in four judged that new technology had had a significant impact on their processes of production. In this field also the percentage was lowest in Britain at only 16 per cent as against, for example, 37 per cent in Belgium.

Workers' attitudes

Workers' resistance to change is often given as a major part of the explanation of Britain's relatively poor record in installing, or at least in using, new technology to achieve high rates of economic growth. In <u>The Wasting of the British Economy</u>, Professor Sydney Pollard maintains that British unions have always been more resistant than others to technical change, and he suggests that even if the trade unions accept the new technology they will do so only after bitter struggles: 'Any innovation becomes a sitting target for pace-making claims while its progress is held up literally for many years'. (11)

The Finniston Committee reported that it found that the process of consultation, communication and acceptance of changes worked more effectively, and with far less friction, elsewhere than in Britain:

Mutual assumptions of conflicting interests appear to underlie communications between managements and workforce in many British companies; in particular, there seems to be an assumption by many on the shop-floor that innovation is essentially for the benefit of management and hence that is at best it has little bearing upon the improvement of their own welfare or prosperity. Pride in the product being made, and appreciation of the need to make it better than that of competitors appear to have become subordinated to an uninterested attitude to the employing company as simply a source of weekly income. Consequently it is too frequently the case that changes in the products, processes or methods can become the subject of heated bargaining over the immediate material incentives to be offered for the co-operation of those affected, frequently carrying delays in the implementation of changes, retention of obsolete plant and underuse of modern plant and equipment. Many companies and individual managers told us that their time was largely preoccupied with issues such as these. By contrast, on our overseas visits we gained the impression that changes to improve company competitiveness were viewed by employees as supporting the maintenance of real wage increases and were hence welcomed and even demanded.(12)

In their book on The British Economic Disaster, two socialist economists also emphasised the role of worker resistance to change:

The UK working class's strong opposition at factory level thwarted many of capital's attempts to increase productivity. New techniques, involving a sharp increase in the technical composition of capital, were often effectively vetoed by unions which did not want to lose jobs. Where new technology was installed, its effect on productivity was often reduced because unions insisted on maintaining existing operation levels or line speeds. Successful resistance of this sort seems the principal explanation for productivity in the post war UK growing less than three-quarters as fast as the technical composition (of capital), whereas the two grew at an approximately equal rate in other countries.(13)

These views on worker resistance to change have been challenged by other writers on the grounds that it is the official policy of the Trades Union Congress to 'increase the rate at which technological advances are adopted by industry', that 28 unions covering 70 per cent of union membership have expressed similar attitudes, and that case studies of technical change do not support the view that worker resistance to changes has played a significant part in delaying or preventing the introduction of new technologies.

However, several policy statements by trade unions in support of technical change contain the proviso that change should be introduced only with the consent of the workers, and then only if any reduction in the demand for labour can be managed through a reduction in hours. The view of the National Association of Local Government Officers (NALGO), for example, expressed in New Technology - a guide for NALGO negotiators, is that the rapid and uncontrolled introduction of new technologies will add to unemployment, and that reduced working time and shorter working hours with increased leisure time for all who work is the only fair and practicable way of distributing the benefits of technical advances. NALGO's view is that the total number of posts should be maintained 'within the particular employing organization', that particular jobs should not be regarded as the possession of the individual employees who occupy them, and that the 'selling' of jobs through productivity agreements, redundancy payments, or other means such as natural wastage, should never be accepted. Similar views have been expressed by the National Union of Mineworkers, and the strength of their commitment to that view was made plain during the strike in 1984 and 1985. Where in the industrial or service sectors there are powerful unions which adopt such attitudes to technical change, there will be a tendency on the part of employers to delay the investment in new technology or to accept, as the price of industrial peace, 'overmanning' of the new equipment. Few of the case studies have been designed to reveal whether management expectations of worker reactions to technical change have led to delays in decisions to invest in new technologies, and, as pointed out by R.M. Bell, the case studies mentioned above deal with the introduction of technical change and not with the way in which the use of the new technology develops over time. (14)

The benefits and costs of technical change have never been evenly spread, and it is to be expected that groups of workers will attempt to use their bargaining power to increase their benefits and reduce their costs from any technical change affecting them. It is doubtful whether in this the objectives of British workers differ much from workers in other countries. And it is doubtful whether British workers are more inclined to assume that in technical change there is an underlying conflict of interest between employers and workers than are, say, French or Italian workers. If they appear to do so there must be other factors involved, perhaps as much institutional as attitudinal. In terms of its own diagnosis, the Finniston Committee might well have explained workers' attitudes as a consequence of the failure of British managers (and schools of engineering) to comprehend the nature of the engineering dimension. But even if 'bosses tend to get the workers they deserve' it may take a good deal of time for more deserving managers to overcome the legacy of responses to previous managers or commitments to customs such as, for example, the method of setting up new piece-rates and bonuses in the engineering industries. The current chairman of ICI commented in 1979 that:

In my company... we have carried out a number of studies on comparative times spent by managers in factories in the United Kingdom, Germany and the USA which show the enormous amount of additional time which has been spent by managers in the labour relations scene [in the UK] if costly and damaging interruptions of production are to be avoided. (15)

Opinion surveys

There is some useful opinion-survey material on general attitudes to technical change in Britain. In 1980 Opinion Research and Communication (OR and C) questioned a cross-section of people in Great Britain, with the following results:

(i) 30 per cent thought their jobs would be affected by microelectronics, and 66 per cent thought that the microelectronic revolution would cause widespread unemployment - about 40 per cent thought unemployment of 5 million a possibility;
(ii) 83 per cent thought that unemployment would be

particularly severe among older workers;
(iii) 80 per cent thought that microelectronics
would lead to the loss of many job skills;
(iv) 75 per cent thought that the introduction of
the new technology would lead to industrial disputes
and strikes; yet
(v) 75 per cent of employees replied that they would
welcome and co-operate in the change, and only 13
per cent replied that they would oppose the change
to protect their jobs.

In 1981 MORI were commissioned by Technology Week to
conduct an inquiry into the main effects of new
technologies. In a study designed to be
representative of all adults in Great Britain it was
found that the main effects of new technology were
thought to be:

- an increase in unemployment
- the creation of new opportunities for industry
- an increase in the competitive position of
British firms, and
- a reduction in the working week

TABLE 2.3: Opinions on the Effects of New
Technologies

	Science Educated %	Others %	All %
Will:			
increase unemployment	64	56	58
create new opportunities for industry	73	49	55
make companies more productive	69	47	52
make British companies more competitive	64	46	50
reduce hours of work	68	49	53
reduce the number of boring (repetitive) jobs	61	32	39

Even though new technology was expected to increase unemployment, 82 per cent of those who thought that new technologies had affected their jobs also thought that the changes had been for the better, and, as with the OR and C survey, there was little evidence from expressed views that more than a small proportion of workers would oppose technical change to protect their jobs. There were some interesting differences between those 'science educated' and 'others', as shown in Table 2.3. It is clear that the 'science educated' were considerably more optimistic than were 'others' about the consequences of technical change.

In June 1984 OR and C interviewed 1,045 workers over 18, resident in Great Britain and in full-time employment. Sixty per cent of those interviewed believed that high technology would not create as many jobs as it displaced in traditional industries. The majority thought that over the following 3 years new technology would cause a loss of up to one million jobs. This was a less fearful opinion than that expressed in the first OR and C survey in 1980, when job losses of five million were considered possible.

As in earlier surveys, the great majority of employees of all categories supported the introduction of high technology as rapidly as possible. One clue to the positive nature of this response is that the very great majority did not think that their own jobs were at risk. Even unskilled manual workers who were generally less favourable to the new technology took that view. Most respondents admitted that they knew little or nothing about the effects which the microchip revolution might have on their own jobs and lives, and expressed the view that it was the employers' responsibility to educate them on the implications of the microelectronic revolution for their future.

In 1979 MORI carried out a study for the Department of Industry on the views of members of trade unions and General Secretaries who were attending the annual meeting of the Trades Union Congress (TUC). Over 80 per cent of those questioned felt that the government was not doing enough to inform trade unionists of the implications of the new technology. The most frequent response when the interviewees were asked what information they wanted was the request for discussion or explanation of the likely consequences of adopting the new technology. Its likely impact on unemployment was particularly frequently mentioned; and protection of members' interests and livelihoods was of prime concern. They expressed the view that new technology might bring about a reduction in the working week and a reduction in the number of boring and repetitive jobs, but few thought that it would bring higher wages.

Twenty-five per cent of General Secretaries and 14 per cent of delegates thought it would bring no benefits at all to trade union members, and well over 70 per cent thought new technology would bring about increased unemployment. However, only 53 per cent of delegates and 35 per cent of General Secretaries thought unemployment would increase amongst their own members. Over 60 per cent thought new technology would increase the efficiency of industry, make companies more productive, and create new opportunities for industry, but only about half thought it would make British companies more competitive. Half the General Secretaries and 40 per cent of the delegates felt that the government and the TUC should allow new technologies to develop at their own pace. The vast majority of those favouring government intervention wanted a speeding up of the introduction of new technology. There was no support for a Luddite opposition to new technology. On conditions for the introduction of new technology, the delegates were primarily concerned with manning levels and redundancies. Only 25 per cent wanted prior consultation, which was perhaps a surprisingly small percentage.

In this survey there was also a limited comparison of trade union and management attitudes. The survey indicated that managers were less confident that the working week would be reduced either in their companies or in British Industry as a whole, less inclined to forecast an increase in unemployment, more inclined to forecast a reduction in routine and boring jobs, and more optimistic that new technologies would bring new commercial opportunities for industry.

According to the Aspen Institute survey reported in Work and Human Values: An International Report on Jobs in the 1980s and 1990s, the work ethic is rather weaker in the United Kingdom than in the USA, the Federal Republic of Germany, Sweden and Japan (16).

TABLE 2.4: Perceived Work Ethic

	UK (%)	USA (%)	FRG (%)	Sweden (%)
Strong	17	52	26	45
Moderate	48	21	54	44
Weak	31	27	15	7

Source: Work and Human Values, p. 65.

There is indirect support for the Aspen Institute survey in the changes in normal hours of work in the UK. In the 50 years since 1929, hours of work in Britain fell by about 10 per cent more than in France and Germany, despite the considerably smaller increase in output per worker-hour in Britain. In so far as this apparently greater preference for more leisure rather than for more income reduced the growth of the home market, it would have had an indirect restraining influence on technical change.

The main impressions conveyed by this opinion-survey material are that the proportion of workers who oppose technical change is very low, that there is a general recognition that technical change is needed to maintain the competitiveness of British industry, but that workers expect that technical change will create considerable unemployment and, apart from reduction in hours, will not bring them much benefit. This rather pessimistic attitude might explain why firms such as ICI find it necessary to devote a substantial portion of management time to labour relations.

Technology and the environment

During the 1960s there was in most industrial countries a marked increase in concern for environmental issues. Rachel Carson's Silent Spring, published in 1962 to warn of the dangers of the over-use of pesticides, helped to create a more general interest in environmental damage. The evidence of nitrates in drinking water, lead in the atmosphere, mercury in fish, unsightly detergent foam on canals, acid rain, the thalidomide tragedy,

the development of thermonuclear bombs and delivery
systems of great accuracy and range, the pressure
from the nuclear lobby to build nuclear power
stations before the invention of methods to deal
with highly radioactive wastes, and from the
aircraft lobby to build and operate supersonic civil
aircraft - all such developments contributed to
fears that scientists and technologists were
treating all new technologies as 'a good thing' to
which society should adapt itself. The same prime
minister who in the early sixties had proposed to
forge Britain's prosperity in 'the white heat of the
technological revolution', in the mid-seventies
sensed a change of mood in the electorate and voiced
concern that technology was being allowed to 'run
over' people's lives.

The growth of concern for the environment led
in most countries to government measures to protect
the environment. Some of these measures reduced
measured output per unit of input, though they did
not necessarily reduce technical change and may have
increased it. For environmental measures to prevent
or reduce the pollution of the land, water and air,
will often require technical change, and that change
may induce a considerable amount of research to
develop less costly ways of meeting the new
environmental standards. For the US economy, E.F.
Denison estimated that changes in pollution
abatement costs subtracted only 0.02 percentage
points from growth in national income per unit of
output from 1964-9, 0.10 points in 1969-73, 0.23
points in 1973-5, and 0.08 points in 1975-8. (17)
There has not been a similar estimate for the UK,
but it is not likely that the impact on growth was
as much as in the USA.

Between 1978 and 1980 Professor Stephen
Cotgrove conducted surveys of the attitudes of the
general public in Britain, and of particular
subgroups classified as new environmentalists,
traditional nature conservationists, industrialists
and trade union officials. (18) A high proportion
of the new environmentalists were employed in the
personal service professions and creative arts.
Table 2.5 gives the results of his inquiries into
some general objectives. Two of the most
interesting results are the similarity of the views
of the public, industrialists and trade unionists on
the importance of stability in the economy and the
price level, and the very low priorities assigned by
trade unionists to high rates of growth and
industrial democracy.

Table 2.5: Percentage Assigning 'High Priority' to Stated Issues

	Public	Environ-mentalists	Nature conserv-ationists	Indust-rialists	Trade unionists
Maintaining a high rate of economic growth	46	2	38	44	29
Making sure that this country has strong defence forces	41	12	38	18	45
Maintaining a stable economy	70	50	71	70	68
Fighting rising prices	78	42	65	71	68
Giving people more say in important government decisions	42	42	13	42	27
Progressing toward a less impersonal, more humane society	48	71	37	65	57
Seeing that people have more say in how things get decided at work	31	33	7	60	18
Progressing toward a society where ideas are more important than money	35	54	15	43	34
n =	(275)	(423)	(218)	(285)	(289)

Source: Cotgrove, Catastrophe or Cornucopia, p. 48.

Cotgrove's inquiries into judgements on the seriousness of environmental problems revealed that more than one-half of the public, and almost two-thirds of trade union officials, judged that environmental problems were extremely serious or very serious. But an important test of the degree of concern for the environment is the extent to which people are prepared to pay taxes to control pollution, or to sacrifice jobs in cases where there is a conflict between growth and the protection of the environment. Less than 7.5 per cent of the public or the subgroups expressed opposition to the idea of raising taxes for pollution control, and over 60 per cent of the public and all subgroups gave support or strong support. The expressed support of the public and industrialists was similar, but significantly less than for the other subgroups. But only the environmentalists expressed a clear preference for protecting the environment at the expense of jobs.

The strenght of the 'new environmentalist' challenge to economic growth as an objective of policy has declined since Cotgrove's inquiry. The sharp rise in unemployment in the 1980s has been a major factor in the change. However, even before that increase in unemployment, the impact of the new environmentalist movement on public policy was small. For the most part the public and governments do not regard worthwhile measures to control pollution and protect the environment as threats to economic growth. Whether or not to support the building of more nuclear power plants is a live issue, in part at least because of doubts about whether they are at all essential for economic growth. And whether to introduce much more stringent and expensive measures to reduce the emission of sulphur dioxide from power stations is also a live issue, but doubts on this issue are perhaps due more to lack of knowledge about the sources (and effects) of acid rain than they are about fears for the consequences of a rise in the price of electric power.

Summary

Since the Second World War successive governments have introduced many measures to promote the creation and use of new technologies. Government expenditure on research and development is higher relative to GNP than in the USA, France, Germany and Japan. However, a very high proportion of this expenditure is on military R & D, and this has restricted the range of new technologies created and the prospective contributions of government-financed R & D to the increase in measured economic growth. Governments have also given fiscal encouragement to firms to spend on R & D and to invest in new plant and equipment.

Education has grown in importance as a factor in the generation and effective use of new technologies. Despite many warnings from expert committees, between the 1870s and the 1930s British governments did not introduce the measures required to create either the scale or type of education needed to match economic growth in other industrialised countries. But the effects of the low growth-rates and high levels of unemployment in the twenties and thirties, and of the important role of scientists and engineers during the Second World War, led to a major change in government attitudes to education. As a percentage of GNP, public expenditure on education is now higher than in most other countries, though provisions for the post-school education of manual workers and the distribution of students in higher education between the humanities, sciences and technologies is not ideally suited to the promotion of rapid technical change.

However, in the explanation of the British education system, much more is involved than the attitudes and actions of central governments. The attitudes of teachers in the schools, of lecturers and professors in the colleges and universities, of parents and of employers, must also be brought into account. Staff in the secondary schools, colleges and universities have not been very supportive of proposals for vocational education other than for the traditional learned professions. That attitude has been reinforced by the high proportion of employers who display anti-intellectual attitudes and have been very critical of policies designed to increase retention rates in schools and participation rates in higher education.

Most of the parents who encourage their children to proceed to higher education appear to accept the predominant attitudes of staff in the formal education system. However, they, like their children, are influenced by the salaries and status of particular occupations. If, for example, the salaries of engineers were higher, the proportion of students enrolling in the engineering schools would rise relative to enrolments in science. It is not clear why, if, as is frequently claimed, there is a shortage of engineers needed for the creation and diffusion of new technologies, the salaries of engineers have not been higher. The explanation of the Finniston Committee is that because the universities have treated engineering as a subordinate branch of science and not drawn into engineering studies a sufficient proportion of able and energetic young people, engineers in industry have not been very productive. They have not displayed sufficient skill in devising technical and economic solutions to practical problems, and management has therefore continued to treat them as mere purveyors of technical services. This may not be a complete explanation, but does appear to be an important part of it, and the government's decision to create an Engineering Council is designed to prevent industry and higher education from reinforcing each other's weaknesses.

One of the weaknesses of management in Britain is the proportion of managers who maintain the traditions of empiricism and of learning by experience which were established during the first 100 years of industrialisation, but became progressively unsuccessful once academic science and organised R & D grew in importance, as they did from the 1870s onward. This attitude is responsible for a reluctance of many firms to promote change, for the low overall expenditure of British firms on industrial R & D, for the low priority given to getting information on new technologies introduced elsewhere and on likely future developments in technology, and for the reluctance to adopt, or comprehend, systems approaches to problems. It has also contributed to the marked reluctance of many firms to recruit the scientists or technologists needed to introduce and operate new technologies and management systems, on grounds that their recruitment would upset salary relativities and be unfair to staff who had spent many years in the firm.

Fear of worker resistance to technical change is often blamed for the reluctance of many firms to make changes in technology until faced with the threat of bankruptcy, and worker resistance to technical change is often blamed for the frequent failure of many firms which pioneer new technologies to achieve subsequent incremental increases in productivity which many firms in other countries achieve. It would be very surprising if workers who thought their jobs or skills were threatened by new technologies, or trade unions who feared a substantial loss of membership and industrial clout, did not press for the maintenance of manning levels, and use or threaten to use their collective powers. There have been various case studies of technical change which led the research workers involved to conclude that fear of worker resistance to change was not a significant factor, but few case-study projects have been designed to test the role of workers' attitudes and trade union pressures in the apparent reluctance of many British firms to innovate, or to innovate before it was clear that the continued existence of the firm was becoming doubtful, or in the failure of many innovating firms to make adequate learning-curve improvements in productivity.

If worker resistance to change is a bigger problem in Britain than elsewhere, the explanation could be institutional - derived, for example, from the laws relating to trade unions and industrial associations or the laws and administrative procedures relating to monopolies and restrictive practices, or from the low level of provisions for compensating workers for loss of income caused by technical change and/or for retraining - or it could be attitudinal. There is some survey and statistical evidence that British workers have a less powerful work ethic, and a stronger preference for stability than for a high rate of economic growth (and therefore technical change), than workers in many other industrial countries.

Not all technical change leads to or is designed to lead to economic growth. Some changes in technology are designed to increase military strength, or to improve health services, or to reduce environmental damage. The Cotgrove inquiry between 1978 and 1980 indicated that 41 per cent of the public and 49 per cent of trade unionists assigned a high priority to strong defence forces. In a Gallup poll conducted in 1980, 59 per cent of those questioned thought that expenditure on armaments and defence was too little or about right, and 92 per cent thought that expenditure on the National Health Service was too little or about right. (19) In the Cotgrove inquiry 57 per cent of the public and 67 per cent of trade unionists thought that environmental problems were extremely serious or very serious, though only 22 per cent of the former and 14 per cent of the latter expressed support for environmental measures which threatened jobs.

NOTES

1. See Engineering Our Future, Report of the Committee of Inquiry into the Engineering Profession, Cmnd. 7794, 1980, p. 24.

2. Cabinet Office, Annual Review of Government Funded R&D 1984, HMSO, 1984.

3. The OECD Observer, March 1985.

4. S.J. Prais, 'Vocational Qualifications of the Labour Force in Britain and Germany' in National Institute Economic Review, November 1981.

5. Ibid.

6. Education and Training for New Technologies volume 1, Report of the House of Lords Select Committee on Science and Technology, HMSO, 1984.

7. See also British Institute of Management, The British Manager in Profile, Management Survey Report, No. 51.

8. Journal of the Royal Society of Arts, April 1985, 329-37. See also D.C. Coleman, 'Gentlemen and Players', Economic History Review, Series 2, 26, 1973.

9. Engineering Our Future, p. 31.

10. See R. Dore, 'Financial Structures and the Long-term View', in Policy Studies, volume 6 part 1, July 1985, pp. 10-29.

11. S. Pollard, The Wasting of the British Economy, London: Croom Helm, 1982.

12. Engineering our Future.

13. A. Glyn and J. Harrison, The British Economic Disaster, London: Pluto Press, 1980, p. 50.

14. R.M. Bell, 'The Behaviour of Labour, Technical Change, and the Competitive Weakness of British Manufacturing' in Bruce Williams (ed.), Knowns and Unknowns in Technical Change, London: The Technical Change Centre, 1985.

15. 'The British Manager', Management Review and Digest, vol. 6, No. 1, 1979.

16. D. Yankelovich and others, September 1983.

17. E.F. Denison, Accounting for Slower Economic Growth, Washington, DC: Brookings, 1979, pp. 71-2.

18. Stephen Cotgrove, Catastrophe or Cornucopia, Chichester: John Wiley & Sons, 1982.

19. N. Webb and R. Wybrow (eds), The Gallup Report: What YOU Said in 1980, London: Sphere Books, 1981.

Chapter Three

CHANGING ATTITUDES TOWARDS TECHNOLOGY IN THE UNITED
STATES.
Dorothy Nelkin. (1)

The public acceptance of new technologies in the
United States today must be considered in the
context of the 1960s and 70s, when many Americans
involved in grass-roots consumer and environmental
movements questioned the definition of progress and
the role of technology in serving the needs of
society. The challenges of this period - expressed
in attitude surveys as well as in public
demonstrations, litigation, and other forms of
protest - led to a number of institutional and
political changes directed towards winning greater
public acceptance of controversial technologies.
These include techniques to assess the social
implications of technology, and efforts to involve
citizens more directly in the decision-making
processes of those public agencies responsible for
regulating technological change. Soon these
participatory and assessment efforts themselves
became the focus of dispute, viewed less as reforms
than as impediments to progress. In the
conservative political context of the mid-1980s,
technocratic strategies and image-building efforts
are increasingly employed to develop and sustain
public acceptance of technology. These efforts
have, to date, produced mixed results. While there
is less opposition to specific projects, and indeed
widespread support for the development of high
technology, there is also a growing preoccupation
with the moral implications of technology and its
effect on individuals' rights.

Public Attitudes Towards Technology
The late 1960s marked the beginning of a period of
public concern in the United States about the social
and environmental implications of technology.

46

United States of America

During this period we saw a remarkable amount of
litigation, demonstrations opposing the siting of
technological facilities, and active efforts to
establish rules and standards to protect public
health and the physical environment. Technologies
of speed and power - airports and nuclear power
plants - provoked antagonism as local communities
protested against noise and environmental
disruption. The risks of science-based
technologies, such as recombination DNA, became
sources of public debate. The potential
side-effects of drugs and food additives evoked
public alarm. Activists - often within the
scientific community - challenged the legitimacy of
the experts and the authorities responsible for
decisions about technology, and called for greater
public control over technological choices.
Neighbourhood groups demanded a voice in the
location of power plants and airports. Patients
demanded greater control over their medication.
Consumers raised questions about the validity of
data that supported government regulations bearing
on public health and safety. Citizens sought to
participate in shaping the rules and standards that
would govern the use of new technologies.
'Demystification' of expertise, 'public accountability' of scientists, 'citizen participation' were
popular slogans in many public disputes. (1)
     The emergence of doubt about the role of
technology as a manifestation of progress has often
been identified with the 1962 publication of Rachel
Carson's Silent Spring, the book which introduced
millions of readers to the environmental effects of
pesticides. During the 1960s the fear of radiation
fallout from the testing of nuclear devices also
provoked active public concern. The continued flow
of news about environmental crises - the Santa
Barbara oil spill, the Dugway sheep kill, incidents
of herbicide and pesticide pollution - sustained the
concern.
     By 1972 some three hundred books on
environmental problems had been published in the
United States dealing with air, water and noise
pollution, civilian nuclear power, pesticides and
industrial wastes. (2) Some, like Barry Commoner's
best-selling book The Closing Circle (1971),
specifically maintained that the environmental
crisis was created by technological innovations -
for example, electrical generating facilities, the
increased use of automobiles, and phosphate
fertilisers. The technological factor, he claimed,

47

accounts for 40 per cent to 85 per cent of the increase in environmental impact. This view became extremely influential during the 1970s, when voluntary associations, organised around the environmental and social impacts of technological development, proliferated. They ranged from national organisations (such as the Sierra Club and Friends of the Earth) to local groups (the Cayuga Lake Preservation Society), from pragmatic research and litigation organisations (the National Resources Defence Council) to more ideological groups (the Clamshell Alliance). In effect, the 1970s saw the development of a social movement mobilised nationally around public concern about the impact of technological change, and organised to influence the growth and regulation of technology through action in the legislatures, the courts and the streets.

Yet public attitude surveys during the 1970s also suggested considerable ambivalance about technology. Optimistic expectations about the benefits of science and technology mixed with fear of their undesirable consequences. In a 1976 survey, only a small majority (52 per cent) of respondents believed that science and technology had produced more good than harm. They saw the greatest benefits in the advances in medicine (81 per cent). In other areas, such as housing, environment, energy and communication, science and technology were reported as beneficial by only 10 to 14 per cent of the public. Longitudinal data assessing public attitudes towards the people who run institutions indicated that between 1966 and 1976 the loss of confidence in the scientific community kept pace with the loss of confidence in other institutions, and declined from 56 per cent to 43 per cent. (3)

In 1979 the National Science Board sponsored a major survey of American public attitudes towards science and technology (the survey involved a base of 1,635 interviews in a national probability sample of the adult population in the contiguous 48 states). (4) While public enthusiasm about technology had clearly declined since the late 1950s, the survey found that attitudes were nevertheless generally favourable. Seventy per cent of respondents believed that the benefits from research outweighed the costs and 46 per cent attributed America's prestige and influence to 'technological know-how'. However, ambivalence persisted: a significant proportion of the public viewed some aspect of science and technology in negative terms (changing our lives too fast - 53 per

cent; breaking down people's ideas of right and
wrong - 37 per cent). The public looked to science
and technology for solutions to many problems
concerning health, energy and the control of natural
disasters. Ninety per cent of the respondents
anticipated that cancer, cheap energy and earthquake
prediction would be resolved within 20 years.
However, several technologies were viewed with
concern; in the case of nuclear power, 78 per cent
saw potential harm and 62 per cent opposed the
location of a nuclear plant in their neighbourhood.
Food additives were perceived as harmful by 80 per
cent of the public, though 60 per cent saw their
benefits as well.

This survey differentiated the 'attentive' from
the general or 'non-attentive' public. About 18 per
cent of the general population were evaluated as
attentive, that is interested and knowledgeable
about science and technology, a percentage that had
doubled since 1957. This group was generally more
favourably inclined towards science and technology
and more optimistic about the possibility of
resolving problems through technological advances.
They were, however, twice as likely as the
non-attentive public to become involved in specific
controversies.

A somewhat less optimistic picture of public
attitudes emerged from surveys during the same
period by the firm Yankelovitch, Skelley and Wright.
(5) Pollster Daniel Yankelovitch describes the
general sense of 'foreboding' about technology and
the measurable erosion in the unqualified belief in
technology as an instrument of growth and progress
that had marked the period immediately after the
Second World War. Finding that younger,
better-educated people had the least confidence in
technology, he refers to the sense of
'disillusionment' among educated youth.

Surveys in the 1970s on attitudes towards
technological risk also reflected a growing public
concern about the potentially harmful consequences
of technology. A series of surveys by the
University of Michigan on 'Quality of Employment'
found greatly increased awareness among workers of
work-related risks. (6) Between 1969 and 1977 those
who said that they were exposed to hazards on the
job increased from 38 per cent to 70 per cent.

A Harris Poll in 1980 asked whether people were
subject to more or less risk today than 20 years
ago: 78 per cent of the public (1,488 respondents)
believed more, 6 per cent less, and 14 per cent the

same. Among top corporate executives (401 respondents), the same question elicited a different response: only 38 per cent believed there was more risk, 36 per cent less risk, 24 per cent the same amount. This survey also asked these groups to agree or disagree with the statement 'Society has only perceived the tip of the iceberg with regards to the risks associated with modern technology', 62 per cent agreed, 28 per cent disagreed. Among the executives, 19 per cent agreed and 78 per cent disagreed. (7) The disparity in the attitudes of different sectors shown in Table 3.1 is a striking indication of the difficulty of resolving risk disputes.

Table 3.1: Today's Risk Compared to That of 20 Years Ago

Q: Thinking about the actual amount of risk facing our society, would you say that people are subject to more risk today than they were 20 years ago, less risk today, or about the same amount of risk today as 20 years ago?

(no of respondents)	Top corporate executives (401) %	Investors/ lenders (104) %	Congress (47) %	Federal regulators (47) %	Public (1488) %
More risk	38	60	55	43	78
Less risk	36	13	26	13	6
Same amount	24	26	19	40	14
Not sure	1	1	--	4	2

Q: Society has only perceived the tip of the iceberg with regard to the risks associated with modern technology.

(no of respondents)	Top corporate executives (401) %	Investors/ lenders (104) %	Congress (47) %	Federal regulators (47) %	Public (1488) %
Agree	19	20	47	38	62
Disagree	78	71	51	60	28
Not sure	3	9	2	2	10

Source: Lou Harris Poll (1980).

United States of America

Psychological surveys on the acceptability of risks consistently suggest that people are most concerned about risks that are involuntary, uncertain, unfamiliar and potentially catastrophic. These of course are the characteristics of the risks associated with technologies such as nuclear power and toxic chemicals - the most visible and widely reported issue in the press. (8) But concerns about the impact of technology extend well beyond the fear of risk. The burgeoning of electronic technology in the early 1970s, especially the development of computerised data banks, prompted debates about the possible abuse of privacy, the loss of civil liberties, the increasingly impersonal character of medical care, the routine character of work, the threat of unemployment, and the possibilities of centralised political and social control.

Sources of Public Ambivalence

A number of factors contributed to public ambivalence about technology in the United States during the 1960s and 70s. Some were directly related to the increasingly visible side effects of specific technologies; others reflected certain changes in the American political process that fostered concern about the accountability of the experts and bureaucrats responsible for decisions about technological change.

Several dramatic and costly incidents during this period called public attention to the health and environmental impact of new technologies. Among them were the Santa Barbara oil spill in 1969; a series of fires and minor incidents in nuclear plants, followed in 1979 by the Three Mile Island accident; and repeated indications of occupational hazards from exposure to toxic chemicals. Such incidents, increasingly frequent and widely publicised in the media, could not be dismissed as aberrations or unique events. They demonstrated the potential dangers of specific technologies, reduced the credibility of industry and government as strong, efficient protectors of the public interest, and enhanced the credibility of environmental and other protest groups.

Indications of public concern often expressed in dramatic demonstrations were also widely covered by the media, further catalysing public sensitivity as concerned individuals realised that they were not alone in their worries about technological change.

To some degree disillusionment about technology at this time reflected disappointed expectations. Atomic energy and the space programme had brought science and technology from the obscurity of the laboratory to the forefront of American consciousness, creating optimistic expectations – the so called 'moon ghetto' syndrome – that technology would solve social problems. (9) It was soon apparent that technological changes not only failed to solve problems but contributed to them as well. Barry Commoner's dictum – 'There is no such thing as a free lunch', became a popular slogan. The policy importance of these new attitudes was reflected in the much publicised debate over the SST, and the subsequent Congressional decision in 1971 to kill this technological project after it had consumed nearly a billion dollars for development and design.

The controversies of the 1970s reflected more than simple concern about environmental impact of technology. They also expressed the general political anxiety about the effect of technological change on the vaguely defined set of values known collectively as 'democracy'. To many critics, new technological developments indicated the blurring of distinctions between public and private interests and the increased dominance of technical expertise. Government, intended to be an independent and representative force that could control vested interests, was seen as having a stake in the private sector development of major and costly technological areas. Nuclear power became a symbol of these critics' concern.

Technological decisions had encouraged the transfer of public policy from the arena of politics to that of expertise. During the 1960s scientists and engineers became increasingly active in policy decisions through advisory boards, special commissions and consulting groups. Federal government employment of scientists grew by 49 per cent between 1960 and 1970, while the total federal government employment grew by only 30 per cent during the same period. Many of these scientists worked on military projects.

The Vietnam War raised public consciousness about the politics of science and the role of scientists in major public decisions. Critics claimed to see a growing institutionalisation of an 'intellectual technocracy', a shift of power to knowledge elites who derived their authority from the cultural emphasis on efficiency and technological progress, and who were not directly accountable to the public. This concern, an important factor in many technological disputes, reflected the broad tendency to question authority in the late 1960s and 70s. Opposition to technology often became a surrogate for such general political concerns.

Finally, public opposition also reflected the growth of 'single issue politics' in the United States. American mainstream politics has tended to avoid extreme positions, so that political party representatives share a common outlook on many issues. This has aggravated peripheral groups, encouraging them to focus on single issues rather than party positions. Specific technologies often become 'issues', the focus of political demands from groups who feel that their views are not represented in mainstream politics.

Institutional and Political Response

The technological controversies of the 1970s succeeded in delaying or obstructing numerous projects, and the public pressure to control the health and environmental risks of technology encouraged costly regulation. Many government officials and industry observers worried that the United States was becoming a 'risk aversive society', and that public fear of technology would have a significant effect on future progress and technological growth. This concern, compounded by growing political mistrust, evoked a policy response. During the 1970s the effort to enhance public acceptance of technology took three forms: the development of organisations and techniques to assess and evaluate the social impacts of technology; the creation of programmes to increase public participation, that is, to engage the citizen more directly in decisions about technology; and an effort by the promoters of technologies such as nuclear power to sell their product.

The 1970s saw the development of several 'technology assessment' initiatives. Techniques of forecasting and assessment were designed to anticipate prospective technologies and to predict

their potential consequences. Computer-based
information systems, social indicators, modelling,
gaming and simulation, Delphi, and programme-
evaluation were all elaborate methods intended to
reduce uncertainty and to guide more rational
planning of technological change. (10)

In the late 1960s a subcommittee of the United
States Congress coined the terms Technology
Assessment as a 'method of analysis that
systematically appraises the nature, significance,
status and merit of a technological process'.
Essentially an extended version of cost-benefit
analysis, technology assessments were expected to
evaluate the full range of possible effects of
technologies in order to guide public policy and to
make anticipatory social decisions that would
minimise undesirable side effects. Congress created
the Office of Technology Assessment (OTA) in order
to provide the legislative branch with independent
analyses of technical matters and to assist in
decisions about technological priorities. Its
location in the Congress reflects the desire to
increase legislative (i.e. public) control over the
execution and implementation of technology. (11)
However, while intended to allay public concerns
through more rational assessment, OTA in fact
provoked considerable political debate. It was soon
clear that assessments must take place within
well-defined boundaries that excluded political,
i.e. controversial, considerations.

While OTA continued to assess new technologies
on Congressional request, governmental agencies and
technological industries mounted a related effort to
allay the growing public concerns about risk. Their
efforts to quantify risk and to develop models to
rationalise risk decisions, began in response to the
public reaction to specific technologies such as
nuclear power. Risk assessment has since grown into
a mammoth business. The assessment and management
of risk has been estimated to cost over $200 billion
per year, or 15 per cent of the GNP.(12) Analytic
approaches, based on engineering and economic
models, were developed and encouraged by legislation
requiring that the risks of technological choices be
weighed against their social benefits. The
assumption underlying much of this work was that
better information about risks would lead to greater
public acceptability. (13)

The second response to public concerns about
technology was reflected in the requirements for
public participation as a means to incorporate

citizen values into decisions about technology. All
major environmental laws of the 1970s - the Airport
and Airways Development Act, the Federal Water
Pollution Control Act, the National Environmental
Protection Act, the Energy Reorganisation Act -
contained requirements for direct participation in
administrative-agency decisions. Participatory
reforms were mostly intended to expand the
information available to the public. However, some
procedural reforms sought to allow public
representatives to take part in the development of
policies. And some agencies, such as the Food and
Drug Administration, put consumer representatives on
their advisory panels.

Buttressing such reforms, the Freedom of
Information Act extended public access to
information about technologies, and so too did the
requirements that agencies circulate their
Environmental Impact Statements prior to undertaking
environmentally significant projects or to licensing
new technologies. Citizen access to the courts was
also expanded during this period with the extension
of the legal doctrine of standing which determines
who has the right to be heard in court. In
addition, the 1970s saw the development of some
participatory experiments. The National Science
Foundation (NSF) created a Science for the Citizens
Program to encourage public interests in science and
thereby to better distribute technical expertise to
citizens' groups seeking to influence technological
decisions.(14)

Inevitably, increased public information and
broader participation created a more attentive
public, and just as inevitably, public values often
clashed with goals of efficient technological
growth. Popular slogans suggested the dilemmas that
followed greater citizen involvement: 'Environment
Versus Jobs', 'Profits Versus Health', 'Local Costs
Versus Regional Benefits', 'Individual Choice Versus
the Public Interest', and even 'People Versus
Penguins'. Developers feared that participation
would encourage endless demands from self-appointed
public representatives who would obstruct
decision-making, cause costly delays, and preclude
the long-range perspectives necessary for effective
technological planning. Tension developed between
the ideal of participation as a source of legitimacy
and its practical consequences. Tracing the history
of participatory and assessment measures through the
1970s, one observes a continued struggle to limit
the range of issues debated in a public forum, to

manipulate the terms of public discourse, and to control the boundaries of participation.

Changes in the 1980s

In establishing policies of deregulation, the Reagan Administration assumed that economic concerns were shifting public priorities to greater support of unobstructed technological change. Abandoning the participatory reforms of the 1970s as interfering with industry's right to manage its own affairs, the Reagan Administration has tried to depoliticise technological choices by increasing the use of analytic techniques as a basis for regulatory decisions. In addition, the Administration has engaged in a variety of image-building activities intended to foster public support for technological progress, especially in the area of 'high technology'.

George Keyworth, President Reagan's former science advisor (with reference to a decision about toxic waste disposal) saw the use of analytical techniques as a means 'to reduce the excessive burden of federal regulation by improving the rational basis on which those regulations are made. (15) It was assumed that decisions, scientifically made on the basis of technical criteria, would be less subject to debate. The goal was to create a rationale for reduced public control of the technological sector while neutralising public dissent.

The ground for technical policy analysis had been cultivated during the Carter Administration on the conviction that analytic procedures would help to resolve the environmental and energy crises by providing a scientific base for political and value decisions. More recently, analytic techniques have become mandatory by law and executive order. For example, in the dispute over the regulation of benzene, the courts demanded evidence of significant risk suggesting the need to perform quantitative risk assessments before regulating carcinogens. Similarly, Executive Order 12291 on Regulatory Impact Analysis required quantitative analysis of the costs and benefits of proposed new or revised regulations. Also, a number of bills have been circulating in Congress to increase the scientific and technical basis of political decisions. The Risk Analysis and Demonstration Act of 1982 included the requirement that all federal agencies concerned with decisions relating to health and environment

United States of America

should have a capability for technical-risk
analysis. The Regulatory Reform Act of 1983
required technical evaluation of differential
regional impacts, judgements of the success of
regulation, and documentation of all scientific
information used to develop regulation. (16)
Although scientific uncertainty often precludes
rigorous analysis, quantitative risk-benefit
assessment has become an accepted basis for
decisions, on the assumption that it will lead to
fewer and more effective regulations.

While quantitative techniques are viewed as a
means to depoliticise technological decisions, even
some of those who created the techniques, are
concerned that they are seriously flawed. A report
on the appropriate role of technical analysis for
public sector decision-making, (prepared by the
Environmental Policy Analysis Unit of the Department
of Energy) documents the tendency to use
sophisticated techniques that are not mature enough
to handle the situation that, in fact, arise. (17)
Increased use of such techniques, claimed the
authors, would 'please scientists and analysts;
future environmental decisions would be made the
'right' way because the 'right' amount of data would
be gathered to conduct the risk analysis'. However,
the report suggests that even if such techniques
were perfected and even if perfect information were
available, the public does not necessarily think the
same as analysts about complex questions of
political choice.

To buttress its effort depoliticise
technological choice, the Reagan Administration has
also sought to create a public consensus about the
need for unfettered technological growth. A great
deal of its public relations effort is directed to
convincing the public about the relationship between
new technology and the economic and employment
advantages of an expanded industrial base. The
media has helped to advance this position through
its coverage of 'high-technology'.(18) Journalists,
for example, have described high-technology as 'the
key component to our future economic base', 'a force
for revolutionary change', 'the route to full
employment'. Articles on stock prices, product
sales, computer camps, career choices and local
economy, all refer to high-technology as a 'new
frontier'. Images of war, battle, struggle,

revolution, prevail. 'We need high-level
co-ordination to catch up and confront.' 'The
technological battle with the Japanese is really an
industrial equivalent to the East/West arms race.'
The media's coverage of technology is, in effect, a
call to arms, an appeal for public mobilisation to
support re-industrialisation and technological
progress.

Analysis of the media also illustrates the
widespread use of rhetorical devices in advertising
and press releases intended to improve the public
image of technology. Controversial technologies are
associated with jobs and opposition to technology
with crisis or disaster: 'We will freeze in the
dark', threatened the nuclear industry. 'There is
no life without chemicals' claims a chemical
company's ad. Pejorative terms such as 'radical',
'left wing', 'extremist', 'luddite', 'chemophobia',
'cancerphobia' debunk those who oppose technologies.
Environmental leaders are dismissed as 'out of touch
with the rank-and-file', representing 'the selfish
interests of the leisured class' or 'a minority
fringe'.

The remarkably lively concern about science
literacy in the United States, expressed in
governmental, industrial, and scientific circles,
can be viewed as still another effort to influence
public attitudes about technology. Concern about
educational reform has focused mainly on technology
and computer literacy as the basis of the future
prosperity and international dominance of the United
States. A report from the National Science
Foundation expressed the goals of the science
education movement:

Improved preparation of all citizens in the fields
of mathematics, science and technology is essential
to the nation's economic strength, military
security, commitment to the democratic ideal of an
informed and participatory citizenry, and
leadership in mathematics, science and technology.
(19)

Much of the discussion about the importance of
science education assumes that an educated public is
more likely to accept the conclusions of scientific
experts and to support technological change.

The media has picked up and disseminated these
themes. A New York Times editorial in 1983 claimed
that 'the battle for the future is being waged in
the classrooms and America is losing. Science and

mathematics, once the backbone of education, is now the soft underbelly.' The blame for America's educational problems is often placed on the '60s mentality' with its priorities of social relevance: 'The public programs of the 1970s, drained school coffers'. In 1982 Reader's Guide listed 28 articles on the problems of science education, mostly focusing on computer skills and the need to develop these skills as an economic resource. 'The quality of science and mathematical education must be boosted if the state is to have workers trained for the jobs of the future.' (20)

While advocating greater science and technology literacy, the Reagan Administration has also systematically sought, in various ways, to restrict public information. A major arena of conflict in the 1980s has focused on the 'right to know'. Disputes are taking place over the labelling of chemicals and the responsibility of industries to inform their workers, over the limits of trade secrecy, over the extension of military classification of information and the restrictions imposed by export controls, and over access to information under the Freedom of Information Act. In each of these areas, government policy has moved in the direction of greater restriction on the free flow of public information about technological developments.

Attitudes in the 1980s

The effect of these policies on public attitudes toward technology is not yet clear. But observations of controversies suggest certain changes, both in the nature of concern about technology and in the way it is expressed.

Overt dissent, that is active opposition to new technological facilities, is far less visible than in the 1960s in the sense that there are fewer demonstrations and public protests, fewer 'media events'. Most of the demonstrations of the 1970s had been organised over the siting of nuclear power plants, and the fact that no new plants are under construction has effectively removed this target. Concern about technology in the 1980s has come to focus mainly on two issues: acid rain and toxic waste disposal. The discovery of toxic waste disposal sites at Love Canal and Times Beach, Missouri attracted a great deal of media coverage and local mobilisation in the neighbourhoods of toxic dumps. However, this issue failed to involve a national constituency. Similarly, acid rain preoccupies committed environmentalists and has raised some public concern about the side effects of technology, but this concern has not evoked public protest. Today there is no equivalent to the anti-nuclear-power movement of the 1970s. Nuclear power had a very symbolic character in the United States that brought together many groups - women, Native American students, professionals - with very different agendas. No other technology has had quite the same mobilising effect.

Indeed, issues which were once the focus of protest, such as airport noise or the siting of Liquid Natural Gas terminals and other potentially hazardous facilities, are now virtually ignored. In the early 1970s the relationship between the military and university research mobilised active protest; today this relationship is developing once again, but it is no longer a public dispute. With the contraction of the nuclear-power programme because of reduced energy demands and high construction costs, this issue has also virtually disappeared from the protest agenda. The Star Wars debate over the proposal for a Strategic Defense Initiative has evoked hardly any overt public interest, suggesting perhaps a tendency toward technological fatalism. A few voices are questioning the implications of the computer revolution - its effect on employment, on the nature of office work, on the balance of industrial and university commitments. They have no constituency. Indeed, the scale and speed of acceptance of - indeed, demand for - computers at home and at work is remarkable. Perceived as a source of employment and an opportunity for entrepreneurs, computer science has become the most popular major field in American universities. Public attitudes towards the risks of technology remain ambivalent. Qualitative research suggests that many people are fearful and confused about technology when faced with unpalatable choices. Citizens living near toxic dumps feel that they must choose between their health and their homes. (21) Workers talk about the choice between their jobs and their health. (22) Those labour unions which had been active in occupational health issues in the chemical and high technology industries in the 1970s have turned back to bread and butter issue in the 1980s. Confronted with complex technical information about health risks, many people feel a sense of impotence. As a chemical worker put it: 'The company will give us information if it is required by law, but it is like handing a Stone Age man a rubber grip for his golf club. What the hell do we do with it?.' (23)

In changing the direction of environmental policy, the Reagan Administration interpreted the renewed interest in technological progress as support for relaxing environmental controls. But Reagan clearly misread the strength of pro-environmental attitudes in the United States. The membership of environmental organisations has grown in the 1980s. During the first three years of the Reagan Administration the Sierra Club grew from 180,000 to 335,000 members. The economic problems at the end of the 1970s and the efforts to equate employment with unfettered technological development failed to divert the consistent support for environmental protection. (24) The public has continued to demand public controls over industrial priorities in order to protect public health. The 1980 Gallup Polls found that a large majority of the public believed that government regulation was worth the extra costs and that economic growth was compatible with high environmental standards. Similarly, a 1982 poll by ABC News and the Washington Post found that most people did not want the government to relax anti-pollution laws.

The NSF followed up its 1979 public attitude survey with another in 1982. Little had changed; the public ambivalence about technology was still striking. While three-quarters of the respondents believed that problems could be solved by better technology; the same percentage also believed that science and technology 'often get out of hand, threatening society instead of serving it'.(25)

The most recent survey of public attitudes comes from California, a state that has traditionally led the nation in shaping public views. In 1984 the Field Institute found that Californians were positive in appraising technological advances, especially in areas of high technology which were felt to create new jobs and new tax revenues. However, 70 per cent of respondents were concerned about the use of toxic chemicals in these high technology industries. While 87 per cent of the respondents agreed that modern life was better because of the wonders of scientific progress, and 58 per cent believed that with fewer technological developments our standard of living would decline, some 70 per cent believed that technological developments were accompanied by considerable risks. By a three-to-one margin people believed that because of these risks more regulation may be required.(26)

Perhaps the most significant change in the American public's response to technology in the 1980s is the growing concern with its moral and religious implications. This reflects the considerable importance of religion in American life. According to a 1984 Gallup Poll, 90 per cent of Americans believe in God, 70 per cent belong to a church, and 60 per cent attend services at least once a month. Sixty per cent say they are more interested in religion than they were five years ago.(27) The NSF survey found that 67 per cent of Americans felt that deep religious belief was a major factor contributing to America's greatness as a nation. In a poll of the confidence of the public in the leadership of major institutions, organised religion fared only slightly less well than medicine and science in inspiring public confidence. It fared far better than political or legal institutions.

The strength of religiosity in the United States has brought issues such as abortion, the teaching of evolution, prayer in the schools and animal rights to the political agenda. Increasingly, public acceptance of science and technology seems to rest less on technical than on moral authority, less on questions of social need than on the protection of moral 'rights'. For example, animal-rights advocates are objecting on moral and ethical grounds to the use of animals in biomedical research. They argue that animals have inherent rights to a full life and that it is 'immoral' and 'unethical' to exploit them, whatever benefits might accrue to humans. The movement, springing up throughout the country, is estimated to involve millions of supporters. It is revealing of their moralistic attitude that these groups no longer talk in terms of animal welfare, but rather of animal 'rights'.

Similarly, moral issues increasingly pervade the discourse on technological and environmental risk - a discourse also permeated with the moral language of 'rights': the right to know, the right to health, the right to a clean environment.

Religious values are also influencing attitudes towards technological applications current research. For example, reproductive technologies such as in vitro fertilisation that are associated with abortion, or techniques of genetic manipulation that challenge the traditional definition of 'personhood', have become the most recent focus of legal and policy disputes. As research in biology touches on human evolution and the nature of human life, it too is vulnerable to religion-based objections. Well over half of the American public believes that scientists should not conduct studies directed to creating new forms of life.

Politically today, moral and religious values appear to be playing a more important role in the United States than in any other Western country. In a nation historically concerned with the relationships of the individual to the state, the preoccupation with individual rights is to be expected. And in a nation with a strongly embedded fundamentalist tradition, the periodic revival of religious values is not surprising. However, using these values as the main criteria for the acceptability of science and technology is limiting - far more limiting than the public concerns that had inspired social and environmental movements of the last decade.

Notes

1. I appreciate the extensive comments provided by Robert Cameron Mitchell.

2. For case studies of these disputes over technology see Dorothy Nelkin (ed.), Controversy: The Politics of Technical Decisions, Beverly Hills: Sage Publications, rev. ed 1984.

3. For a bibliography see L.K. Caldwell and Siddiqui, Environmental Policy, Law and Administrative Environmental Studies Program, Indiana University, Bloomington, Indiana, 1974.

4. Georgine Pion and Mark Lipsey, 'Public Attitudes Toward Science and Technology', Public Opinion Quarterly, 45, 1981, 303-16.

5. 'Public Attitudes Toward Science and Technology' in National Science Board, Science Indicators, 1980, USGPO 1981; and Kenneth Prewitt, 'The Public and Science Policy', Science, Technology and Human Values, Spring 1982, 5-14.

6. Daniel Yankelovitch, 'Changing Public Attitudes to Science and the Quality of Life', Science, Technology and Human Values, 7, no. 39, Spring 1982, 123-9.

7. Robert P. Quinn and Graham Staines, The 1977 Quality of Employment Surveys, Ann Arbor: Institute for Social Research, University of Michigan, 1979.

8. Louis Harris Poll, 1980.

9. Baruch Fischoff, Sarah Lichtenstein et al., Acceptable Risk, New York: Cambridge University Press, 1981.

10. Richard Nelson, The Moon and the Ghetto, New York: W.W. Norton, 1977.

11. Peter House and Roger Shull, The Rush to Policy, Washington, D.C.: USGPO, 1984.

12. For a review of assessment procedures and a history of the Office of Technology Assessment, see F. Hetman, Society and the Assessment of Technology, Paris: OECD, 1977.

13. Decision Research: Improving the Societal Management of Risk, National Science Foundation Project Report, June 1979, pp.2-4.

14. For a review of the literature, see Dorothy Nelkin, 'On the Social and Political Acceptability of Risk', Impact of Science on Society, 4, 1983, 225-31.

15. For a review of participatory efforts in the US, see Dorothy Nelkin, 'Science and Technology Policy and the Democratic Process' in National Science Foundation Five Year Outlook, NSF 80-30, 1980, 483-92.

16. George Keyworth, US Congress, House Committee on Energy and Commerce, Subcommittee on Commerce, Transportation and Tourism, 98th Congress, 1st session, 17 March 1983.

17. See discussion in Chapter 6 of David Dickson, The New Politics of Science, New York: Pantheon, 1984.

18. House and Shull, The Rush to Policy.

19. Dorothy Nelkin, Science, Technology and the Press, New York: W.H. Freeman (forthcoming).

20. National Science Foundation, <u>Science and Engineering Education for the 1980s and Beyond</u>, Washington, DC, October 1980.

21. Dorothy Nelkin, <u>Science, Technology and the Press</u>.

22. Adelina Levine, <u>Love Canal: Science, Politics and People</u>, Lexington, MA: D.C. Heath, 1982.

23. Dorothy Nelkin and Michael S. Brown, <u>Workers at Risk: Voices From the Workplace</u>, Chicago: University of Chicago Press, 1984.

24. Quoted in Michael S. Brown, 'The Right to Know: Hazard Information and the Control of Occupational Health Risks', PhD thesis, Cornell University, 1984.

25. Robert Cameron Mitchell, 'Public Opinion and Environmental Politics in the 1970s and 1980s' in Norman Vig and Michael Kraft (eds), <u>Environmental Policy in the 1980s</u>, Washington, DC: Congressional Quarterly Press, 1984.

26. National Science Foundation, <u>Science Indicators 1982</u>, Washington, DC: USGPO, 1983.

27. Field Institute Release 1232, 26 January 1984.

28. George Gallup, Jr, <u>1984 Religion in America</u>, Report of Gallup Poll, 1984.

Chapter Four

JAPANESE ATTITUDES TOWARDS SCIENCE AND TECHNOLOGY

Sigeki Nisihira and Ronald Dore

The Historical Background

One ought not to be too surprised at Japan's recent successes in the science and technology field. Where else in Asia in the 1790s would one have found doctors struggling to learn Dutch in order to master the only Western anatomy book they could get hold of? One might have found princes in other places gazing in wonder at a machine which generated static electricity and caused a spark to leap across the void. But where else would one have found a retired statesman writing in his diary this sort of put-down:

People take enormous delight in playing with this machine. We go too far, I think, in marvelling at these machines that originate in Western countries. What foolishness it is, after all, to gape wide-eyed at this, and yet not to wonder at the way a piece of amber picks up light pieces of fluff when you rub it; it is the same fire-spirit at work. ... (1)

Or the same statesman, on another occasion, recording how his scientific curiosity had led him to sit in the garden all day with a bamboo stick down well attached to a float, to determine once and for all whether there were tides in wells. (2)

True, there were moments in the 1870s when the local populace was less than wholly welcoming to innovations. The appearance of telegraph wires in some areas was a cause of riots which, prefectural governments reported, were caused by peasant ignorance and suspicion. The peasants connected the new contraptions with the new conscription laws called the 'blood tax', and thought were going

66

to be used to transmit the extracted blood of their children to dye red blankets in Tokyo.(3)

Was this wholly true, or were the rioting peasants simply expressing entirely rational objections to having their sons taken away to fight in wars which they saw as none of their business - and simply taking out their fury on anything which was the government's - schools and telegraph poles included? Modern historians explain the 'misled by false rumours' stories as a typical expression of the contempt shown by samurai administrators for the vulgar masses, and their unwillingness to allow that these masses might have honourably rational motives. (4) Given what one knows from many countries about the views of contemporary administrators on the anti-nuclear protests of contemporary peasants, this is a not wholly implausible story.

Certainly, historians have not found instances in Japan's last century of the sort of resistance to science-based innovations which springs from fear of the unknown, threats to established religion, superstition or conservatism of the bone-headed kind. From the tremendous campaigns 80 years ago against the Ashio copper mine, whose smelter's effluents ruined vast areas of rice crops and left whole villages to starve (5), through to the great Minamata and Niigata and Smon pollution scandals of the 1970s, (6) protesters have generally had strong personal-interest reasons or public-spirited denunciation-of-injustice reasons for their protests. There have been no sectarian resisters to vaccination or blood transfusion. There have been no Leavises in Japanese academia. Herman Kahn sells more in translation than Rachel Carson. The recent campaigns against local nuclear power stations do, as elsewhere, utilise fears of effects like radiation which are both incomprehensible and insidiously invisible. They build up resentment of the remote, faceless, heartless, calculating scientist offering take-it-or-leave-it expertise. But those who seek to keep these sentiments alive, like the strontium-90-in-your-milk campaign leaders who finally got us a test-ban treaty in the 1960s, do have reasoned political purposes in view, and not any general antipathy to science and technology as such. (In atom-bombed Japan 'nuclear never' is the last-ditch citadel of all who oppose Japan's rearmament, and the vague impression that 'nuclear is indivisible: think of Hiroshima' clearly suits their book.) Even the students still breathing defiance against the police and the world airlines from inside their fortress towers at Narita airport are not opposed to jumbo-jet technology in itself: only to its being used (according to some obscure logic, more especially there than anywhere else) in the service of world capitalism and US imperialism.

And in recent years the degree of public attention devoted to science and technology has intensified by leaps and bounds. So it has everywhere, of course, as the 'microprocessor revolution' catalyses public awareness of the effects of long-standing trends - the acceleration of the process of innovation, the increasing interpenetration of national economies and intensification of international trade competition, and the increasing importance of innovation and product quality in that competition. But the effect seems greater in Japan than in most other places, for one particular reason.

National Goals

'Catching up with the West' has for over a century been defined as the nation's dominant goal - so defined in the speeches of leaders, in the national textbooks used in primary schools, and in the implied questions built into newspaper reports on foreign countries: 'How do we rate compared with them?' 'how well do they think of our films/ motorbikes/politics/food/poetry, etc?'. To achieve the goal of catching up the Japanese were prepared to learn walking before they tried running. They had the humility to sit - deferentially, though often smarting with resentment - at the feet of, often arrogant, foreigners who had knowledge they needed - in the first 20 years of the new regime inviting large numbers of them to teach in Japanese colleges at salaries comparable with that of the Japanese Prime Minister. They were honest imitators - delighted, of course, when a Noguchi made a genuine advance in immunology or a Toyota invented a loom that the great Platts of Lancashire wanted a licence to copy, but for the most part directing their brain power to the job of absorbing others' technology. So much so that they still, themselves, more than half believe the 'nation of mere imitators' stereotype which the outside world came to attach to them. When the Institute of Statistical Mathematics did its 1983 survey and asked respondents to select from a list of adjectives which might be used to characterise the

Japanese people, any ones they thought appropriate, only 11 per cent of Japanese respondents chose 'creative originality' - compared with 69 per cent who chose 'hard-working'; 47 per cent 'punctilious about manners'; 42 per cent 'kindly'; and 30 per cent 'idealistic'. That 11 per cent was still higher than the 7-8 per cent recorded in the four previous surveys, 1958-73. (7) (Note 7 gives detailed references to all the surveys referred to in this chapter.)

Until recently 'catching up with the West', in the sense of joining the pioneers at the frontiers of knowledge, was not an end in itself. For the last quarter of the nineteenth century the purpose was so to impress foreigners with the modernity of the country - with the nation's ability to run railways and hospitals and diplomatic receptions, to build a capital like London and to establish courts and parliaments - that Japan could negotiate its way out from under the humiliating 'unequal treaties' and end the extraterritorial zones over which, foreigners insisted, Japanese police and courts were not civilised enough to have jurisdiction. The achievement of equality in this sense was soon followed by victory in the Russo-Japanese war, and the sudden realisation that perhaps Japan really could compete in the imperialism league and take the Asian empires of the European powers back into Asian hands where they belonged. Better science and more technology were necessary for the strength needed to pursue that goal. They were equally necessary for the alternative peaceful strategy which for a while after 1918 competed with the goal of Imperial expansion - the strategy of joining Japan's 1914-18 allies as equal partners in the joint regulation of a world which had outlawed aggression. Neither internal nor external factors gave the latter alternative much of a chance.

Defeat in 1945 ended all thoughts of such military ambitions for at least two generations. But it did not end the 'catching up' - or rather now the 'catching up/surpassing' - preoccupation. It simply altered the way the media, politicians' speeches and bath-house conversation defined the dimensions in which the race was being run. For a while, in what was, after all, the first nation to renounce war in its Constitution, there was a strong sense that the best thing to be a shining example of was a peace-loving democracy. Then, in the 1950s, the economy began to revive. By 1964, when Japan finally regained full world citizenship and readmission to all international organisations, she was simultaneously impressing the world by the efficiency with which Tokyo organised the Olympics, and achieving economic growth rates of a level which made everyone else envious. For a decade the Japanese enjoyed a sense of being world leaders in the economic growth-rate stakes. The oil crisis and the subsequent gearing down of growth, together with the impressive rise, as growth-rate leaders, of the other Asian Newly Industrialising Countries (NICs), somewhat took the shine off that glow of satisfaction. But obvious alternative definitions of the competition for international prestige soon offered themselves. The American sociologist author of Japan as Number One(8)(over a million copies of the Japanese translation sold) told them that they ought to be proud of their social cohesion, their peaceful readiness to compromise, their ability to run industrial policy, their mature attitudes to authority, their caring concern for deviants and unfortunates - and their ability to excel in industrial research. This was just about the time that haado (hardware, material technology, material values) and sofuto (software, social organisation, quality-of-life values) were coming into wide use in everyday vocabulary. The media attention given to the government's annual Science and Technology White Paper brought an awareness that by the end of the 1970s Japan had a larger share of foreign patents registered in the US than had any other nation - over one quarter of the total. Perhaps the 'imitator' stereotype was wrong after all? Perhaps they could compete in the biggest league? By the mid-1980s (by 1985, on recent trends, the Japanese will have a larger share of all patents registered in the UK than will UK citizens) there is a general world consensus that in some industries - notably electronics - the race is between the US and Japan, with Europe nowhere in the picture. And - if one can still, in this rapidly changing world, use the idioms of the 1960s - neither the Americans nor the Japanese now have any doubt (and their assertions mutually reinforce each others') that science and technology is where 'number oneness' is at. And 'number oneness' counts in Japan, as it does in any country experiencing a change in its position in the international pecking order - as the British are becoming conscious as they slip inexorably down the ratings.

There have been a number of public opinion surveys in Japan reflecting this preoccupation. In

1979 a Yomiuri poll found that 53 per cent thought that Japanese science and technology had 'caught up' with that of 'Western advanced countries', 27 per cent thought it had not, and 7 per cent thought it was ahead. A later survey would probably show a different balance. The 1983 'national character survey' of the Institute of Statistical Mathematics showed a general increase in national self-confidence. Ever since 1953 they have been asking the simple and provocative question: 'do you think the Japanese people are superior to Westerners, or inferior?' Thirty-five per cent of the sample have consistently rejected the possibility of generalising or said that there is no difference, and there have been a (diminishing) number of 'don't knows'. But among the others the superior/inferior balance has gradually shifted from 2/28 in 1953 to 53/8 in 1983. Another 1983 survey by Yomiuri was directly comparative. It asked about the relative prowess of the Americans and the Japanese in specific fields of technology. Seventy-five per cent conceded, and only 9 per cent disputed, American superiority in space rocketry. But on computers 43 per cent voted for Japan (25 per cent for the US) and on machine tools 59 per cent for Japan as leader as against 16 per cent for the US. 'What do you think you can most be proud of as a citizen of your country?' is a question which was used in a 1983 international survey organised by the Youth Problems Bureau of the Japanese Prime Minister's office. The samples were of 18-24 year olds. The proportion answering 'our science and technology' was 46 per cent in Japan, second to the 49 per cent who said 'our history and cultural inheritance'. There were parallel studies in a number of other countries, and of all the national samples Japan had the second highest proportion choosing the science answer. (In the highest, the USA, it was 68 per cent; Germany, 44 per cent; UK, 41 per cent; Sweden, 36 per cent - see Note 7 for a detailed reference to the survey.)

But all this 'rivalry consciousness' does not, it should be emphasised, make the Japanese secretively nationalistic. Perhaps because they now have confidence in their capacities, they turn out to be more in favour of co-operation in science and technology than are other national populations. Yomiuri used Gallup International for a survey which asked about the way forward in the development of high technology such as robots and computers. In the British and American samples 60 per cent were

for going it alone, 30 per cent for international
co-operation. In the Japanese sample the balance
was strikingly different - 30 and 51 per cent
respectively. The other two countries were France,
41 (go it alone) as against 47 per cent; West
Germany, 46 against 37 per cent.

The Media Today

Whether fuelled by rivalry consciousness or hopes of
international co-operation, or just concern about
domestic effects, a heightened awareness of
developments in science and technology is reflected
clearly enough in the media. The December 1983
index of the leading, and by no means elite,
newspaper (6 million circulation), Asahi, shows
nearly 50 references to high technology, new
products or computers (including a feature article
on 'high technology and the Japanese garden' and
another headlined 'Japanese high technology: a
threat to the future of the Japanese-American
security system'). The index for the corresponding
month 10 years earlier, full of references to oil
and energy, has only a single reference in these
categories - to a government White Paper on the
future of computers.

How much these articles are read, of course, is
another matter. In a 1976 government survey, 15 per
cent said that they were very interested in reading
news about science and technology, and another 47
per cent said they were 'fairly interested'. Seven
per cent said they had specialist books and
magazines, and 8 per cent that they went to
scientific exhibitions and films. The only more
recent comparable survey (1981 Mainichi) asked
specifically about news items on computers; 12 per
cent and 44 per cent respectively said that they
'frequently' or 'occasionally' read them.

Certainly, for the book buyer, there is no
shortage of things to read on science and
technology. Here is a list of titles copied down
during a by no means exhaustive search of one corner
of a central Tokyo bookshop in the spring of 1984 -
admittedly a bookshop round the corner from the
Ministry of International Trade and Industry and
probably not, therefore, typical of Japan's
half-million bookshops.

Industries of the Future: the need to get it right
The Realities of Venture Business
The Computer Data Era

Technology, Lifeblood of the Nation: The End of
the 'Trade-as-the-life-blood' Era
The New Media: A Reader
Warning Bells for Japan, the 'Quasi Great Power'
New Frontier Enterprises
The Challenge of Frontier Technology
Managements Strategies for the New Media Era
How to make out in the Information Society
Towards a Networked Society
Frontier Technology: A Bedside Reader
Informatisation Leaps Ahead
The New Alchemists of Silicon Valley
No World for the Technological Innocent!
The New Media: a Dictionary of Specialist Terms
The Japan Economic Times High Tec Dictionary

The fact of the boom is claer; its
manifestations protean. The Japanese reader
certainly has the means to keep himself well
informed. Above all, perhaps, the last item
deserves mention as a godsend to those who really
want to understand some of the more arcane sections
of their newspapers. Opened at random, a typical
pair of its 400 pages yields lucid explanations of:
STOL planes; subscription TV; SUS316; separative
work units; total company automation (TCA); TDMA
(time division multiplex access); tracking data-
relay satellites; Train a Grande Vitesse (TGV);
tumor necrosis factor; take-over bid. (9)
'New media', which recurs in some of these
titles, is explained in a recent (31 August 1984)
Financial Times article.

'New media' is the current buzz word in Japan;
'OA' (office automation) has been bypassed as the
Japanese electronics industry perpares for a
service economy based on home banking, home
shopping, video-text information services,
advanced telecommunications services and local
community cable.

Hi-tech hype has, of course, become a standard
service device of governments and business
communities trying to introduce an element of
buoyancy into an economy which is otherwise screwed
down by deflationary macroeconomic policies.
Certainly, the buoyancy of the Japaneses economy
owes a good deal to the way in which concern with
new technology translates into investment and into
a responsive market for new gadgetry. The rapidity
with which automatically opening sliding doors

spread to every little corner store a decade ago
has few parallels in other countries, even
countries where the temperature-control properties
of such doors offered equivalent advantages. As
consumer polls show, however, the Japanese public
is not indiscriminate in its love of novelty. A
Mainichi poll in 1981 about the then-new video-
discs found only 5 per cent who intended to get
one: 57 per cent who did not want one, and 36 per
cent who were going to wait and see. A 1981 survey
(Mainichi) asked about the prospects for
telecommuting. (That is, employees - or contract
workers - whose work consists entirely of on-line
keyboard manipulation, working from home with a
terminal linked to an office.) Thirty per cent
liked the idea; 66 per cent did not.

The Computer

Computers and current attitudes towards them have
been the subject of numerous other surveys. In
1970 (an NHK survey) only 5 per cent of respondents
(9 per cent of those at work) had anything to do
with computers directly or indirectly. In 1981 (a
Mainichi survey) 20 per cent were using computers
at work or at play and another 37 per cent
expressed an interest in having a personal
computer. When offered a choice of words to
characterise their mental associations with
computers, the majority chose words with a positive
connotation - speedy (52 per cent), convenient (47
per cent), progress (44 per cent), accuracy (36 per
cent). The words with negative connotations chosen
were: expensive (17 per cent), difficult (14 per
cent), crime (11 per cent), breakdowns (10 per
cent), inconvenience (1 per cent).

Another (government) survey, aslo in 1981,
produced the pattern of assessments of the impact
of computers on society shown in Table 4.1.

Japan

Table 4.1: Assessment of the Impact of Computers (in per cent)

	Agree	Hesitantly Agree	Disagree	Other
Indispensible to modern society	59	28	7	6
Have made life more convenient	62	27	4	7
Wether beneficial or not depends on who uses them	44	26	11	19
They threaten to invade privacy	20	22	37	21
They make one feel vaguely uneasy	30	26	34	10

The overall impression is positive, but with a good deal of discriminating awareness of the dangers to be guarded against. In another, Yomiuri, survey (1981), 42 per cent agreed and 37 per cent disagreed that 'information is getting centrally controlled by computers and privacy will be invaded', though only 32 per cent said in a Mainichi survey the same year that they 'felt uneasy about the invasion of privacy with the increasing use of computers' (and 64 per cent denied it). 'Invasion of privacy' was also a reason given by 32 per cent of the respondents in the latter survey to a question about the government's decision to have a single 'citizen number' for all administrative purposes - the so-called 'number on every back'. Twenty-seven per cent objected on the grounds of its strengthening government control: 74 per cent in total were against the system and only 17 per cent in favour - for the saving in administrative costs (14 per cent) and/or the saving of citizens' time and trouble.

The question arises: how far are these assessments based on accurate knowledge of what computers do? The 1970 NHK survey provides some data on knowledge at that date. Respondents were offered nine activities and asked which of them involved computers. In order of the frequency with which they were chosen they were: calculations in banks and offices (51 per cent); spaceships (50 per cent); election forecast (50 per cent); train-ticket booking (44 per cent); factory automation (39 per cent); filing in government offices (37 per cent);

university research (29 per cent); paying telephone
bills (22 per cent); live baseball TV broadcasts (18
per cent).

The same technique of offering a multiple
choice of replies was used for the question: what
can computers do better than human beings?
Sixty-three per cent chose 'accurate calculations'.
The other items were 'large volume memory' (48 per
cent); 'large volume calculations' (46 per cent);
'immediate responses to any questions' (29 per
cent); 'invariably gives correct answers' (22 per
cent); 'operates even without instructions' (13 per
cent). Ten per cent thought at that time that
computers could 'produce original ideas'.

On the direct impact of computers on the people
using them, the Ministry of Labour carried out a
large-scale survey in November 1983 covering white
collar workers in about 6,000 establishments
employing more than 100 workers. Sixty per cent of
the women and 47 per cent of the men were using
somekind of computer or word processor. Half
thought the automation of the office had no great
difference to the ease or difficulty of the work,
and of the rest, the balance was in favour (32 per
cent easier; 19 per cent more difficult). Of the
women working with VDUs, 71 per cent complained that
they got tired eyes; and 59 per cent stiff necks.
One question probed the extent to which workers were
confident of keeping up with the new technology, and
as shown in Table 4.2, a quarter of the women were
not.

Table 4.2: Do you think you will be able to keep up
with the new technologies? (in per cent)

	Male	Female	Total
Quite confident	20	6	15
Will manage somehow	64	69	66
Not very confident	15	24	18
Other	1	1	1
Total	100	100	100

Images of Science and Technology

Attitudes towards specific innovations like
computers are presumabley influenced by certain,

probably ill-articulated, general attitudes towards
science and technology. Some surveys have tried to
get at these; one by the government which asks for
associations of science and technology with one or
other of a set of adjective pairs, and two
successive ones by Yomiuri which required choice
between contrasting general characterisations.
These are shown in Table 4.3.

Table 4.3: Images of 'Science and Technology'
(answers in per cent with Don't Knows excluded)

Government 1976

Hope	66	Despair	5
Satisfaction	30	Dissatisfaction	28
Amazement	38	Perfectly according to expectation	30
Excitement	14	Coolness	41
Peace of Mind	15	Anxiety	53
Clarity	8	Hard to understand	56

Yomiuri 1979 and, in parentheses, 1981

Bright prospects	72(74)	Far from bright prospects	14(15)
Familiar in daily life	37(36)	Remote from daily life	42(44)
Serves peace	56(52)	Leads to war	22(30)
Brings progress	73(73)	Leads to abuses	11(15)

These general sentiments do seem to fluctuate
with the mood of the times. (The growing world
concern with nuclear weapons, coming in particular
from the installation of Cruise and Pershing
missiles, probably explains the one significant
shift in the last set of figures on the war/peace
question.) An NHK 1970 survey brought out many more
who endorsed favourable characteristics of science
and technology (everyday convenience, better
economic and managerial planning, relief from
drudgery) than unfavourable ones (central control,
having to work harder, being controlled by
computers, etc.) But in 1973, the year before
preoccupation with the oil crisis and inflation took
over, survey results showed theeffect of several
years of concern with environmental pollution
(during which some of the world's strictest control
standards were imposed) and of public talk about
'due attention to social expenditure to correct

unbalanced emphasis on economic growth'. Fifty-five
per cent were negative about the impact of science
and technology: 5 per cent worried about
unemployment; 15 per cent about pollution, and 35
per cent had a general sense that the 'warm-blooded
human-ness' of life was being destroyed. (The 32
per cent who gave positive replies instanced 'work
easier' (17 per cent); 'more comfortable living' (9
per cent); 'more leisure' (6 per cent).)

Three years later, in 1978, the Science and
Technology Agency found 55 per cent agreeing that
the ill-effects of science and technology outweighed
the good, while 36 per cent thought the pace of
scientific change too fast (30 per cent 'about
right' and the rest didn't know.) As to what,
precisely, brought the ill effects, respondents were
asked to comment on three particular phenomena shown
in Table 4.4.

Table 4.4: Effects of Science and Technology (in per
cent)

	Good effects predominate	Ill effects predominate	Effects Ambiguous
Diffusion of automobiles	34	28	38
Diffusion of alkaline detergents	25	31	34
Rising standards of consumption	25	30	45

The 'warm-blooded human-ness' or 'richness of human
feeling' (ningensei) theme is prominent in many of
these surveys. The national-character survey found
between 1953 and 1973 a growing number (from 30 to
50 per cent) agreeing the statement: 'with the
development of science and technology, life becomes
more convenient but at the same time a lot of human
feeling is lost'. This was, of course, one cliché
version of the 'economic animal' theme, the charge
of 'too much emphasis on economic, not enough on
social development', which was a main motif of late
1960s and early 1970s politics. The popularity of
this theme has fallen off since 1973. Only 43 per
cent endorsed it in 1978, and 47 per cent in 1983.
Other reflections of this concern may be seen in
Table 4.5, containing the results, first of a
Yomiuri-Gallup joint survey, three times repeated in
Japan, and of a 1981 Mainichi survey. Both asked

about the effects of computers and robots.

Table 4.5: Effects of Computers and Robots (multiple answers allowed)

| | Japan | | USA |
	1968	1981	1983	1983
Life becomes more convenient	22	21	12	24
Work gets easier	13	13	11	20
Jobs are destroyed	12	17	34	38
We become slaves of machines: loss of 'human-ness'	36	39	37	19
Other, or no answer	17	10	6	4
Totals	100	100	100	105

Effects on	Positive		Negative		Other
The urge to work	Intensified	9	Weakened	58	33
Personal relationships	Enriched	11	Impovorished	60	29
Creativity	Enriched	25	Impoverished	50	25
Ways of thinking	More variety	31	More uniformity	45	24

Employment

The concern about jobs, Lable 4.5 suggests, increased suddenly in 1983. Even earlier, direct questions had shown some underlying concern. A 1970 Mainichi survey found 45 per cent who thought their job might eventually be taken over by computers or machines, and the same percentage who did not. In 1981 (in a survey done by Mainichi again), there was still a majority in favour of the introduction of robots into factories (50 as opposed to 44 per cent) but 33 per cent gave 'increased unemployment' as a reason for opposition, and only 1 per cent gave a probable eventual increase in jobs as a reason for being in favour. The Yomiuri survey in the same year used a translation of a European Community survey which had asked: 'here are certain kinds of fears which are sometimes expressed about the future of the world we live in. For each one I would like you to tell me if it is something which really concerns you or worries you or not.' In response to 'increase in unemployment as a consequence of the contraction of jobs', almost the same number said they were not worried - 29 per cent in Japan, 28 per cent in Europe. Those who said yes, they were worried, numbered 67 per cent in Europe, 59 per cent in Japan. As is usual the Japanese proportion of 'don't knows' was the higher one.

The Medical Field

The menical field is one which brings out in most populations a most complex mix of hope, fear of the unknown, and sense of potential invasion of individual integrity. As for hope, 'Cancer and other intractable diseases' headed the list of fields in which the respondents to a 1976 survey saw science and technology as offering prospects of significant advances. (The others were earthquakes and typhoons, 37 per cent; pollution, 37 per cent; agriculture and food, 25 per cent; new energy sources, 24 per cent.)

The evolution of attitudes towards test-tube babies is of interest in this context. When the first in vitro fertilisation succeeded in Britain in 1979, 35 per cent of Japanese respondents were opposed to it on the grounds that children were a gift from heaven and one ought not to meddle in the process. By the time the first successful Japanese birth took place in 1982, 47 per cent held this view, (Both Yomiuri surveys.) Another 33 per cent were opposed because the moral, legal and religious problems which it posed nad not been settled. Those who were opposed because of possible medical dangers had been 24 per cent in 1979 and fell to 17 per cent in 1982. But altogether the proportion who welcomed it as good news for childless couples fell from 24 to 17 per cent. On the other hand, a later (1984) Mainichi survey, confined to wives under 50, found 62 per cent who thought that it was a good thing that childless couples could now have babies through test-tube fertilisation, and only 32 per cent opposed.

As for the 'fear of the unknown' aspects, the Japanese seem somehow less prone to them than Europeans, to judge from another question in the Yomiuri-European Community Survey quoted earlier, as shown in Table 4.6.

Table 4.6: Are You Concerned About the Risk that the Use of Some New Medical or Pharmaceutical Discoveries May Severely Affect the Human Personality? (in per cent)

		Yes	No	Don't know
Japan	(1981)	41	43	16
European Countries	(1979)	53	38	9

As compared with Canadians, Japanese are rather less prepared to register as willing donars of organs after their death - 23 per cent in the 1982 Yomiuri survey and another 18 per cent who would do so to benefit friends and relatives compared with 53 and 59 per cent who had signed or were willing to sign donar cards in successive (1978 and 1983) surveys of the Canadian Institute of Public Opinion. Only 12 per cent of the Japanese sample rejected the idea outright, however; 42 per cent said they 'would have to see at the time'.

Nuclear Energy

Japan was the country about which - perhaps even in which - the term 'nuclear allergy' was invented to describe an antipathy towards all things associated with nuclear fission, whether or not rational assessments of the risks they involve have or can be made. Hiroshima and Nagasaki remain potent symbolic weapons in the armoury of those who advocate Japan's distancing itself from the US and its nuclear weapons - powerful enough to have caused successive governments to maintain a non-nuclear stance, declaring that its US ally must refrain from bringing nuclear weapons on to Japanese soil or into Japanese harbours. (And the well-grounded suspicion that American aircraft carriers do not observe this rule causes periodic political uproar.)
    When the first nuclear-powered ship was built, it had the misfortune to leak a little radiation on its first trials, possibly thereby endangering the mass of fishermen's boats which were buzzing it in protest about its very existence. Thereafter for 10 years, only massive central government subsidies could persuade the mayor of a quiet harbour town to give it mothballed shelter, and although it has finally moved out, it still has not had proper trials.

But the vociferous minority is a minority, and the majority who incline in favour of the development of nuclear power stations does not seem to be a particularly volatile majority, if the impact of particular incidents is traced through the opinion polls. Three Mile Island was, of course, reported with great prominence in Japan, and the controversy in the press was slow to die down. So, to a much greater extent was a radiation leakage at the Tsuruga power plant, involving the release of irradiated sludge into the bay and exposure of 100 employees to radiation. The impact was reinforced in this case by the subsequent piece-by-piece revelation of the design faults and operating errors which had been responsible. But a poll 6 months later showed only a very slight shift against nuclear power. There is, one might say, a state of accepted and distanced polarisation on the issue. Those who are pro-nuclear are a majority and that majority support for nuclear power seems pretty stable, not much disturbed by newsworthy incidents like Three Mile Island. The number who express anxiety about the safety of nuclear power stations does seem to have increased between 1981 and 1984, as a result, perhaps, of changes in US opinion, but this does not seem to have much affected positions taken on whether or not the nuclear-power programme should go ahead. There is nothing particularly Japanese, of course, in the discrepancy between the numbers in favour of nuclear power in principle, and the numbers willing to have a nuclear power station in their neighbourhoods.

Facing Forwards

The overall impression which emerges from these surveys is of a people who are facing forwards, and doing so with a certain confidence. The deep structure of attitudes on these matters is always hard to get at, and nowhere, perhaps, more than in Japan where 18 million reader households are shared between a mere three major newspapers almost identical in style and content, and in the frequency of their resort to the vague and solemn cliché (e.g., 'science and technology bring great benefits, but can lead to a loss of "warm-blooded human-ness"'). The Japanese are very prone to know what they ought to think about science and technology, though more likely to have diverse and independent views the more concrete the issues.

There are, as suggested at the beginning of this chapter, historical ingredients in the generally positive attitudes towards science and technology. It may also be that, for the reasons enumerated earlier, the Japanese are more aware than many other peoples of the international implications of developments in science and technology - of the effect on trade competitiveness and international prestige. But for the most part, as in other populations, attitudes are primarily determined by the impact of innovations on personal circumstances, and it is reasonable to suppose that the generally positive attitudes revealed by Japanese surveys reflect the cicrumstances of a society with a high overall growth rate where more people find themselves in growing sectors than in declining ones. Stagnant societies where innovation is commonly part, not of expansion but of a cost-cutting exercise in process improvement, may well be less disposed to welcome the changes which new science and new technology may bring.

Notes

1. S. Matsudaira, Taikan zakki (Jottings of a Retired Official, (1793-1800) in Kokumin Tosho Kabushiki Kaisha, (ed.), Nihon zuihitsu zenshu (Collection of Miscellaneous Writings), Tokyo 1927-30, vol.14, p.169.

2. Ibid., vol.14, p.214.

3. E.H. Norman, Japan's Emergence as a Modern State, New York, IPR, 1940, p.73.

4. S. Kidota, 'Ishinki no nomin-ikki ('Peasant Revolts after the Restoration'), Iwanami Shoten, Nihon Rekishi, vol.15, Tokyo, 1962, p.203.

5. T. Hayashi, Tanaka Shozo, Tokyo, Nigatsusha, 1974; K. Strong, Ox Against the Storm, Tenterden, Kent, Paul Norbury, 1977.

6. See, e.g., M. McKean, Environment Protest and Citizen Politics in Japan, Berkeley: University of California Press, 1981.

7. The surveys quoted in this paper include:

   (i) The regular sample surveys conducted by:
       The NHK (The Japan Broadcasting

Corporation);
The Public Opinion Division of the Prime Minister's Office;
The Mainichi newspaper;
The Yomiuri newspaper;
The Asahi newspaper.
The first two are published in the monthly official publication:
Gekkan Yoron Chosa.

(ii) A seven-times repeated quinquennial survey of the national character Tokei Suri Kenkyujo (Institute of Statistical Mathematics), Kokuminsei no kenkyu: Dai-nanakai Zenkoku Chosa (A study of national character; Seventh national survey), Tokyo, 1984.

(iii) A special international survey of youth sponsored by the Youth Development Head-quarters of the Prime Minister's Office, English summary published by the Head-quarters, Japanese Youth, 1984.

8. E.F. Vogel, Japan as Number One: Lessons for America, Cambridge, Mass., Harvard UP, 1979.

9. Nihon Keizai Shimbunsha, Nikkei Hai-tekku Jiten. With great charm, the entry on 'take-over bids', after explaining briefly the nature of tender offers, time limits etc. in the US and UK, continues: 'The system was also introduced in our country with the 1961 revision of the Securities Law, but the idea of taking over a company by the power of money was too dry (dorai) to have achieved full acceptance here.' [The word for 'take-over' is that also used for hijacking an aircraft.]

Chapter Five

PUBLIC ACCEPTANCE OF NEW TECHNOLOGIES IN THE
FEDERAL REPUBLIC OF GERMANY
Thomas Peter, Peter Mann and Georg Thurn

Acceptance of Technology as Topic and Problem

In common with the experience of most Western
industrialised nations, the public in the Federal
Republic of Germany (FRG) became increasingly
concerned with the topic and problem of technology
acceptance during the course of the 1970s. It is
true to say that the problem of the acceptance of
technologies, with their accompanying risks, has a
much longer tradition, as have the varied measures
of individual social subsystems in securing
technology acceptance. What is new and
qualitatively different, however, are the objective
risk potentials of modern technologies, their
subjective perception by the members of society
and, as a consequence, higher demands on risk
management in society as a whole.
Since the 1970s the new technologies have
become more and more characterised by their
tendency towards global pervasiveness and
irreversibility, and by their growing potential for
catastrophe. With regard to the changed perception
of new technologies, in broad sections of the
population there is a more mature consciousness of
the social risks of new technologies, and a
sensitivity towards the aims and paths of
technological progress, which no longer bear
qualitative comparison with previously existing
critical attitudes towards technology. Finally,
another new factor is a raised level of expectation
of risk management and demands on policy-making,
especially the necessity to adopt preventive
measures rather than react incrementally to each
situation. Thus, in comparison with traditional
technologies, it is much more difficult to procure
the acceptance of new technologies, and this in

turn intensifies the 'pressure of tasks and problems'.(1)

The silent consensus which accompanied technological progress in the 1950s and 60s has been disturbed, especially by the conflicts over nuclear energy in the early 70s. Since then public debate has been characterised by a growing awareness of risk and an increasing politicisation regarding the risks of technology. In the course of the debate it has become clear that technology is no longer unquestioningly accepted, the ability of scientists and experts to solve problems is critically examined, and doubt is cast over the legitimacy of political decisions on matters of technology policy.

Generally, this basic characterisation also fits the present situation. The debate has, however, lost much of its bitterness. Whereas the conflicts over nuclear energy in the 1970s reached proportions that some observers have compared to conditions of civil war, and that could well be interpreted as a 'shaking up of the system', the current situation in the debate on the acceptance of modern information technologies appear to be less tense. No opposition has emerged that has to be taken seriously, although there are discernible waves of unease as information technology continues to diffuse through society. Microelectronics has awakened very high expectations. It is seen as the common task of the business world, science and the state to propagate its use, upon which, it is maintained, depend economic vitality and social stability.

It would, however, be premature and superficial to conclude that the problem of acceptance has lost its explosive force; it will, as before, remain an important factor. First, a situation of social upheaval comparable in its effects to the crisis of acceptance which arose over nuclear technology cannot definitely be ruled out. Secondly, the results of surveys and research into attitudes point to a widespread scepticism regarding technology. Thirdly, experience has shown that the acceptance of technologies is determined by the structure of the technology on the one hand and by the social environment on the other. Technologies are never perceived as purely technical systems, rather always as sociotechnical ones; in the social dimensions of technology can be found the basic reasons for any difficulties in acceptance. This is especially true in the case of

information technologies, as will be shown later.
Risk-assessment research leads us to a similar
conclusion. The affected public not only evaluates
technologies and their real or perceived risks, but
also pays particular attention to those bodies and
institutions (together with the relevant underlying
value-systems) that are associated with the
technology concerned.

Given a reasonably stable social and economic
framework, a socially responsible and politically
legitimate approach to new technologies, especially
on the part of their promoters, is the deciding
factor in how those technologies are accepted by
the public at large. Experience in the area of
nuclear power in the 1970s has shown that there
were sizable structural deficiencies in this
respect. There is no reason to assume that these
deficiencies have been fundamentally removed from
the new field of information technologies.
Contrary to first impressions, the acceptance of
technology will continue to belong to the catalogue
of problems facing modern industrial societies, and
thus also facing the FRG. For the foreseeable
future, public sensitivity to technological risks
seems not to be reversible. The 'protest at
innovation' has come to acquire a quantitatively
significant and, above all, qualitatively
impressive, dimension in the form of a
'progressive, educated movement taking the
offensive on social issues'.(2) Furthermore, the
acceptance of technology cannot be ignored as a
topic of public debate and as a social problem
because it is closely knit with other problematic
areas in society.

We shall return to this point at the end of
this essay, where we shall deal with problem areas
where technical innovations and structural problems
in society have combined to produce crisis
symptoms. We shall begin by trying to characterise
the 'climate of acceptance' in the FRG. After a
discussion of the results of opinion polls and
attitude surveys, the various actors on the stage
of technology policy will be introduced, together
with their programmes and strategies; following
that we shall briefly examine the role of the
education system and the influence of the mass
media. Finally, we shall address the development
of the problem of acceptance in the course of the
past two decades of West German History. Here we
will first outline of the debate on nuclear energy
and the associated acceptance crisis; following

that, the debate on the introduction of information technologies will be analysed, both with regard to its present state and with an eye to future problems. And in order to convey an understanding of the current discussion and the stage of public consciousness regarding the problems of acceptance in the Federal Republic of Germany, we have - in an Annex - attempted both to obtain an overview of the various meanings and usages of the term 'acceptance', and to summarise the tendencies of social science research into acceptance.

The 'Climate of Acceptance' in the Federal Republic of Germany

Technology and New Technologies: Attitudes and Opinions

In contrast to the situation in other Western countries, the population in the FRG seems to regard technology with great scepticism. An example is afforded by the fact that only 55 per cent of Germans agree that more should be done to secure the advance of technology (USA 78 per cent; England, France, Spain, Italy 60 per cent and above; but Netherlands and Denmark 32 per cent and 34 per cent respectively). The influential opinion pollster Elisabeth Noelle-Neumann draws our attention to the fact that the 'large majority' are not 'in favour of progress at any price'. Survey data, she concludes,

illustrate very clearly how the technological euphoria of previous decades, the unquestioning acceptance of technology during the 'economic miracle', has now in the 1980s, but with its beginnings already discernible in the 1970s, turned into a basically soberer but also more responsible critique of the social, human, economic and ecological consequences of technical development.(3)

The problems associated with a climate of public opinion that generally regards technology very sceptically, and with an allegedly 'irrational mistrust of experts' are interpreted by Noelle-Neumann, and not by her alone, as a danger to the 'security of the future of an industrial nation strongly dependent on its exports'.(4)

In evaluating attitudinal patterns towards technology, the remarkable tendency seems to be to regard such identified problems of acceptance as inhibiting innovation, or even as a danger to the functioning of society as a whole. This kind of interpretation is quite often presented by politicians and representatives of economic interest groups, but also by a number of social scientists. We have the impression, however, that in many cases such a view is based on an understanding by which certain sceptical attitudes directed at the negative consequences of individual technologies are directly identified with general opposition to technological progress. One reason for this conclusion seems to be an underlying tendency to equate the factual behaviour of the protest movement against nuclear energy in the 1970s with the voiced opinions expressing sceptical attitudes towards information technologies in the 80s.

Therefore, it is appropriate first of all to remind oneself of the available empirical data, from which a diversified, but at times also quite diffused, reading of the nation's mood can be obtained.

Drawing the results from different surveys together, cannot, of course, lead to a complete and indisputable view of the present situation. It requires an act of judgement to select and interpret a number of significant data from a large body of surveys differing in their approaches and methods and undertaken at different times. A particular problem lies in the fact that results of time-series studies are not available, with a few exceptions which are rather limited in scope.

Despite these and other restrictions and without discussing the underlying conceptual and methodological problems, we shall nevertheless attempt to characterise public attitudes on new technologies by quoting survey data. The general picture looks like this:

• Despite the belief in progress no longer being intact, there is still a basic faith in the ability of science to solve problems.(5)

• Despite public scepticism with regard to technology, there is a widespread acceptance of technologies on an individual basis: that is, the benefits of technological progress in everyday life, for example, the home, are undisputed.

In attempting to draw all available results

together, one quickly realises how implausible it is to talk of 'rejection of technology', or of a 'bias that is nowhere as strongly present as in the Federal Republic'.(6) We can rather draw other conclusions from the same empirical data, if we accept another paradigm as the basis for our interpretation. Thus, it is true to say that there is a considerable awareness of the fact that the use of technologies can bring with it threats to human existence. But would not the absence of such an awareness be very surprising, given the well-known strains imposed on society by technology?

One can also point to conclusions of attitude surveys which confirm a great faith in the capacity of science to solve problems; and also a large measure of confidence in the opportunities offered by technology to solve problems and bring lasting benefits both now and in the future. Taking all these things together, it is legitimate to speak, at the level of attitudes, of considerable openness towards technological change even in the face of possible dangers. There is, at least, no sign of 'technological nostalgia'.

This interpretation can be supported by a series of further points, of which only a few can be mentioned here.

1. Those who maintain that there is a growing hostility towards technology usually base their claims on a recent Allensbach survey, according to which 72 per cent of the population believed in 1966 that technology was a 'blessing', while in 1984 'only' 31 per cent shared this view. Granted that there has indeed been a very surprising shift in opinion, this could also be interpreted as a normal rational social learning process. For all that, 54 per cent of the population tend to the plausible view that technology can be partly a curse and partly a blessing, and only 11 per cent are of the firm opinion that it is a curse. In another survey 56 per cent of the population are characterised as 'approving by and large of technology', while a mere 20 per cent felt 'ill at ease' with technology. If one further considers that one-third of the population still believes that technology is a blessing, despite the enormous potential threat posed by it, then it is hard to understand how one could take the expressions of scepticism as evidence of 'hostility' towards technology.(7)

2. The significance of technology for securing

the future is fully realised by all sectors of the
population. 'Prosperity' and 'progress' are
positively associated with technology, particularly
by 'opinion leaders' in the FRG. It is interesting
to note furthermore that the 'large enterprises
with their research and development departments'
and the 'universities with teaching and research'
are credited with being able to do something for
this progress (according to 73 per cent and 59 per
cent of the population respectively).(8)
3. If we look at the attitudes of youth, we find
that in 1980 a mere 26 per cent tended to be
sceptical in their attitudes to technology, 46 per
cent(!) tended to be positive, whereas 29 per cent
did not know. Elsewhere, we read that one in
five(!) young people were hostile to technology.(9)
Concerning nuclear energy, a noticeably high
proportion of 18-25-year-olds are in favour.(10)
It is generally valid to say that young people are
very interested in questions of technology, that
they would like to acquire adequate knowledge and
corresponding skills, and that they perceive
technical incompetence as a disadvantage. Beyond
this, a large measure of curiosity about technology
goes hand in hand with complete acceptance of
everyday technologies, especially in the realm of
consumer goods. There is also a critical attitude
to the possibility of negative consequences which
cannot be denied. However, the term 'hostility
towards technology' does not seem to be
appropriate, and even the label 'sceptical of
technology' seems a little exaggerated when
attached to the young.(11)
4. In 1981 33 per cent of the population held the
view that 'accidents were practically impossible'
with nuclear energy. Immediately after the
accident at Harrisburg, 13 per cent still 'excluded
the possibility of such accidents', and as many as
46 per cent 'regarded this as highly unlikely'.(12)
This very high readiness to entertain risk - 'after
all that we know today' - is even more clearly seen
in the fact that, in 1979 a good half of the
population maintained that risks had to be accepted
in order to ensure the adequate provision of energy
and to continue technological development.
Finally, in order to put the significance of the
fear of risk and the willingness to entertain risks
into perspective, we may mention the oil crisis of
1973: at that time 70 per cent of the population
were in favour of the expansion of nuclear
energy.(13)

5.   Depending on the data quoted, at least half
the   population   are in favour of   nuclear   energy.
Those   who oppose it believe for the most part that
nuclear power will prevail.   In 1981,   only 13 per
cent   believed   that   nuclear   power   should   be
completely   abandoned   and existing power   stations
dismantled.(14)   These data indicate a   basis   for
acceptance which, if not optimal, is at least quite
favourable,   especially when one considers that all
existing   fears of risk are completely overshadowed
by   the   fear of economic   crisis,   which   prevents
existing   fears from having a widespread effect   on
attitudinal   patterns   to the building   of   nuclear
power stations.   For example,   in 1980 62 per cent
believed that long-term jobs were being created   by
nuclear energy; only 17 per cent did not agree.
6.   Surveys   on   attitudes   to computers   also
indicate   positive trends.   Computers are seen   as
making scientific progress possible, lightening the
burden   of work,   and   strengthening   international
competitiveness. A relative majority (34 per cent)
expects   a   prevalence   of   positive   consequences.
Computers   are   hardly   associated with   threats   to
health,   security,   or safety,   or leisure time.(15)
The   younger   the interviewees,   the greater   their
readiness to work with computers (e.g.   42 per cent
of 14-19-year-olds).(16)
7.   In an analysis of perception and acceptance of
technological   risks,   Renn   has   pointed   out   that
there   is   a   'surprising   positive   rating'   of
automation at the place of work.

Very positive attitudes towards modern machines at
the place of work seem to be prevailing,   not only
for   future generations and for society at   large,
but   also   the   personal   well-being   of   the
individual.   Automation as a humanisation of work
is more firmly rooted in the public   consciousness
than the possible negative effects of redundancies
and conditioning by new work procedures determined
by machine.(17)

8.   Surveys at workplaces appear to confirm this.
They   do indeed indicate fears of losing   jobs,   of
alienation,   and of too much control.(18)   But they
also show that as technology spreads,   barriers are
removed,   the job situation is favourably seen, and
there is no desire to return to a pre-technological
stage.   The majority of working people think   that
technological change has,   on the whole, made their
work   more   responsible,   more   interesting   and

cleaner: the same goes for those sceptical of technology. The most prevalent feeling seems to be that the individual himself will enjoy the benefits of new technologies and will be spared the disadvantages.(19)
9. A recently completed research project, based on a survey in the metal industry, presents findings which suggest that technological change has increased motivation to work, and that morale does not sink but is restructured, 'probably resulting in greater productive power'.(20)

The foregoing comments are certainly not intended to dispute either the existence of ambivalent attitudes among the population of the Federal Republic or the threats posed by technology. On the contrary, the sceptical ratings of modern technology, and its actual and potential risks, are, and will remain, of political significance. But it should be said that in a description of ambivalence to technology one should not just simply emphasise the negative and sceptical attitudes. It is well known that social scientists themselves take part in the process of defining social problems - and here we are speaking of the social problems of scepticism towards technology - so they should refrain from exercising their power as 'definers' to propagate one-sided messages, thereby helping to construct defective and misleading social policy models.

Programmes and Strategies of the Main Actors in Technology Policy

In addition to analysing the results of public opinion surveys, the 'climate' of public acceptance of new technologies can be further characterised by looking at the most important actors in the arena of technology policy. The significance of the political and economic elites we are concerned with here is undeniable, as it is government, political parties, employers and trade unions who have opportunities to exercise power and influence over whether, and what kind of, policy decisions are made regarding research and technology. Furthermore, they also have the best chances of influencing the public, of using or blocking the various channels of communication, of encouraging or discouraging wider participation or action, and of deciding when to initiate and (especially) conclude in dialogue.
In the context of this study, the following

observations are limited to references to stated
programmes and policies - addressed by the relevant
groups to their own members and the public at large
- and do not claim to provide a thorough analysis
of technology policy decisions and resulting modes
of behaviour. Instead, the following remarks are
intended only to illustrate how political elites
legitimate the process of the pervasion of society
by technology and how their views of the acceptance
issue can be characterised.

If one reviews the programmatic statements of
the various groups involved in policy-making, a
preliminary assessment would be that the situation
is less deeply divided than is the case with
'public opinion'. It rather seems that there is a
basic acceptance of technology by the majority of
the bodies concerned. This consensus only begins
to crumble a little, and then undramatically, when
we look at the question of 'how' new technologies
are to be introduced, especially at the speed and
extent of that introduction.

All the deliberations on strategy by the
government of the Federal Republic and its
statements on new technologies, especially
microelectronics, can be reduced to the core
premise of the 'technological imperative', that is,
that the increased diffusion of modern technologies
is absolutely essential because of the Federal
Republic's dependence on its ability to export.
Beside biotechnology and materials technology, of
particular importance are the production and
information sectors (information technology,
communications technology, and microelectronics).
As a consequence of this 'technological imperative'
there arises the need for a modernisation strategy
for the national economy, oriented towards the
world market, with a specific component, namely a
technology policy, to encourage new technologies,
to which all other deliberations have then to be
subordinated. Accordingly, the Annual Economic
Report (Jahreswirtschaftsbericht) for 1982 says
that 'enough jobs can be retained and new ones
created only through the modernisation and
expansion of the production potential'. Such an
understanding of technology policy as a policy of
modernisation of the national economy emphasises
the economic aspect of the acceptance problem.
According to this, acceptance means willingness on
the part of the parties concerned to invest and
innovate, expectations of growing sales, demand
created by increased purchasing power, appropriate

training and further education programmes, active research and development and a general climate in society favourable to technology. This has not always been the case. The former socialist-liberal coalition propagated and practised at first a 'socially oriented' technology policy from the beginning of the 1970s onwards, which found its expression, among other ways, in a programme of research and promotion known as 'the humanisation of work'. In terms of the programme, 'socially oriented' meant bringing the two global aims of 'productivity' and 'humanity' into harmony within the framework of economic rationalisation programmes. To this end, it was intended to involve the trade unions in rationalisation programmes in firms and industries by means of 'discursive cooperative alliances and social learning processes'.(21) With respect to the question of acceptance, a more sensitive approach of governmental research and technology policy was revealed in the recognition of the necessity of justifying policies in the face of growing public scepticism. Moreover, with the establishment of a 'public debate on nuclear energy', and a 'dialogue on technology policy', an attempt was being made to compensate for perceived shortcomings in the legitimisation and execution of research and technology policy by means of corporate problem-solving mechanisms.(22)

However, already by the late 1970s analyses of the structure of research and technology policy showed that it was basically 'economic policy', with a stated priority of 'securing and raising the general standard of performance of the economy', whereas the 'declared aim of "increasing knowledge of the opportunities and risks presented by technologies" is merely assigned to a legitimating function'.(23) This observation is being increasingly made today, and correspondingly the question of acceptance and measures to ensure acceptance have dropped back to a lower position in the hierarchy of values of research and technology policy as a whole. Certainly, we find many statements emphasising the overriding importance of social acceptance of modern technologies; there is, however, no significant parallel in practice. This can also be formulated in another way: the issue of acceptance is primarily viewed as a matter of barriers to acceptance in the spheres of economy and organisation, which can be removed by means of financial and technocratic measures.

Today the government of the Federal Republic 'sees no reason to overdramatise the fact that people have become more sceptical about technology',(24) but rather stresses the need to 'show greater acceptance of technology in our own country'.(25) To this end, the following are suggested as 'necessary measures':

• overcoming the rejection of technology among the younger generation through 'instruction' on the necessity of a free-market economy and competition at school and in training programmes;

• countering 'ideological currents which threaten our future' with a 'policy of clear and honest information' pointing out that 'there is no alternative to technical progress in our country';

• making young people familiar as early as possible with the possibilities offered by modern technologies.(26)

From the government's point of view the encouragement of acceptance is a problem of therapy: healing the 'split relationship of Germans' to the new technologies. 'While we are just talking about information technology, other countries are launching products on the market, gathering working experience, and testing acceptance by the users,' states the Federal Minister for Research.(27) From this idea of the relation between the diffusion of a technology and the debate on its introduction and consequences, the task of defining technology assessment can be derived. Of this the Minister has said:

I am very much in favour of technology assessment: this is not least to be seen in the fact that I recently set up a department to deal with this area. However, technology assessment must not become the eye of a needle or an instrument for hindering the development of technology; but we need it in order to be able to develop humanely designed and socially acceptable new technologies.(28)

It is no wonder that the government's current policies in the area of new technologies, and its understanding of acceptance, meet the approval of the employers, and, indeed, conform almost entirely with their view. This is inevitable, given the priority of implementing new technologies 'as a necessary prerequisite for the German economy to retain its leading position and to remain competitive and as an essential source of economic growth. If one were not to implement them' - the journal Der Arbeitgeber (The Employer) continues - 'or if one were just to delay their implementation assuming that one could thereby exclude suspected negative consequences, there would be serious consequences for the economic and social future of our country that we could not answer for'.(29) So the measures employers put forward to promote acceptance (and the significance placed on these measures) is not far from the official government view, in so far as it is a matter of 're-establishing the consensus where it had been eroded and overcoming accumulated fears'.(30)

The dominant block of policy-makers in technology who perceive technical progress and acceptance in the ways outlined above is growing, if one looks at the large political parties. Apart from a few minor details that differ, the starting point of all deliberations is the recognition of the necessity of introducing new technologies in order to ensure international competitiveness. As an example of this, we can compare the economic programmes of the SPD and CDU which arrive at almost identical statements. The 'Stuttgart Guidelines for the Eighties' of the CDU say that unemployment can be reduced only 'if we accept modern technologies'. The SPD says: 'An industrial policy relying on modernisation can only be realised if those immediately concerned accept it. Modernisation and structural change cannot be carried through against the will of those affected'.(31)

However, the two views do begin to drift apart on the question of 'how' the 'social consensus', acknowledged by all as necessary, is to be accomplished. The sensitive area where clear differences in the strategies of the two big parties emerge is the SPD's emphasis on the strong and active role of the unions in the process of modernisation, and on the basis of extended co-determination. This is an indication that there are indeed controversial stands in this debate. But they revolve not around whether to introduce technologies, just on how to do it.

This is illustrated by the position of the unions, whom one would anyway most expect to oppose technology on the grounds that it may threaten jobs. Underlying their policies is a basic affirmation of technological progress - an attitude which has a historical tradition in the German trade union movement, and which has often been described as 'technological gullibility'. But it is also true to say that recently there has been a much clearer criticism than before of the current direction of technological progress, which 'neglects the requirements of working people' and only considers the 'one-sided interests of industry, instead of, for example, promoting new technologies which result in socially useful and compatible products'.(32)

A review of the ways in which the unions have reacted to the consequences of rationalisation in production and administration since the end of the 1970s, reveals a shift in their priorities from the securing of assets and the obtaining of wages settlements, to negotiation over improved working conditions. In view of high redundancy rates and the higher demands on the worker brought about by the introduction of new technologies (accompanied by the jobs themselves often becoming empty of meaning) the new focus on quality in formulating demands and in negotiating agreements attempts to keep the negative consequences in check and to compensate for them (by guarding against new strains imposed by new kinds of work; with regulations about breaks; with free shifts; and with agreements governing retirement and shorter working week, at first for older employees and subsequently for the whole labour force).

In this way, the problems resulting from new technologies are cut up into more manageable proportions by many regulations of this kind. Thus, large-scale disputes have arisen only 'where the negative social consequences of technological progress affected the core of the union members' - the skilled workers - as the disputes in the printing, metal and steel industries have shown.(33) In these cases, the most points were the effective down-grading of qualifications, especially of those holding medium or high qualifications, and the question of redundancies in groups which were highly unionised. The struggles of the unions resulted in an alleviation of the negative consequences of technological progress; their demands at the beginning of the disputes could, however, not be met, especially not in a time of economic recession and high unemployment. So, in effect, the interests of the employers in a modernisation and rationalisation drive have finally asserted themselves.

The fact that the interests of the state, the economy and the workers have been and continue largely to be identical has to be viewed against the background of these experiences, and together with the conviction of the unions that opportunities for development and growth in the Federal Republic, and the assurance of continued international competitiveness (in the benefits of which the employees can then also participate) can be secured by a dynamic modernisation of production.

These examples also show clearly the logic of trade union policies on technology. On the basis of an underlying affirmation of technological progress as the economy's driving force, and of the conviction that the process of automation can be 'shaped', the trade unions' strategy - aiming at the mitigation of the inevitable hardships for workers - is not directed against the principle of further rationalisation. It is one of reacting to circumstances rather than taking the offensive. From this it follows that deliberations by the unions to pursue the modernisation of the economic system, not as 'a rationalisation from above' have as yet borne no fruit. Central to this model of communication and interaction is the desire to play a participatory role already at the stage of choosing the technology, the result of which would be so-called 'technology agreements'. This is a comparatively long-term aim. Until now it has remained a stated goal rather than one that has been realised, which is probably because the concept of the unions influencing investment decisions by firms is irreconcilable with the currently practised principles underlying the market economy.

In reviewing the programmes of the various bodies involved in policy-making, one could say that the discreet charm of microelectronics has captivated everyone - rather different to the very cool relationship to nuclear technology. Every policy programme acknowledges the pervasion of society by information technology: the acceptance of new change by the 'modernisation cartel' comprised of state, economy and unions (34) seems at the moment to be assured.

As this summary leads us in the same direction as our discussion of the results of opinion and attitude surveys, it will be of interest to check the validity of our conclusions by looking at the role of the educational system and the mass media.

The Role of the Education System:

In the debate on the acceptance and acceptability of new technologies, there is general agreement in the Federal Republic that the education system has a key role. There are two main aspects to this. First, it is expected of the education and training system that it will create the preconditions in terms of 'technical knowledge' and 'skills' to enable a broad section of the population, and young people in particular, to work with modern technologies. Secondly, and going beyond the first aspect, this concern with new technologies in the classroom and in training is designed to encourage a sensitivity to the role of technologies in modern life. It should in particular overcome awe in the face of the unknown, and correspondingly, promote readiness to work with new technologies.

Simultaneous with these insights, there is also agreement that the Federal German education system has not yet reacted adequately to the challenge posed by information technology. Thus, for example, the Commission of Enquiry of the German Parliament (Bundestag) on 'New Information and Communications Technologies' can confirm a 'pressure to conform', and at the same time a 'need for re-orientation' of the entire system of training and further education at school and at work.(35) In the Commission's view, a flexible reaction is essential in that only through a modernised education system will it be possible to guarantee the continued diffusion of information technologies and thus, too, the continued economic vitality and competitiveness of the Federal Republic. The Commission's Report formulates its objectives for the education system in general terms, as follows:

- improving basic technical knowledge,
- increasing social acceptance of technology,
- provision of additional qualifications in technology,
- enhancing professional flexibility and mobility,
- increasing interest in and readiness for further education.

The necessity for innovation in education in the Federal Republic has been acknowledged on all sides: 70 per cent of those in employment are already in need of more or less explicit knowledge of the field of information and communications technology, many jobs which require training are affected by the consequences of microelectronics, especially in the commercial professions. However, each year about a million young people leave school of whom only a very small proportion have been given an insight into information science.

In contrast to the economic system, administration and industry, the education system has not yet· systematically taken on the challenge of the new technologies; pupils and apprentices are not being adequately prepared for the world of work. There are still considerable obstacles on the road to an 'appropriate' education system, the facts, for example, that teacher training has hitherto been lacking, that there are in particular no teacher training courses in information science, and that schools do not possess computers or other items of hardware, nor teaching materials or software to an adequate level. Neither are appropriate teaching methods available. A general objection levelled at the education system - over and above that it is unable to turn out qualified young people - is that the latter are encouraged to adopt a 'hostility towards technology', which can then lead to a shortage of qualified personnel coming up through the ranks, especially in engineering and science. The following are seen as causes of this suspected uncoupling of the education system from the world of work: over-emphasis on abstract and theoretical teaching, a lowering of standards, particularly in the grammar schools, a lack of willingness on the part of schools to teach matters of relevance to the world of employment, and a failure to promote an appreciation of the demands of the economy and of the necessity for technology.

Recognising these obstacles, the state has decided on a series of measures which are designed to place the education system in a position to react to the challenges of new technologies:

· The Federal government and the Länder (states) governments have decided upon an 'outline concept for education in information technology'. It foresees a basic grounding in 'informatics' for all pupils aged between 10 and 14 years, to be incorporated into already existing subjects. This comparatively early universal 'basic qualification' is particularly important in the education system of the Federal Republic because 60 per cent of any one school year leave at the end of compulsory secondary schooling. A more thorough education in information technology in the form of information science is proposed for the higher classes of the secondary schools. The main foci of an education centred on information technology are computer-aided drawing, design and construction, the programming of machines and manufacturing processes, and integrated data and text processing.

• The Federal Institute of Vocational Education has set up research projects and model experiments to improve vocational education with regard to new technologies, and others are planned. In the area of 'information technology and vocational education', however, the number of model experiments is still very small.

• In 1984 the Federal Ministry of Research and Technology began a model experiment, 'New Technologies in Vocational Education', which is to include changes in qualifications in vocational education. Various projects in conjunction with firms and training centres and in vocational schools are promoted by the Federal Lander-Commission for Education and Research Promotion.

• Wide-ranging changes are being planned in educational regulations and vocational school training and are being prepared in model experiments, which aim particularly to accommodate microelectronics in electrical, metal, commercial and administrative professions.

• At the universities courses in information science and electrical engineering are to be expanded. In addition, students of both other technical subjects and non-technical and non-scientific subjects are to be given the opportunity to become acquainted with the relevant technologies, especially computers.

• Within the 'German Research Network', certain selected local networks are to be equipped with personal computers on an experimental basis. So-called computer pools are to be set up by the state governments and supplied with software by the universities, so as to provide students with the opportunity to borrow personal computers.

• New models in further vocational education are to be tried out, with particular emphasis on the needs of small and middle-sized companies.

• Work-relevant and work-oriented courses

offered by the adult education services and other providers of further education are to be encouraged and expanded, to reduce fear of the unknown and to afford practice in the home as well as at work.

· In co-operation with the Federal Ministry for Education and Science, the Federal Ministry for Research and Technology has staged a campaign called 'Computers and Education', which has brought together representatives of the different education ministries, German economic life, scientific and technical associations, and research institutions, in order to support schools and other educational institutions.

Of particular interest in this connection are the various activities of the economic sector concerned with exercising influence on the education sector. Thus, a 'Society for the Promotion of Computers in Education' has been founded by two major associations of the machine construction and electrical engineering industries, to activate and co-ordinate fund-raising activities for educational institutions. This campaign parallels other, existing activities of the information industry which have as their aims the equipping, free of charge, of schools, adult education institutes and other educational institutions, and the provision of courses of instruction.

Such donations and offers are to be seen as being as much a part of an economic market strategy as measures designed to promote acceptance. For example, Nixdorf donated around DM 10 million to install 250 computers in vocational schools in the Federal Republic. One can suppose the following reasons for this to be:

1. a direct market interest on the part of the firms who are eager to make use of the available funds; even if the initial investments are made in the form of a donation there remain subsequently enormous markets, chiefly in the fields of software and in replacing equipment rapidly outdated by succeeding generations of equipment;
2. an indirect market potential, brought about by making future customers with great spending power accustomed to one's own brand, if not actually committed to it;
3. and not least, an argument which is not without self-interest in the long term, but which is philanthropic in the short term: the education of a

generation prepared for information technology and who will later be available as qualified workers.

Because of the Federal structure of the Republic, which leaves responsibility for education to the states (Länder), there is no direct centralised process to overcome the above-mentioned obstacles, and probably for that reason such a process has not emerged in a way comparable with that in other countries. Several lines of development, already clearly discernible in England and France, are only vestigially present in the FRG; for example:

• the installation of (national) computers, the diffusion of (national) educational software, and the setting up of training centres and programmes for additional training for teachers
• rapid introduction of information technology into vocational training, in particular the preparation of less-well-qualified school leavers for future fields of work involving information technology
• promotional measures for 'disadvantaged' groups, such as unemployed young people and women
• a 'general consciousness-raising' of the population, for example by means of combining youth leisure activities with computer education, software especially designed for youth leisure, and holiday centres where the use of microcomputers can be learnt.(38)

Measures to adapt the education system to technical progress, in terms of organisational intensity and funds allocated, are in the Federal Republic not yet as clearly emerging as they are in England and France. But it will only be a matter of time before a grand coalition of bodies from the state, scientific and economic spheres forms, albeit with different motives, but with the common basic aim of introducing a set of measures to change the education system in such a way that it meets the challenges of the new technologies and can play a more positive role than hitherto as an acceptance factor.

The Influence of the Mass Media:
The question of whether and how the acceptance of new technologies is influenced by the mass media is to be viewed against the fact that communication research appears as yet not to have established whether the mass media are a major influence on the

formation of opinions and attitudes. Equally little understood are the channels and mechanisms by means of which the messages of the media influence or change the disposition of the recipients. In general, the influence of the media should not be overestimated. Attitudinal patterns, and even more so patterns of behaviour, are seen to offer considerable resistance to information dissemination by the media. It is therefore probable that attitudes and opinions cannot be 'created' by the media, the effect of the media rather consists in strengthening or weakening attitudinal patterns which already exist.

These general conclusions can be confirmed in the context of our topic. Let us examine two exemplary studies on the influence of the mass media on the attitude of the population of new technologies.

A survey on the way reporting in the media on nuclear energy is perceived (39) shows first of all that the frequent reporting of accidents in power stations and of the debates about the introduction of nuclear energy stick strongly in the public memory. It seems certain that far more negative topics than positive ones are noticed. But at the same time more than 50 per cent of those asked in this survey believed the reporting of the media to be well balanced, about a quarter perceived it to err on the negative side, and about a fifth saw it to err on the positive side. This particular survey on the perceived tendencies of the mass media confirms general findings from communications research that positive and negative aspects of reporting are generally regarded as being given equal weight. To differentiate this finding a little, it can be said that opponents of nuclear energy are somewhat less convinced that reporting is balanced. Similarly, over 70 per cent of those who are undecided on the nuclear energy issue are of the opinion that reporting is balanced, only 16 per cent of those in favour are of the opinion that their positive attitude is reflected in the media.

This individual case study draws the conclusion that reporting can have an important influence on the perception of the problems associated with nuclear energy. However, its impact on the individual's attitude is still only small. Instead, it seems that the perception of media reporting is determined by rather stable fixed attitudes to the theme or object in question. For example, opponents of nuclear energy, having a basically negative attitude, either have no cognitive perception of the positive aspects in a report by the media, or they declare it to be unfair reporting. Those who are in favour and those who are undecided perceive both positive and negative aspects corresponding to their attitudes. It could be said, as a very cautious generalisation, that a decisive and direct influence by the mass media on the attitudinal patterns regarding nuclear energy is probably out of the question. Attitudes which have once become fixed can be changed very little by information and judgements pointing to the contrary conclusion. The recipient tends to protect himself against contrary information through selection, re-interpretation of content, or by refusal to accept the 'transmitter', in such a way that his own opinion and attitude can remain stable.

In a survey of reporting by the mass media on the computer, similar results were recorded. (40) Fifty-five per cent of those included in the survey rated reporting as partly positive, partly negative; 28 per cent regarded it as positive, and 17 per cent as negative. This general assessment can be differentiated by mentioning that opponents of computers assess reporting in the media as being far more negative than those who are in favour of computers. Furthermore, it appears to be valid to state that in this field also the recipients select, interpret and suppress information from the media according to their attitude. That is to avoid so-called 'cognitive dissonance', the perception of the content of the media conforms to opinions that have already been formed. From this the conclusion may be drawn that the reporting in the media contributes functionally to a stabilisation of judgements on the computer · which were already present.

From communications research in general, and from some specific investigations into the influencing of attitude formation by the mass media in the FRG in particular, we conclude that only little weight can be lent to appraisals that assign an overriding influence on attitudes to technology to the way in which a topic is reported, and which therefore accord priority to the mass media factor as a main one in creating acceptance. Rather, we will recognise the fact that the effect of the mass media is very limited and that its significance or function will have to be seen in terms of strengthening or weakening social structural or individual factors that are already present. Thus, for the constitution or attitudes, and also for the question of the acceptability of individual technologies, the mass media have only the status of 'intervening variables'.

The influence on the formation of the opinions of the so-called average citizen, which the mass media certainly have, has to be seen in conjunction with social structural factors and processes of change, which have a greater effect on the formation of attitudes. For example, it is to be assumed that an improvement in living standards, growth in the proportion of those who have spent longer in the education system, and an increase in cognitive skills and in available objective information, all determine and limit the mechanisms and the extent of the effects of the media.

The Development of the Acceptance Question

Nuclear energy: social polarisation and acceptance crisis: A crisis in the acceptance of nuclear energy technology, and of technology in general, arose only in the later stages of the introduction of nuclear energy in the FRG, at the beginning of the 1970s. Against the backcloth of the oil crisis of 1973, the economic recession of 1974-5 and the disputes surrounding the planned construction of the nuclear power station at Wyhl, a public debate gradually grew, manifesting not only a growing consciousness of risk and of the threat to the environment, but also a specific aversion to nuclear energy technology and to the way it was being introduced. In particular, doubts were cast on the legitimacy of the authorities concerned to make decisions, and on the authoritativeness of the experts concerned.

Patterns in attitudes towards technology, experts and politics in general, ranging from sceptical to hostile, began to form at about this time. The socio-political controversies arising from this and lasting through the entire decade were, however, much further-reaching and amounted to a very basic questioning of the future path of industrial society in the Federal Republic. Reactions were provoked from almost all participants, which caused a very confused debate on 'ecology /technology /politics /public opinion /democracy /resources /energy'. (41) The explosive nature of the negative attitudes among the public towards technology in general, and nuclear energy technology in particular, resulted from the fact that the political culture in the FRG now became characterised by the emergence of the new social movements and the formation of 'citizens' initiatives' groups throughout the country.

From very localised beginnings movements developed 'in a process of generalisation which mobilised people on a regional and then on a national level', a process which reached its height in 1976-7 and then again in 1979-80.(42) Within these social movements forces for resistance formed alongside forces favouring retreat. These forces were at first active on an extra-parliamentary level, articulating their political needs and goals in public, and trying to incorporate their concerns into institutionalised decision making processes. Legal forms of dispute played a special role here, that is, people tried exhaustively to exploit the opportunities for resistance offered by the legal system.

The stabilisation of the organisation of the civic action movement, through the founding of the BBU (Bundesverband Burgerinitiativen und Umweltschutz), the formation of a 'Green' party, which is contradictory as it is, into politics, and not least the peace movement which came into being with surprising speed, are important indicators of the fact that the political scene in the Federal Republic of Germany has changed radically.(43)

The established political-administrative system and the party system did not seem to be in a position to resolve the conflict, let alone assimilate the extra-parliamentary opposition, using the current forms of problem-solving (even if one could speak of the process of bringing the 'Greens' into parliament as such a development). In general, however, one can note distinct deficiencies of the established democratic processes,(44) the magnitude of which should not be underestimated, especially regarding its long-term effects. There resulted an inability to reconcile divergent interests with more than symbolic policies. Communication, where it took place at all, was dominated by those with the greatest power. Although the state, quoting the principles of democracy, declared itself in favour of making information publicly available and of encouraging discussion and dialogue with citizens and, indeed, itself partially initiated these processes (albeit always rather late in the day), it was at the same time very clear that there was no chance of reconsidering the decision already taken in favour of nuclear energy.(45) 'Under such conditions it also proved impossible to allot more than a peripheral role in the energy programme to renewable energy systems. Such promotional measures had only a very limited effect with regard to creating political legitimacy.'(46)

The crisis over the acceptance of nuclear energy demonstrates clearly that the interplay of communication and interaction accompanying its introduction was characterised by structural defects that gave rise to sometimes very fierce debates, and were therefore also partially responsible for the lack of acceptance. Recent literature on the processes of discussion and decision-making in the development of nuclear energy show how an important principle of fair debate was ignored when a discussion of various options regarding reactor types broke down, which also meant that learning processes were interrupted. 'Open discussion suffered further in the area of safety questions regarding nuclear energy, and so one grew accustomed to being presented with "faits accomplis" as far as the construction of nuclear power stations is concerned.'(47) As the development of nuclear energy proceeded, a process which existed only in an embryonic form never really got under way. This was a broad-based public discussion, a system of control by parliamentary democracy, public participation and a commensurate policy of education, all aimed at preparing society for the nuclear age. Looking at the process by which nuclear technology gradually assumed its ascendancy, defects in communication can be found there too, as for example in the relations of science and industry: the nuclear research community and the nuclear power industry often talked at cross-purposes or were even opposed to each other.

Further structural deficiencies can be seen in the fact that at first the arena of nuclear energy was dominated almost entirely by the actors in the ministerial administration and in industry. An especially typical example here is the very low significance accorded to the parties and their forum, parliament, as opposed to the executive, which had the upper hand throughout almost the entire process of discussion and decision-making.(48) A study of the decision-making processes in the field of nuclear energy has shown that communication and interaction took place more or less exclusively between the ministry concerned and the organisations and firms commissioned to carry out the project. Information was released only selectively outside this circle, if at all. Ministry decisions were never modified by the other people or bodies concerned, and in particular no watchdog role was played by parliament.

Alongside the fact that other important interest groups were excluded was the inability to anticipate problems of acceptance, let alone the capacity to deal with them. Moreover, long-term economic, ecological and social consequences were for a long time not topics for discussion. This was attributable not only to lack of material rationality on the part of politics and industry, but also to a lack of experience and awareness on the part of potential opposition groups and the public at large. Thus, nuclear energy policies long remained in complete isolation from problems of acceptance. A genuine acceptance crisis arose only at the beginning of the seventies, when the problems which had hitherto been systematically ignored intensified to a situation of conflict. Among these problems could be counted: the social and economic consequences of a compulsory extension of the nuclear power station programme, questions of safety technology and its effects, the strain imposed on ecology, the differences between the decision-making channels at federal, Lander and local authority levels, regional planning policy, and finally the lack of transparency in decision-making processes.(49) The suppression of these problems and the resulting costs, the absence of political rationality, and the inadequacies of administrative organisation coincided with a growing awareness of risk and an increasing willingness of various groups in society to take action. The executive, administration and the energy sector of the economy were then no longer immune to criticism or alternative proposals.

The beginnings of this acceptance crisis were characterised primarily by the technical and scientific aspects of nuclear energy (especially its safety), and by opposition to the research policy of the federal government. Later, it became clear that a growing part of the population had begun deeply to mistrust the whole political system. Thus, symbolic gestures by the government or measures aimed solely at increasing the safety of power stations were no longer accepted as an adequate response to the political dimension of the issue.

This contradictory situation and the already mentioned problems of the costs of the intensified introduction of nuclear energy finally resulted in conflicts which 'made themselves felt as a shock to the political system'.(50) Attempts to deal with the problem by means of a 'fair dialogue' failed because of the structural defects in communication and interaction - both long standing and recently emerged. The experiment, initiated by the state, of a public dialogue on nuclear energy, did nothing to fulfil the demands for rational discourse - it was probably not designed to. It comes as no surprise that the 'use of such methods which are more or less of a manipulative nature' led to a heightening of the conflict and to a reinforcement of the barriers to communication.(51) This statement is not used to claim that this problem, which has come to be more than a conflict about nuclear energy, having broadened out to become a more generalised conflict over political authority and values, could have been solved by a more functional communications structure alone. But that was an essential if not wholly self-adequate, prerequisite if the crisis of acceptance of nuclear energy was to be defused.

Be that as it may, it is a fact that at the beginning of the 1980s (after an occasionally very violent debate in society, particularly in 1976-7) the process of confrontation came to a standstill. This showed itself in many ways: the economic and political necessity of expanding the nuclear energy programme continued to be upheld, though those who influence policy-making in industry and politics and who had been made to feel insecure remained relatively reticent. Between 1975 and 1979 no new nuclear power station was authorised and only two were commissioned. Furthermore, there was a careful attempt to re-orientate energy policy. But we cannot see this causally as a consequence of the opposition to nuclear energy, or as a 'learning process' by the coalition of politics, bureaucracy and industry. The changed basic economic situation was probably more instrumental in effecting this restraint. Apart from that, in the mid-1980s two developments have arisen which likewise point away from the hypothesis of 'success' or of a 'learning process'. On the one hand there is once again little delay in getting the go-ahead for the building of new power stations; and on the other there is a strengthening of long-standing aspirations to reduce opportunities for the public exercise of influence so as to speed up authorisation procedures.

The anti-nuclear movement, which has never been a unified block either in its social composition or in its ideological direction, is now showing signs of becoming fragmented and less capable of action, especially when it is a question of uniting to oppose the existing system. Finally, there no longer seems to be such an urgent need for the further expansion of nuclear energy to ensure economic growth. However, it is not possible to say that the conflict has been resolved, especially as 'the problem of waste disposal, which has now come to be the sticking point in the debate on acceptance, has no satisfactory solution'.(52) Surveys now show, moreover, that the public's main concern over the risks of nuclear energy is the question of atomic waste disposal.(53)

All in all, the stagnation of communication in the field of nuclear energy is an expression of the failure of all attempts to arrive at a social consensus by way of rational argument. If one examined the history of nuclear energy and its acceptance and measured this up against a model of a successful learning process, one would notice a high degree of irrationality, authoritarian decision-making and lack of transparency. Many studies on the processes of decision-making in the nuclear technology sector have concluded that this took a course which was not planned in advance and which showed a complete lack of rationality on the part of those participating: this is reflected in the acceptance situation. Measured against a democratic society's demand that problems be tackled in a way that is open and visible to all, and which is oriented towards achieving consensus, the history of the efforts to encourage acceptance of decisions regarding future energy policy has to be seen as a history of failure.

Reflections of the acceptance climate and the stagnation in communication are to be found in several empirical studies on nuclear energy which look at obstacles to communication, the lack of public participation, and stereotypes in perception. From these studies it becomes clear that all concerned are aware that public participation is not genuine participation: the public are of the opinion that decisions have long since been taken, whereas politicians believe that the public wants to have a say only in order to be able to sabotage decisions. A Battelle report on the attitudes of the Federal Republic's population towards nuclear energy has thus identified as a prerequisite for the acceptance of political decisions about technology that planners and politicians should recognise the 'way that the public sees things as social reality' and should attempt to 'see their modes of interpretation, which they confront the public, more clearly themselves'.(54)

Although today the disputes centred around nuclear energy seem more or less to have come to a standstill, the analysis of the acceptance crisis of nuclear power is still of far more than merely historical interest. For, if it is correct to say that

social values are at the root of the widespread scepticism towards nuclear energy which transcends the issue of nuclear energy itself and that therefore there are processes of social change underlying the acceptance issue which will slowly gain political recognition(55)

then one cannot completely exclude the possibility that in future the question of acceptance of new technologies could again lead to a challenge of the political system. If 'non-responsiveness'(56) continues to characterise the actions of politics and administration, as shown in the disputes over the acceptance of nuclear energy, then it should not be precluded as a possibility that latent scepticism towards technology will emerge in new forms and constellations. A coalition of 'Greens', parts of the SDP's left wing and the unions, supported by grass-roots movements, could then become a 'brake on modernisation'.

Information Technologies: Social Consensus and the Assurance of Acceptance?

The history of the acceptance issue in the case of information technology cannot be reported on in the same way as in the case of nuclear energy. Not only is the period of public and scientific discussion of this issue still too short, but there are no such lines of development and conflict, or such clear fronts and polarisations in the debate.

Given the structure of this technology, the fact that it is able to pervade many sectors of society and has a multitude of applications, which are moreover constantly changing, then it follows that it is difficult to draw a full picture of its consequences or a clear 'acceptance profile'. Information technology is, of course, not an undisputed area, but we cannot speak of an acceptance crisis comparable to that over the issue of nuclear technology. The fact that there are no such great differences over information technology is due to a series of factors, both internal and external, which allow of a provisional description as follows:

• There seems to be basic acceptance of this new technology in spite of the general scepticism encountered in opinion polls. Further, the question under discussion is not whether to introduce, but how. The consensus on the need for information technology, explicit or implicit, extends to almost all groups in society, and even embraces opponents.
• Accordingly, there is no real conception of an 'alternative' to the use of this technology. On the level of the firm, for example, the dispute does not take the shape of a struggle for an alternative form of rationalisation, but of participation in new production concepts.
• In contrast to the massive and undifferentiated nature of nuclear technology, there is a wide range of possibilities of planning, shaping and using information technology, which awaken a large number of expectations. This suggests that this technology has an instrumental character which opens up opportunities for supervising the way it is introduced and used.
• There is no dramatic reaction to information technology. Even the strongest negative myth, which sees microelectronics as a 'job-killer' seems merely to raise a few fears, but does not cause upheaval. Rather, these fears are countered by the positive effects hoped for and also by a resigned attitude which says that there are no real alternatives in the face of pervasion by information technology of the world of work and society at large.
• Despite the facts that both the positive and the negative sides are clearly perceived, there is no resulting polarisation and no emotionalism emerges in public opinion. The computer 'is not as

yet a significant creator of fear'.(57)

* The handling of the social consequences of information technology is managed by involved parties well-versed in the strategies of conflict management - trade unions, employers and government. There is no protest group that could answer those accustomed to having the upper hand in discussion, among whom the unions are characterised by a basically positive attitude and a concept of collaboration which is both defensive and corporate.

* The market offers a wide range of opportunities to test things out and therefore allows the possibility of social adaptation and for learning processes to take place.

* Contrary to the case of nuclear energy, information technologies provoke a wide spectrum of different reactions on the part of those who are affected. On the one hand we have different kinds of information technologies with different forms and uses and thus different reactions. On the other hand, moreover, we find differing reactions with respect to one and the same kind of information technology. Therefore, one cannot identify homogeneous groups which could potentially form small protest groups or even a popular movement. Taking as example the acceptance of new technologies within a firm reaction will depend on whether the individual stands to gain or lose by rationalisation or just tolerates it, and on whether we are looking at marginal groups or at the unemployed.

So we see a very wide applicability of this technology and at the same time many ways of organising and monitoring its use, at least from the point of view of those involved in its operation. As the 'openness' and 'flexibility' of technology are seen as reality and not as fiction, there is a generally favourable attitude towards acceptance.

All those involved accordingly share the views that the application and consequences of information technology follow no natural course of development, and that the structure of the technology does not enforce any specific changes (for example, in the nature of work), but rather that the process of automation is an open-ended learning process which can be shaped by specific policies and influenced by co-operation between politics, industry, science and general public.

On the level of the firm the characteristics hitherto formulated in general terms take on concrete form: attitudes and opinions at the place of work show an awareness of problems associated with new technologies but no signs of rejecting them.

The trade unions, while basically approving the new technologies, oppose the 'break-through mentality' of management. Their own position, however, represents a rather defensive strategy aiming at "social control" of technological development. This concept is characterized by notions of protecting jobs and at maintaining the level of qualified jobs for skilled workers. Using the traditional and largely institutionalised channels of dialogue and co-operation, the trade unions also try to include their own aims in the process of implementing new technologies. But this is not to be seen as an attempt to close minds to the necessity, explicitly recognised by the unions, for structural change in the economy, nor as an attempt actively to support opposition to technology.

From the structure of the situation, it follows that such factors as oppose the speedy and frictionless implementation of a technology are seen as constituting problems of acceptance. The state and industry, especially, see problems of obtaining and recruiting qualified personnel as being relevant to the acceptance issue, and propagate measures to counteract them. Other aspects which are taken as constituting barriers to acceptance are: hesitancy on the part of the user to innovate (which results in a widening gap between the technological possibilities theoretically available and those actually in use, for example in the area of office communication); insufficient awareness on the part of the producers of the need for their products; lack of conception of their possible uses; and products which are not 'user-friendly'.

We do not in the least want to deny the importance of these factors for technological innovation. Being concerned with the acceptance issue, however, we expect problems of a different nature to play a role in the medium and long term. In the following section we shall touch upon some of these possible future problems of acceptance, concentrating on the production sector.

Information Technologies: Possible Future Problems

of Acceptance: The result of the foregoing seems to be that the introduction of information technologies is proceeding without any problems of acceptance worthy of mention. In the economic sphere, however, important problem areas can be seen which are just beginning to emerge, as regards: participation; constellations of interest and conflict; power and control; and strategies of action.

The fact that these increasingly relevant problem areas are only reluctantly acknowledged may be due to the 'incubation period'(58) of new technologies, which has only now made it possible to see and evaluate the whole range of consequences which had hitherto remained concealed. This is significant chiefly in that here it is a matter of coming to terms with new technologies at the socio-cultural level. It concerns the substitutive behaviour on individual and collective levels which is essential if technological progress is to be rendered assimilable, that is the adaptation or modification of existing patterns of perception, behaviour and values. These social barriers to acceptance - as opposed to the acceptance problems which can be resolved by economic and organisational measures - fall within the sphere of the social sciences, within a public research and technology policy, resting on a basic notion of acceptance which does not exhaust itself in merely economic and organisational perspectives.

Participation. In the FRG the situation in firms is characterised by there being a number of legally prescribed, as well as a number of informally practised, regulatory mechanisms which aim at allowing workers to have a voice in decisions on the introduction of new technologies and their use. In addition to the traditional and legally enshrined procedures, new ones are increasingly being sought, which often bypass the previous exercises of influence. For this reason, and also because traditional procedural mechanisms are no longer regarded as satisfactory or adequate to the demands of information technology, the result will be an increase in the tensions that are already present in that managements are more oriented towards efficiency and the unions more towards the possibilities of participation. The desire for participation becomes a problem of acceptance at the point where it goes beyond the hitherto practised forms of 'user participation' as an early

warning system for management, and begins instead
to aim at

the participation of those involved with a view to
taking into consideration interests hitherto
neglected. The demand for participation in the
development of systems and in the process of
putting them into practice is not only a result of
the value systems of our democratic and social
constitutional state. It is also a means of
assuring applications of technology that are more
geared to human needs.(59)

In this sense, the following factors are decisive
in determining the quality of participation as an
integral part of the implementation of a
technology:

• the extent to which the various bodies involved
are taken into consideration when technologies are
designed;
• how their defined interests and chief values are
articulated and realised;
• how mechanisms for the balancing of interests
are set up;
• whether there are any accompanying measures at
the level of the firm and above.

These aspects are also relevant to the problem
of acceptance in that they afford insights into the
previous processes of mechanisation, into the
'processing' of social aspects of the various
groups involved, and into difficulties and
opportunities anticipated in the future. The
question of the granting or withholding of
acceptance will change and intensify with the
various 'qualitative' repercussions deriving from
the use of new technologies (demands for
flexibility in the re-definition of jobs; the
confidentiality of personal information; demands
for adaptations of an ergonomic nature; comfortable
and convenient machine-operation design; changes in
working conditions and in work procedures through
an intensification of work; the devaluation of
and/or an increased requirement for qualified
personnel, etc.), not forgetting the effects of the
rate of redundancy; in other words, this
development affects employees as users of the
technologies in a new way. They are directly
confronted by massive change and react to increased
burdens and threats individually through unrest and

a refusal to accept the changes, or via those bodies representing their interests (works councils, trade unions) with demands for protection against the negative effects of rationalisation, regulations governing breaks, etc.

This development has by no means been concluded. It will continue to entail many changes: for administration, for the nature of work, for family life and leisure, for the expansion and spread of knowledge generally - and these changes will probably be greater than those experienced hitherto.(60)

Constellations of Interest and Conflict. On the basis of numerous empirical studies, sociological theses have interpreted the mechanisation of production, and of the sectors preceding and following it is a socio-technical network of communication and interaction, in the development of which changes take place in the balance of power and in the nature of conflicts, or attempts are made to ward off these changes. In the course of this, new configurations are constantly emerging in the distribution of interests and influence and in the constitution of conflicts.

It is to be assumed that problems of acceptance will arise where new technologies and the way they are integrated into the existing constellation of power bring advantages or disadvantages of varying degrees to the parties involved. Because the application of technologies is thus a 'social process guided by interests', the guarantee of acceptance can be given only when a balance can be guaranteed between the various interests'.(61)

New technologies can pose a threat to the social status and qualifications of the skilled workforce, especially in the case of a rapid introduction of rationalisation technology. Such a situation can provoke severe opposition. A prominent case in point is the printing industry, where the machine compositors particularly affected took action. The skilled workers in machine tool construction, on the other hand, have been able to deflect the threat to their status by the individual adoption of alternatives offered by the firm, through which they could avoid dequalification and being placed in a lower income bracket. This resulted in an acceptance of new technologies, as in the areas of individual production and maintenance, where acceptance was assured because the position of the skilled workers was not threatened, but was even upgraded.

The general picture of the shifts that take place in firms is that a small class of privileged skilled workers in supervisory roles is opposed to an ever growing class of unqualified, semi-skilled personnel. A levelling out in the status of the work of skilled and semi-skilled workers is often discernible in cases of thorough rationalisation, so that new groupings and coalitions form, which then fight against speedy and all-pervasive rationalisation.(62) Other groupings can also be observed, where certain factions (for example, in the field of maintenance foremen and chief maintenance men) profit from technical progress and are therefore prepared to go along with it.(63)

In the second place a change in 'constellations of behaviour within firms'(64) is noted which has consequences for acceptance in the administration sector as well. By this is meant the existence of a specific form of 'pluralism of interests'(65) in firms, which is based on a varied distribution of formal and practical competences, of actual opportunities for exercising influence, and of interests. This 'constellation of partial interests, competences and opportunities of influence, and the resulting conflicts, alliances and competition'(66) can be as important (and sometimes more important) a factor as shortfalls in the employees' qualifications or ability to adapt, as studies on the introduction of word processing have shown. Because of a close link between the established structure of interests and acceptance difficulties, it can happen in individual cases that new technological systems are hesitatingly introduced.

A third point of significance for the acceptance of new technologies is the extent to which the interests of employers and unions are identical. For example, at Volkswagen, although the unions reckon that one-quarter of all jobs will be lost in the long term, a system of co-operation operates in decision-making about the rationalising measures to be taken. Acceptance has thus hitherto been procurable 'by externalising costs and excluding marginal groups among the employees', and by an orientation towards social technical solutions which are negotiated on a central level while avoiding mobilisation and politicisation of the basis.(67) The success of this consensus-finding system in an individual case also indicates, however, that such a model is not implemented everywhere in the manufacturing sector - with sharper conflicts of interest and a partial refusal of acceptance as a result. But as rationalisation progresses, and against a background which is probably worsening, even at Volkswagen the question is raised as to what extent this concept is still practicable, as 'the core work force [is] itself threatened' and 'unrest, conflicts, and loss of legitimacy are most likely to come from the skilled workers'.(68)

A fourth consideration is that studies of rationalisation in various sectors have shown that the distribution of power within the firm shifts in favour of the management because of greater controllability of production brought about by technology.(69) That does not always result in further reactions from those affected, but is rather a 'shift in power without a power struggle'.(70) The main explanation is that the workforce has an interest in the continued existence of the firm - especially in crisis-ridden sectors like shipbuilding - which leads to a remarkable degree of technology acceptance on the part of the employees: the unions and the works council themselves press for the introduction of technology. This fundamental acceptance is supported by the fact that potential conflicts are neutralised through keeping on a large pool of skilled workers although this is technically no longer necessary to such an extent, and the creation of an attractive wage structure as a safety net.

A fifth aspect is that it has become clear from studies of technological innovation in the chemical industries that a certain type of 'worker with a character determined by his work milieu' contributes greatly towards an acceptance of new technologies, in so far as he fits into the management's rationalisation concept. As already seen in previous waves of innovation, skilled workers in the chemical industry have shown themselves very willing to adapt, a tendency which continued even when the industry went over to full automation of supervision and control of processes with all the consequences which that entails. At least in the chemical industry, it appears possible to cushion this situation by means of a specific wages policy and by avoiding redundancies. In the maintenance sector of the chemical industry, too, the degree of acceptance is determined by the 'dominant type' of worker. There could be difficulties with a worker with a more traditional attitude to his job, and who is either obliged to adapt to new technologies or whose qualifications are downgraded. On the other hand, the 'individualistic technocratic' type, denoting younger persons who are career-oriented and in favour of innovation, is regarded as the 'promoter of full automation'.(71)

A final example of configurations of interests and conflicts between a (usually smaller) group of those who 'profit' from rationalisation and a (usually larger) group of those who 'lose', can be seen when one takes the factors of length of service and chances for promotion into account, together with their changing significance.

According to the principle of seniority, a worker's status increases with age and length of service, as a growth of experience is assumed. This principle affords a calculable security. Knowledge acquired through experience is rendered largely superfluous by computer-controlled systems. Instead, a willingness to innovation is demanded that can hardly be expected from employees who are oriented towards the principle of seniority. For this reason, technical innovation is rated positively more by the younger employees and those with lower functions, as these are not yet thinking in terms of the seniority principle.(72)

Power and Control. In connection with the changing constellation of interests and conflicts, it is all the more important to study changes in the mechanisms of power and control. Of special relevance in this respect are the shifts in the distribution of planning and supervisory knowledge brought about by information and communications technology. A glance at the use in the work of engineers and technicians reveals shifts not only in the manoeuvring space of the dependent employees but also in the opportunities for control and monitoring by the management. The result is increased manoeuvrability and, simultaneously, a tighter, more centralised, control of work, given the present state of technical development together with appropriate planning within the firm.(73) Such observations offer themselves in industrial production in preparation and planning and in workshop control, in the service sector in the processing of applications and claims (e.g. in insurance), and in industrial management in the computer-aided support of marketing.

In any case, it can be assumed that new technologies will bring about clear changes in the accustomed structures of interaction and communication, for example, between supervisors and workforce, which have traditionally been based on knowledge of the person and his/her peculiar characteristics. By means of information technology internal directives tend to become more anonymous, centralised and objective. Thus, the extent of personal transactions is reduced in favour of behavioural patterns ostensibly deriving from a neutral technology.

Control and power again seem to move closer together as a result of certain characteristics of the new technologies. Because of the increased opportunities for behavioural control as well as the uncertain nature of the use and evaluation of data stored in computers there is a growing feeling of abstract threat in many sectors. As a result of computer-aided process control and/or computerised techniques of personal control with their practically unlimited possibilities for gathering, storing, combining and interpreting data, it has become virtually impossible for those concerned to foresee the consequences of any 'deviant behaviour' regarding rules and regulations in the firm.(74)

In the service sector it is especially evident that, besides saving labour and capital, the new information technologies make new possibilities for control available, which support hierarchically organised systems in exercising power. This fact is undoubtedly another reason, besides the criterion of economic efficiency, for the readiness of those involved in decision-making to introduce new information technologies. That is to say, 'the interest in upholding and maintaining hierarchies in these organisations'.(75) On the other hand, a potential for resistance begins to form and acceptance begins to be withheld in a number of ways as those affected, especially works councils and trade union members, grow increasingly aware of the possible consequences of new control mechanisms.

Experience to date with the use of 'personnel information systems' and 'management information systems' as 'new management technology' shows a particularly embattled case of the application of information technology.(76) The attempt by management to introduce these information systems, especially in large-scale enterprises, has resulted in much resistance from the unions, and much publicised disputes.

These, and other, consequences, are of decisive significance for the question of acceptance, not only because established and hitherto consensus-guaranteeing opportunities for power and control in the interplay of operative forces are changed, but also because these changes are realised, anticipated and incorporated into appropriate strategies by the relevant group or body. These strategies are then put into practice on the introduction of such technologies, and on the attempt being made to secure a measure of influence (maybe under new terms), and they can generate conflicts.

Strategies of Action. In industrial sociology and the sociology of technology the change of the nature and form of jobs has for a long time been deduced from the structural principles of new technologies. But recently research has turned away from this technological determinism towards concepts of the 'policies of the firm'. At the basis of this perspective lies the idea that technologies can be regulated and steered. This theoretical insight finds its practical complement in union policies within companies.

Taking this insight as their starting point, many studies have tried to show that the ways of acting and reacting displayed by the groups relevant in a company, and the strategies and philosophies of rationalisation underlying them, are factors of significance. Thus, the question of technology acceptance acquires a new dimension. In the process of the introduction of new technologies special importance is assigned to the world views, ideologies and political concepts of individuals and groups in a company's day-to-day life that gain in importance in so far as these factors can act as barriers to acceptance, and thus as brakes on innovation. Where, for example, the logic of the 'social technologies' of management meets with the unions ' 'control strategies', and with the concept of collective bargaining, not just for higher wages but also for improved working conditions, there will always be material for conflict. In order to create a situation of acceptance which goes beyond mere 'tolerance', it will be important to note to what extent the strategies of argument and self-legitimisation employed by the parties involved preclude the development of a consensus.

In connection with these strategies of the parties immediately involved at company level, it is also worth noting the 'Humanisation of Work' programme promoted by the Federal Minister for Research and Technology. This is of importance to the whole question of acceptance in so far as this programme presents those components in the government's modernisation concept which are designed to try and achieve consensus in society on new technologies, and which at the same time therefore represent the state's justification of its actions.(77) This part of the government's technology policy is thus a contribution to the promotion of public acceptance of new technologies.

Technology Acceptance in the Context of Structural Problems in Society

Over and above the foregoing treatment of possible future problems of acceptance in a social subsector, namely the production sector, it is necessary for a global understanding of acceptance as a factor in modernisation, to obtain some grasp of its significance for society as a whole. The reason for this is that the acceptance issue has to be seen in connection with the fact that technological innovations are always structurally interlocked with social issues. Technological innovation and structural problems in society can thus together produce crisis symptoms, which then generate problems of acceptance.

This is of especial relevance, and will remain important, in the following contexts: the ability of the political system to regulate and justify itself; the strain on the environment and the consciousness of ecological crisis; changing values in modern industrial societies; and structural changes in employment and existential changes in the nature of work.

In the Federal Republic, debate about the consequences of technology for the environment, the economy and employment, has gone beyond the exclusive circles of science and politics. To the extent that technology has become the subject of public discussion, problems of control and legitimisation have emerged in a number of social subsystems which seem to be overburdened by the rapid introduction of new technologies. The political decision-making system, the legal system and the cultural system are fully stretched. But in addition,

the concept of the social compatibility of technology points to limits on regulation: the plausible justification and its social acceptance are important factors in introducing new technologies. It is no longer the interaction of politics and science that constitutes the sticking point in techno-political decisions: it is the debate between experts and citizens, in court or in the street, that has become the centre of the decision-making process in technology policy.(78)

The ability to obtain a consensus among differing interests plays an important role, along with the ability to process information, in putting technological innovations into practice, and this can be seen as a process of social modernisation. If it is the undisputed aim of the present political system to press ahead with modernisation, there will be no way round developing long-term strategies aimed at achieving consensus. It is, however, possible that government and administration are structurally incapable of, or at least have considerable structural difficulties in, managing this task for longer than just the short term. As the history of nuclear energy shows, problems of overload and difficulties in providing justification can arise, which can then lead to loss of public confidence if these difficulties and problems are not only objectively present but are also perceived by the public. These problems are not merely due to deficiencies in formal rationality of politics and administration, but also arise from a divergence in the values of politics and administration on the one hand and the critical public and the other. If confidence in the system is lost, this has negative repercussions on the capacity of the system to solve difficulties and so makes the whole process of introduction less manageable.
A consciousness of ecological crisis, which has become an important factor in the acceptance issue, has resulted from a public perception of the burden imposed by technological progress on the psychology and physiology of the human species.(79)

Today, technologies have to justify themselves in terms of compatibility with the environment, and the chances that they will be implemented without too many hitches increase in proportion to their environmental compatibility, and to their ability to replace other technologies which are hostile to the environment.

But the definition of what is a burden on the environment depends to a large extent on the perceptions of the main actors in this area. This can be seen in assessments of technologies. Figures which are quoted here are to a certain degree arbitrary that is, they are socially defined, or 'politicised'. This can be seen, for example, in an international comparison of standards of quality set down for the environment which determine, in apparently objective figures, the maximum concentration of harmful substances acceptable. The differences in admissible emission levels in different European countries suggest that 'there are significant cultural differences in the degree to which humans, animals and ecosystems can be subjected to strain', and can be attributed to the fact that these norms are 'compromise values', incorporating technological and economic considerations and influenced by differing risk assessments, by various disciplines and, not least, by political interests.(80)

Social communication and behaviour with regard to new technologies have as their reference point the value-system inherent in new technologies and the institutions that embody these values. Thus, politics and industry are under pressure not only to introduce efficient measures, but also to be transparent in their decision-making, as well as being clear about the value-making underlying their actions. Where 90 per cent of the population believe that politicians do not do enough to protect the environment, this pressure can no longer be relieved by symbolic policies or by merely establishing new advisory capacities.

The process of <u>shifts in the hierarchy of social values</u> leads to difficulties in acceptance where those value-criteria on which the technology is based are incompatible with those of the groups affected. For example, in the case of nuclear energy the criterion of profitability/utility as observed by industry and politics was for a long time placed above that of safety. As public sensitivity to nuclear energy grew, utility continued to be the dominant criterion, but safety became almost as important. Many sectors of the population rated safety above utility, with the result that the legitimating strategies employed by politics and industry, which were still oriented towards profitability, were no longer so acceptable.

Today, sceptical attitudes towards technology can reach explosive proportions when they correlate with elements of a change in values, as has been confirmed by research. Without wanting to enter into the debate on the change from materialistic to post-materialistic values, we should mention here: several trends in society that are of interest to us in particular, a turning away from traditional bourgeois standards, a decrease in orientation on a career, a tendency to adopt expressive leisure activities, and a growing trend towards unconventional political behaviour. One hypothesis deriving from this is that lack of technology acceptance fits into a complex of attitudes and modes of behaviour which are critical of industrial civilisation, and which could therefore become a problem for the officially adopted path of development in a modern industrial society.

It is expected of the information and communications technologies based on microelectronics that they will achieve a fast and dramatic restructuring of work of hitherto unknown proportions. Technological innovations and the social organisation of work lead to a re-allocation of downgraded work and highly qualified work, and at the same time create an enormous number of potential redundancies, the extent of which cannot accurately be assessed. But it can already be said that in the production sector a comparatively small class of skilled workers will use the opportunities provided by new technologies, whereas the majority, and especially women, will run the risk of unemployment and the downgrading of their jobs.

An empirically proven tendency is the 'split of the world of employment into a stable, highly modernised core sector and 'declining industries', and an increasing separation of those with jobs and the long-term unemployed'.(81) In the service sector the introduction of electronic data processing in offices and specific information technologies in engineering has led to a polarisation into high and low-qualified activities, and it has recently been observed that those areas of work which have been 'upgraded' are both more exacting, and subject to greater control and increasing pressure for high performance.(82)

It is as yet not discernible how rationalisation will affect the working population as a whole with regard to the 'aspect of job loss' and the 'aspect of the re-deployment of those remaining in employment',(83) or with regard to the division of the industrial sector into 'sunset' and 'sunrise' industries. It is, however, clear that technological innovation will create a new arrangement in each sector concerned, and consequently a redistribution of opportunities and risks.

Finally, in the sphere of work, there will be a revaluation of traditional work values, and this will have consequences for leisure too: 'diligence', for example, will lose its significance to the extent that in the industrial production sector human work activity is no longer directly involved in the production processes but is rather increasingly confined to a monitoring and checking activity. On the other hand, in those sectors of the economy where there is an apparent re-definition of Taylorist principles, a new value could be put on human labour, and this could eventually prevail. Increasing automation could lead to a complete re-definition of tasks, from which the higher-graded workers in production and the skilled workers would profit, at the expense of other groups.

The tertiary sector too can profit from the opportunities presented by upgraded work together with a possible revival of old virtues and the creation of new ones. Where technologies and their organisation stabilise or strengthen work motivation, there will be ready acceptance of the technology, and this will lead to an acceleration of the 'informatisation' of the service sector. But the question here also will be whether there will not emerge a group of those who lose out in the rationalisation process, whose existence will bring out the dark side of the new technologies and challenge their acceptance and acceptability.

Job losses, redundancy, re-definition of jobs, changes in the system of training and qualifications, change in the significance of traditional virtues, all demand both individual and collective adaptation, in the course of which problems of acceptance have already emerged, and this will continue to happen. Attitudes and modes of behaviour related to the question of acceptance range from various forms of refusal, and 'opting

129

out' to active co-operation in the process of technological innovation. It cannot be assumed that the one or other tendency will prevail; there will be fair spread over the whole spectrum, with a great deal of variation. This very differentiated reaction will require enterprises, union and state to act in new ways, and to include the problem of the acceptance of specific uses of technological innovation in their deliberations.

In the foregoing discussion the complex situation formed by the interconnection of problems of acceptance and other structural problems in society could only be touched upon. The social sciences should make increased efforts to form a basis for society and politics to come to a better understanding of the conditions- for the creation of a technology, the circumstances surrounding its use, and its consequences. This is especially valid because it seems that Jürgen Reese's observation made back in 1980 still holds true today: 'The Federal Republic of Germany is energetically promoting . the production and diffusion of information technology, without acquiring the necessary information to ensure an effective monitoring of this process'.(84) As Renate Mayntz has recently shown in an overview of sociological research in the FRG, there is

still a significant discrepancy between the recognised challenge which the impact of developments in information technology poses to social science (and to society), on the one hand, and the actual research, whether government-sponsored or coming spontaneously from the scientific community, on the other hand.(85)

ANNEX

ACCEPTANCE AS A SUBJECT OF SCIENTIFIC RESEARCH

On the Nation of 'Acceptance'in Various Disciplines

Although it has long been a topic of public debate and a subject of scientific investigation in the Federal Republic, there is nothing like a general consensus on the notion of acceptance of new technologies. In the academic disciplines a whole series of different notions of acceptance has emerged, with approach, method and motivation being the factors making for variety. Presenting an overview is far from easy, due to the large number of interpretations of the term. This is even more the case since the frontiers between the disciplines involved in the investigation of acceptance are relatively fluid, and it is seldom that within a given discipline there is a uniform interpretation of the notion of acceptance. A further complication is the fact that in interdisciplinary projects attempts are made to integrate aspects of various acceptance notions into one comprehensive concept.

Recognising these difficulties, we will nevertheless attempt to give an overview of the various notions of acceptance in the FRG, and at the same time try to assign them approximately to the various directions that research is taking.

A first notion of acceptance can be found in market and diffusion research, where acceptance is generally understood as the extent to which a technology has penetrated into a market and is actually in use. It is described in terms of the speed and density of its introduction and is measured in terms of the volume of investments, the invested purchasing power, and the amount of time allotted to its use. With such a perspective, research concentrates for example on the 'process of acceptance' of a product as a bearer of innovation, whereby decisions to purchase or the adoption/diffusion of a technical innovation can be understood as an expression of acceptance. Central to this type of 'acceptance research' is the actual acquisition and putting into use of the product by the customer, and there is little attention paid to the degree to which the customer is convinced of the meaningfulness of the innovation and its benefit to society.(86)

Other notions of acceptance focusing on the role of new technologies in the organisation or work deal, for example, with ergonomic aspects and the adequacy of software design of information systems.(87) Going beyond these aspects, other approaches in work science and particularly in industrial management and organisation research have developed more complex concepts of acceptance. Looking at the willingness of firms and organisations to purchase and use these systems, it is also of interest to look at resistance to innovation brought about by factors other than technical ones: for example, personal and social factors. The chief subject of study in this type of acceptance research is the behaviour of individuals in the process of technical (and the accompanying organisational and social) change.(88)

In examining the practical application of new technologies and the difficulties arising in that process, the focus is either on functionally adapted or on resistant behaviour. A situation can be analysed for acceptance where a basic 'tendency towards acceptance or towards questioning actually leads to behaviour which conforms to role expectations or to non-conformist behaviour, as the case may be'.(89) Functional or dysfunctional behaviour on the part of users, seen here above all as a factor on the level of the firm or organisation, can, of course, also be analysed from the point of view of acceptance on the level of the market and the behaviour of buyers.

A notion of acceptance which merely looks at observable user or purchaser behaviour and largely shuts out cognition, emotion and motivation, in a certain sense reduces the acceptance problem to its behaviourist dimensions. Because of this, research tries to include these aspects to arrive at a more complex description of the processes of acceptance. It is to be noted that it is only meaningful to identify acceptance with the actual use of a technology when there is a really 'free' choice whether to adopt it or not. This is seldom the case in production and administration. Therefore, other research approaches have tried to detect the 'grey areas' in apparently functional behaviour which are 'in opposition to [the user's] cognitive, emotional and motivational stance'.(90)

An independent line of enquiry on acceptance, which aims to take these very human concerns into account in order to avoid the growing danger of misjudgement on an individual and national economic level when they are ignored, has now established itself in the investigation of developments in information and communications technology in the office and in administration. It is characterised by the fact that it attempts both to integrate perceptions from several disciplines into a new approach, and to make allowance for the 'social consequences' on the levels of the individual, organisation and society'.(91) This direction in research accordingly sees itself as sociological research accompanying pilot projects in programmes of the Federal Ministry of Research and Technology to promote technology, with the aim of being a 'preventive corrective' of undesirable side effects of technological innovation.(92)

It should be noted here that although this direction in research does aim to analyse the overall societal consequences of new information technologies, it has not as yet solved the problem of measuring these consequences empirically. Therefore, the study of this dimension has been left to other directions and approaches.

The acceptance of new technologies is also studied in research into opinions and attitudes. Acceptance is defined here in somewhat 'climactic' terms as the measure on a positive/negative scale of the attitude to or voiced opinion on technologies. The results are very varied. The studies focus on the attitudinal patterns discernible in the population at large (or in specific groups within it) to technology and technological progress, as well as to specific sets of technologies like nuclear energy and information technologies. Of particular interest here is the public perception of the new information technologies. There have been surveys on public estimation of the expansion of electronic data processing in public administration, and on the rating accorded to computers and other technologies in everyday use.(93) Picking out certain situations, the attitudes and behaviour of operators and consumers of information technologies in industry and administration, as well as of technologies in everyday life (especially the 'new media'), are analysed in accompanying investigations or in studies of the impact of pilot projects like cable television and interactive video text.(94)

Similar to the attempts already described to arrive at a more complex notion of acceptance, we also find attempts at greater differentiation in research into opinions and attitudes. This occurs especially in trying to describe and account for the values, the rational and irrational motives, and the psychological dispositions that form attitudes and direct behaviour and actions when faced with technologies. Primarily dedicated to this enterprise is socio-psychological attitude research, and also cognitive perception and evaluation analysis, which is especially geared to assessing risk acceptance.

Arising from criticism of the individual psychological interpretation of the acceptance problem, new approaches have developed in acceptance research that attempt to link up attitudes towards a technology with a scale of values and social expectations of behaviour that in their turn are largely determined by the basic structure of society. Here, factors like public opinion, social values and the socio-structural position of the body concerned are taken into consideration. The acceptance and acceptability of new technology acquires a macrosociological dimension.(95)

Such a development can be seen in the research on risk acceptance which has hitherto largely concentrated on nuclear energy, and which sees itself today as an advance over that type of risk assessment concerned exclusively with technical risks. The intention is to focus not on the economic or technical side, but on the 'political feasibility' of new technologies. Here, it has gradually come to be recognised that acceptance is not merely a question of complete and correct information, but is above all a 'psycho-social problem of modifying accustomed behaviour and forms of social interaction'.(96) In so far as risk-acceptance research now takes this into account and 'is beginning to abandon the paradigm derived from safety research', it has correspondingly expanded its field of enquiry as a result of a learning process. Where it is concerned with the acceptance crisis of nuclear energy, it begins to see this as connected to problems of the shift in values and of the legitimacy of other political subsystems, thereby providing an attempt to account for the long duration of this crisis. Its attempts at explanation, therefore, tend in the direction of trying to link up individual attitudes and subjective preferences and value systems, which are influenced by technological innovation on the one side, with socio-structural processes in which technical development plays an important role on the other.(97)

In connection with attempts at defining a nation of acceptance which does not exhaust itself in quantifiable opinions and attitudes, approaches are to be noted which emphasise the normative dimension of acceptance. We find terms like 'social acceptance', 'acceptability' and 'social compatibility' being used: the last of these being defined as 'a concordance of technological development with the structure of values present in society as well as the guaranteed keeping open of options regarding the different possible lines of development of society',(98) as opposed to a mere subjective question of personal opinion. Equally, the notion of acceptability contains an explicit reference to values and is distinct from 'acceptance' in that it evaluates the introduction of technology by means of subjective normative acceptability criteria, and accordingly argues for its being worthy of acceptance.(99)

Finally, we have to examine the meaning and significance of acceptance and acceptability in the context of sociological impact research and technology assessment.

The efforts of interdisciplinary impact research as they have developed, for example, in the research institute Gesellschaft fur Mathematik und Datenverarbeitung (Society for Mathematics and Data Processing) since about 1977, are aimed especially at information technologies and their prerequisites and consequences in society.(100) In so far as we are talking about the acceptance of new technologies as an object of research, acceptance research is understood as the 'development, application and evaluation of participatory strategies', that is, as the continued development of 'user participation', in the sense of an 'analysis of needs and the involvement of those affected, with the aim of taking hitherto neglected interests into account' and as a 'means of achieving a human-scale application of technology'.(101) Therefore, acceptance research does not refer to 'the articulated needs of organised interested parties, but to the more rarely articulated needs of the citizen affected by information technology'.(102)

In the case of information technology the aim is practically oriented knowledge of its consequences. To understand this approach, it is important to realise that a practical orientation is not on the same plane as political consultancy. The results are not primarily addressed to policy-makers but rather to those affected by technological change, so that research does not run the risk of merely generating knowledge which serves to maintain the system or to perpetuate the pressure of utility criteria. This approach acknowledges the problem of the refusal by the political system to accept and apply such knowledge and consciously allows for this danger. The hitherto published results from impact research are, however, chiefly characterised by endeavours both to develop the theory and the methodology for such acceptance research, and to put them to some individual tests. Rarely have results been produced which are already geared to being applied in practice, and which especially also meet the demand of being able to anticipate consequences.

In the German variety of technology assessment (Technologiefolgen-Abschätzung), seen as an aid to decision-making on technology,(103) policy research is carried out in the fields of energy, automation, transport and environment.(104)

Up to now, investigations have usually had the character of technically and economically oriented feasibility studies. Only more recently has increased attention been devoted to social consequences, and thus to the problem of acceptance as evidenced by the work of the above-mentioned commissions. Here the activities of the parliamentary commissions should be mentioned, which have been, or are still, concerned with energy policy, information technology and biotechnology. It still has to be said, however, that acceptance and acceptability have not as yet achieved prominent status as of central research interest. Instead 'social acceptance' is rather seen as a 'special problem'.(105) That may be connected with the fact that technology assessment or 'technology appraisal'(106) is still primarily concerned with basic conceptual and methodological problems.

On the Conceptual Development of Acceptance Research

The multitude of notions of acceptance and the number of disciplines employing them in the FRG shows that acceptance research comprises both a number of mixed approaches and interests and a great variety of methods of enquiry. The debate in search of an outline for social science research on acceptance has not yet been able to develop any generally accepted theoretical base which could serve to bring together disparate attempts. The most important questions for the development of a research strategy are as yet still unresolved: these include questions as to the complexity of the notion of acceptance, the temporal framework governing processes of acceptance, the validity of methodologies, and the question of the practical orientation and reliability for making prognoses. Several tendencies may, however, be characterised which can be seen as positive in terms of the further conceptual development of acceptance research:

• There is an increasing number of attempts to work with a notion of acceptance that goes beyond observable behaviour in terms of purchaser decision and intensity of use on the one hand, and voiced opinion and attitude on the other.

• It can be observed that investigations into opinion and attitude are becoming increasingly refined. 'Technology' and 'the public' are no longer seen as uniform phenomena but as being differentiated in many respects.

• Sets of variables are employed that yield far more substantive results, which are much better suited to achieve an elevation of and explanation for the symptoms diagnosed. The opinions and attitudes surveyed are related to, for example, socio-demographic factors or variables such as experience, especially work experience with, for example, computers, or technical knowledge; other parameters are included too, for example, political orientation and value preferences. Furthermore, results are no longer presented in descriptive terms only, but are also interpreted in terms of the various internal variables and then viewed in their relationship with the 'outside world'.

• Even with highly differentiated investigations, one increasingly realises that, for all the empirical soundness with which individual data are collected, their usefulness in more general interpretation is limited, especially with regard to developments or changes over time. We are made aware again and again of the fact that opinion polls can only offer us 'snapshots' of the situation at a specific time and place. There are signs, however, of time series studies being undertaken which allow for a more systematic grip on attitudes and changes in attitudes, which therefore create a wider empirical base.

• With a broadening of the empirical base in mind, efforts are being made to overcome the unsatisfactory situation that

the relative weight of the various factors governing attitude formation cannot as yet be clearly read off from the results of these investigations, and neither have the interdependences which undoubtedly exist between the various influencing factors been empirically recorded. In particular, the feedback processes between public and individual opinion formation have ... still to be analysed.(107)

• In the course of seeking the causes of, and the determining contexts for the acceptance or rejection of technology, the conclusion has been reached that a social phenomenon like technology acceptance cannot be looked at and accounted for in isolation. So various attempts have been made to interpret the phenomenon within a broader framework. Thus, technology acceptance is interpreted as a subphenomenon of larger syndromes, like for example changes in values, social movements, or the critique of industrial civilisation; or else as being related to acceptance problems in other areas of policy, like, for example, functional and structural change of the political administrative system or of political culture in the form of an 'extension determined by socialisation of cognitive and motivational capacity to participate in political decision-making processes'.(108)

These attempts to see problems of acceptance, which are in a sense 'grouped around a technology', in relation to each other, and to analyse them in relation to higher 'systems', point to the inadequacy of empirical investigations into attitudes. They also manifest the disadvantages stemming from imprecise methodology and an inadequate empirical base. So, much remains arbitrary and speculative, though this does not necessarily reduce the heuristic value of such studies, and it should under no circumstances be taken to mean that efforts in this direction should be abandoned.

When aiming at a sufficiently complex understanding of the problem of the acceptance of new technologies, one has to go beyond the instrument of static opinion polls. But it will then, in the final analysis, not be possible to resolve completely the tension between the demand for a sound empirical base on the one hand and for generalising statements on the other, or the question of how the various factors relevant to the problem of acceptance on the micro, meso and macro levels are to be related to each other. Studies on the problem of acceptance, like all studies on the role of technology in society, will continue to move between the two poles of immediate relevance on the one hand and reliable generalisability on the other. As already seen in the field of nuclear energy studies, the long-term problem of the empirical validity of one particular chosen notion of acceptance will remain. This will at least be true when one is not merely interested in static opinion surveys or factual behaviour, or when one is interested in more than one specific form of technology under its restricted conditions of time and place.

The price to be paid for studies which are methodologically and empirically sound is that their results are not readily generalisable; on the other hand, the approaches which go above and beyond empiricism succeed in setting up generalising hypotheses, but the cost of their empirical grounding and immediate practical relevance.

It is clear that the basic set of conceptual and methodological problems of social science research on technology in general, and of acceptance research in particular, have not been exhaustively treated in the foregoing discussion. However, three points emerge which are of importance for further research efforts in acceptance and acceptability: the understanding of acceptance should not restrict itself to observable behaviour or to measurable attitudes; the methodological tools should continue to be refined and the empirical base should be broadened further; the focus of interest should - in spite of methodological uncertainty and lack of empirical evidence - be on the macrosocial dimension of technology acceptance and, therefore, on models which account for its social significance and accompanying social change.

Notes

1. Dierkes, M. and V. Thienen, 'Strategien und Defizite bei der Behandlung technischer Risiken: Ein Problemaufriss', Staatliche Gefahrenabwehr in der Industriegesellschaft, hrsg. von Ulrich Becker (Schriften der Deutschen Sektion des Internationalen Instituts fur Verwaltungs- wissenschaften Bd. 6), Bonn, 1982; pp. 73-91.

2. Renn, O., Wahrnehmung und Akzeptanz technischer Risiken, 6 Bände, Jülich 1981. (= Jülich-Spezial 97 der Kernforschungsanlage Jülich); I, p. 116.

3. Ministerium fur Wissenschaft und Kunst Baden- Württemberg (Hrsg.). Kritik an der Technik und die Zukunft einer Industrienation, Demoskopische Nachwuchsanalyse fur Ingenieurberufe, vorgelegt vom Institut fur Demoskopie Allensbach, Villingen-

West Germany

Schwenningen, 1982; p. 11.

4. Ibid., pp.XXIII and XXV.

5. Dierkes, M., 'Wissenschaft geniesst Vertrauen', Rheinischer Merkur/Christ und Welt, Nr. 48, 26.11.82, p. 17.

6. Beckurts, K.H., 'Die Schlacht kann gewonnen werden: Innovationsstärke und Wettbewerbsfahigkeit der deutschen Industrie',Die Zeit, 7.9.1984, p 44.

7. Archiv des Instituts fur Demoskopie Allensbach, IfD-Umfrage 4048 (1985).

8. Ministerium fur Wissenschaft und Kunst Baden-Wurttemberg (Hrsg.), Kritik an der Technik, p.13.

9. Geissler, R., 'Das Interesse ist gewachsen - aber auch das Gefahrenbewusstsein', Frankfurter Rundschau, 27.4.1983, pp. 14-15.

10. Renn, O. Wahrnehmung und Akzeptanz, III, p. 108.

11. Ziefuss, H., 'Technikfeindlichkeit der Jugend - eine vergebliche Debatte?', Gesellschaft für Arbeit, Technik und Wirtschaft im Unterricht (Hrsg.), in Arbeitslehre zwischen Technikfeidlichkeit und Arbeitslosigkeit, Bad Salzdetfurth und Hildesheim, 1983, pp. 33-55.

12. Renn, O. Wahrnehmung und Akzeptanz, III, p.27.

13. Ibid., p.6.

14. Ministerium fur Wissenschaft und Kunst Baden-Wurttemberg (Hrsg.), Kritik an der Technik, p. 12.

15. Lange, K., Der Computer im Meinungsspektrum der bundesrepublikanischen Bevolkerung: Materialien zur Eindordnung, Differenzierung und Erklärung, St. Augustin, 1983 (=Arbeitspapier der GMD Nr. 64), p. 34.

16. Computer-Image, 'UV Planung und Steuerung,' Winfried Hoffman, 12.12.83, p. 14.

17. Renn, O. Wahrnehmung und Akzeptanz, II, p. 81.

18. Müller-Boling, D., 'Inforantionstechnik aus der Sicht der Benutzer im Wandel des letzten Jahrzehnts', GMD-Spiegel, 3/4, 1983, 35-41, p. 37.

19. Hartmann, M., Rationalisierung im Widerspruch: Ursachen und Folgen der EDV-Modernis-ierung in Industrieverwaltungen, Frankfurt, New York, 1984, p. 282.

20. Schmittchen, G., Neue Technik, Neue Arbeitsmoral: Eine Sozialpyschologische Untersuchung über die Motivation in der Metallindustrie, Köln, 1984, p. 202.

21. Naschold, F., in Bundesminister fur Forschung und Technologie (Hrsg.), Modernisierung der Volkwirtschaft in den achtziger Jahren, Düsseldorf, Wien, 1981.

22. Bruder, W. and Ende, W., 'Forschungs- und Technologiepolitik in der Bundesrepublik Deutschland,' Aus Politik und Zeitgeschichte, B 28, 12.7.1980, 3-31; p. 8.

23. Ibid., pp. 12 and 13; also Bartelt, M. et al. Forschungspolitik, 'Technologiefolgenabschatzungund offentlicher Dialog', Aus Politik und Zeitgeschichte, B 28, 12.7.1980, p. 22-36

24. Bundesbericht Forschung 1984, hrsg. vom Bundesminister für Forschung und Technologie, Bonn, 1984, p. 26.

25. Probst, A., Neue Techniken schaffen neue Möglichkeiten, neue Bedürfnisse, neue Märkteund neue Arbeitsplätze, Bonn, 1984, p. 7.

26. Ibid., p. 7.

27. Reisenhuber, H., Informationstechnologie - eine Gemeinschaftsaufgabe von Wirtschaft, Wissenschaftund Staat, Bonn, 1983; p. 8.

28. Ibid.

29. Kreklau, C., 'Ohne Technik kein Fortschritt', Der Arbeitgeber, Nr. 23/35, 1983, p. 928.

30. Kador, F. J., 'Technischer Fortschritt braucht Akzeptanz', Der Arbeitgeber, Nr. 23/35, 1983, 929.

31. cf. Frankfurter Rundschau 17.3.1984.

32. cf. Frankfurter Rundschau 15.5.1984.

33. Brandt, G., Jacobi, O. and Müller-Jentsch, W.,Anpassung an die Krise: Gewerkschaften in den siebziger Jahren, Ffm. 1982, p. 242.

34. Ibid., p. 81.

35. Zwischenbericht der Enquete-Kommission Neue Informations - und Kommunikationstechniken, Bundestagsdrucksache 9/2442, 28.3.1983, p. 115.

36. Regierungsbericht 'Informationstechnik', BMFT 413-5800-1-2/84, 23.2.84.

37. Kalbhen, U., 'Informationstechnik im Bildungswesen: Überblick über internationale Entwicklungen, Forderansatze und Forderstrategien', Loccumer Protokolle 23/1983, Neue Technologien und Schule, Loccum 1984, pp. 25-50, p. 44.

38. Ibid., pp. 39ff.

39. Renn, O. (1981), Wahrnehmung und Akzeptanz III, pp. 88ff.

40. Lange, K., Der Computer im Meinungsspektrum, and Das Image des Computers in der Bevölkerung, St. Augustin, 1984. (=GMD Studien Nr. 80, März 1984).

41. Battelle-Institut (Hrsg.), Einstellung und Verhalten der Bevölkerung gegenüber verschiedenen Energiegewinnungsarten Bd. 1 und 2, Ffm. 1977; B7.

42. Kitschelt, H., Politik und Energie: Eine vergleichende Untersuchung zu den Energiepolitiken der USA, der BRD, Frankreichs und Schwedens, Ffm 1983, p. 212.

43. Frederichs, G., Bechmann, G. and Gloede, F., Grosstechnologien in der gesellschaftlichen Kontroverse: Ergebnisse einer Bevölkerungsbefragung zu Energiepolitik,Kernenergie und Kohle, Karlsruhe 1983; p. 22.

44. Kitschelt, H., Politik und Energie, p. 392.

45. Koschnitzke, R. and Rolff, H. G. (Hrsg.),

Technologischer Wandel und soziale Verantwortung, Essen, 1980, p.61.

46. Kitschelt, H., Politik und Energie, p. 375.

47. Radkau, J., 'Kernenergie - Entwicklung in der Bundesrepublik: ein Lernprozess?' Geschichte und Gesellschaft, 4, 1978, 195-222, 216.

48. Kitschelt, H., Politik und Energie, pp. 385ff.

49. cf. Kitschelt, H., Kernenergiepolitik: Arena eines-gesellschaftlichen Konflikts, Ffm. 1980.

50. Radkau, J., 'Kernenergie', p. 679.

51. Frederichs, G., and Loben, M., Die Akzeptanzproblematik der Kernenergie. Konsequenzen des grosstechnischen Einsatzes der Kernenergie in der Bundesrepublik Deutschland, Abteilung fur angewandte Systemanalyse des Kernforschungszentrums Karlsruhe (KfU 2705), February 1979, p. 24.

52. Kiersch, G. and v. Oppeln, S., Kernenergiekonflikt in Frankreich und Deutschland Berlin, 1983; pp. 65 ff.

53. Ibid., p. 95.

54. Eisenhart, G., Krebsbach, G. Camilla, Stereotype Wahrnehmung. Ihr Einfluss auf Interaktion und Kommunikation im Bereich politischer Planung, Ffm 1978, p. vi.

55. Frederichs, G., Bechmann, G. and Gloede, F. Grosstechnologien, p. 3.

56. Kitschelt, H., Politik und Energie, p. 384.

57. Lange, K., Der Computer in Meinungsspektrum, p. 34.

58. Kern, H. and Schumann, M., Das Ende der Arbeitsteilung? Rationalisierung in der industriellen Produktion: Bestandsaufnahme Trendbestimmung, 1984.

59. Kalbhen, U., Kruckeberg, F. and Reese, J. (Hrsg.), Gesellschaftliche Auswirkungen der Informationstechnologie: Ein internationaler Vergleich, Ffm, New York, 1980.

60.   Mambrey,   P.   and   Oppermann,   R.   (Hg.)
Beteiligung von Betroffenen bei der Entwicklung von
Imformationssystemen, Ffm, New York, 1983, p. 5.

61.   Kubicek,   H.,   Interessebeerücksichtigung beim
Technikeinsatz   im   Büro-und   Verwaltungsbereich.
Grundgedanken   und   neuere   skandinavische
Entwicklungen, Munchen, Wien, 1980.

62.   Mickler,   O.,   Facharbeit   im   Wandel.
Rationalisierung   im   industriellen   Produktions-
prozess, Ffm, 1981, pp. 215ff.

63.   Malsch, T., Weissbach, H. J. and Fischer, J.,
Organisation   und   Planung   der   industriellen
Instandhaltung, Ffm, New York, 1982.

64.   Weltz,   F.   and   Lullies,   V.,   Innovation im
Buro: Das Beispiel Textverarbeitung,   Ffm,   New
York, 1983.

65.   Ibid., p. 165.

66.   Ibid., p. 156.

67.   Brumlop,   E.   and Jurgen, U., Rationalisation
and   Industrial   Relations   in   the   West   German
Automobile   Industry: A Case Study of Volkswagen,
Berlin   1983 (=IIVG/dp 83-216 des WZB   Berlin),   p.
47.

68.   Ibid., p. 49.

69.   Mickler,   O.   (1981),   Facharbeit im Wandel,
Schumann,   M. and Wittemann, K.P. 'Beherrschung des
Arbeitsprozesses   als   Interessenkonflikt   zwischen
Betrieb   und   Arbeitern   am   Beispiel   der
Industrialisierung im Schiffbaum', Materialien zur
Industriesoziologie,   Sonderheft   24/1982   KZSS,
Opladen, 1982.

70.   Schumann, M. and Wittemann, K.P. 'Beherrschung
des Arbeitsprozesses'.

71.   Kern,   H. and Schumann, M., 'Rationalisierung
und   Arbeitsverhalten   - Ansatz und   erste   Befunde
einer   Folgestudie   zu   "Industrie   arbiet   und
Arbeiterbewusstsein",   Materialien zur Industries-
ozologie, Opladen,   1982 (=Sonderheft 24/1982,
KZSS).

72.   Haider,   E.   and   Rohmert,   W.,
'Arbeitswissenschaftliche   Grundlagen   und
Erkenntnisse   zum   Thema   'Technologieentwicklung,
Rationalisierung und Humanisierung', Technolog-
ieentwicklung, Rationalisierung und Humanisierung,
IAB-Kontaktseminar   1979   am   Institut   fur
Sozialforschung Ffm, Nurnberg 1981 (=Beitr AB Nr.
53   des   Institutus   fur   Arbeitsmarkt   und
Berufsforschung), p. 25.

73.   Seltz, R., Neue betriebliche Machtressourcen
im Wandel des Kontrollsystems durch elektronische
Informations - und Kommunikationstechologien: Eine
theoretische   und empirische Skizze zu 'Kontrolle
im Arbeitsprozess' und 'Arbeitspolitik', Berlin,
1984 (=IIVG/dp 84-202 des WZB, Berlin); pp. 8ff.

74.   Jurgens, U., 'Die Entwicklung von Macht,
Herrschaft und Kontrolle im Betrieb als politischer
Prozess - Eine Problemskizze zur Arbeitspolitik',
Arbeitspolitik,   Opladen,   1984   (=   Leviathan
Sonderheft 5/1983), pp. 58-91, p. 77.

75.   Feser, H-D. and Larm, T., 'Strukturelle
Arbeitslosigkeit, technologischer Wandel und der
Einfluss der Mikroelektronik', Leviathan, 10, 1982,
531-54, p. 549.

76.   Kubicek,   H.,   Sozialtechnologie   des
Managements: Uternehmerische Beteiligungsstrategien
als   Herausforderung   an   die   betriebliche
Interessenvertretung,   Arbeitspapier   83/2,
Universitat Trier FbIV, 1983.

77.   Pohler, W. and Peter, G., Erfahrungen mit dem
Humanisierungsprogramm. Von den Moglichkeiten und
Grenten   einer   sozial   orientierten
Technologiepolitik, Koln, 1982.

78.   Frederichs, G., Bechmann, G. and Gloede, F.
Grosstechnologien. p. 18.

79.   Kessel,   H.   and   Tischler,   W.,
Umweltbewusstsein: Okologische Wertvorstellungen in
westlichen Industrienationen Berlin, 1984.

80.   Weidner, H. and Knoepfel, P., 'Politisierung
technischer Werte', Zparl, 1979, pp. 160-70; p. 160
and p. 163; also Conrad, J. and Krebsbach-Gnath,
C.,   'Zum   gesellschaftlichen   Umgang   mit

technologischen   Risiken',   Zeitschrift   fur
Umweltpolitik, 3, 1980, p. 821-45.

81.   Beckenbach, N.,   'Zukunft der Arbeit und
Beschaftigungskrise. zu   den gesellschaftlichen
Rahmenbedingungen der Neuen Techniken', Prokla, 55,
1984, 22-40, p. 28.

82. Ibid., p. 34.

83.   Kern, H. and Schumann, M. 'Rationalisierung',
p. 19.

84.   Jurgen   Reese in Kalbhen,   U.   et   al.
Gesellschaftliche Auswirkungen, p. 135.

85. R. Mayntz in Szyperski, N., Grochla, E.,
Richter,   U.M. and Weitz, W.P., Assessing the
Impacts   of   Information   Technology,
Braunschweig/Wiesbaden, 1983, p.291.

86.   Schonecker, H.G., Bedienerakzeptanz und
technische   Innovationen: Akzeptanzrelevante
Aspekte   bei   der   Einfurhung   neier
Burotechniksysteme, Munchen, 1980, pp. 82ff.

87.   Reichwald, R., Zur Notwendigkeit der
Akzeptanzforschung bei der Entwicklung neuer
Systeme der Burotechnik, Munchen, 1978.

88. Schonecker, H.G. Bedienerakzeptanz, pp. 88ff.

89.   Kubicek, H., Informationstechnologie und
organisatorische   Regelungen.   Konzeptionelle
Grundlagen   einer   empirischen   Theorie   der
organisatorischen Gestaltung des Benutzerbereichs
in comptuergestutzten Informationssystemen, Berlin,
1975, p.94.

90. Schonecker, H.G. Bedienerakzeptanz, p.88.

91.   Dohl, W., Akezptanz innovativer Techniken in
Buro und Verwaltung, Gottingen, 1983, pp. 111ff.

92. Ibid., p.118; see also Manz, U.,Einordnung der
Akzeptanzforschung   in   das   Programm
sozialwissenschaftlicher   Begleitforschung: Ein
Beitrag zur Anwenderforschung im technisch-
organisatorischen Wandel, Munchen, 1983, pp. 175ff.

93. Computer Image. 'UV Planung und Steuerung' Lange, K. Der Computer in Meinungsspektrum, Lange, K. Das Image des Computers.

94. Bundesminister fur Post - und Fernmeldewesen (Hrgs.), Zwischenberichte zur wissenschaftlichen Begleituntersuchung des Bildschirmtext-Feldversuches im Raum Dusseldorf/Neuss, Dusseldorf 1982; Industriegewerkschaft Metall (Hrsg.), Maschinen wollen sie - uns Menschen nicht; Rationalisierung in der Metallwirtshaft, Ffm, 1983; Meier, B. Die Mikroelektronik: Anthropologische und sozio-okonomische Aspekte der Anwendung einer neuen Technologie, Koln, 1981; Muller-Boling, D., Arbeitszufriedenheit bei automatisierter Datenverarbeitung, Munchen, Wien, 1978; Muller-Boling, D., 'Informationstechnik'; Muller-Boling, D. and Muller, M., ADV-Attituden im zietlichen Wandel: Erste betriebsspezifische Auswertungen einer Langsschnittanalyse, Dortmund, 1983 (=Arbeitsbericht Nr. 4, Universitat Dortmund, Abt. Wirtschafts - und Sozialwissenschaften.) Scheffler, H., 'Bildschirmnutzer: Urteile und Haltungen', IBM Nachrichten Heft, 8, 1983, 21-5; Schellhass, H. and Schonecker, H., Kommunikationstechnik und Anwender Akzeptanzbarrieren, Bedarfsstrukturen, Einsatzbed-ingungen, Munchen, 1983; Schmittchen, G., Neue Technik.

95. Frederichs, G., Bechmann, G. and Gloede, F. Grosstechnologien.

96. Bechmann, G. and Frederichs, G., 'Vom Risikobegriff zur Akzeptanzproblematik in der Kontroverse um die Kernenergie: Probleme und Perspektiven der Akzeptanzforschung', Ms Manuskript, Karlsruhe, 1980, p. 18.

97. Ibid., p. 28.

98. Renn, O. et al., Endgutachen 'Sozialvertrag-lichkeit von Energiesystemen', Masch. Ms., Julich, 1984.

99. Dierkes, M. and Thienen, V., 'Akzeptanz und Akzeptabilitat von Informationstechnologien', Paper des WZB Berlin, 1982.

100. cf. Manz, U. Zur Einordung, pp. 80ff.

101. Kalbhen, U. et al., Gesellschaftliche

West Germany

Auswirkungen, p. 21.

102. Ibid., p. 143.

103. Conrad, F. and Paschen, H., 'Technology Assessment (TA) - Entscheidungshilfe fur Technologiepolitik', Technische Mitteilungen, 75 1980, 5-13.

104. Helle, H.J., Gutachten zum Problemereich Technischer Wandel und dessen Einfluss auf technische, wirtschaftliche und soziale Entwicklungen, Munchen, 1980.

105. Bohret, C. and Franz, P., Technologiefolgenabschatzung: Institutionelle und verfahrensmassige Losungsansatze, Ffm, New York, 1982, p. 51.

106. cf. Munch, E., Renn, O. and Roser, T., Technik auf dem Prufstand: Methode und Massstabe der Technologiebewertung, Essen, 1982.

107. Beker, G., v. Berg, I. and Coenen, R., 'Uberblick uber empirische Ergebnisse zur Akzeptanzproblematik der Kernenergie', Abteilung fur Angewandte Systemanalyse, Kernforschungszentrum Karlsruhe, Mai 1980; (KfK 2976),41.

108. Frederichs, G. and Loben, M. 'Die Akzeptanzproblematik', p. 28.

148

Chapter Six

PUBLIC ACCEPTANCE OF NEW TECHNOLOGIES; FRANCE

I	Government to Citizen	- Jean Paul Moatti
II	Production	- Pierre Rolle
III	From Citizen to Consumer	- Corrine Hermant, Eric Barchechath

Introduction

In reporting the results of any international comparison it is in one's interest to take a few preliminary precautions in order to sidestep any stray ambiguities. As examples, we shall state three such precautionary measures.

The first is to postulate what it is that specifically differentiates one's country from its neighbours. Each nation is sensitive to its 'insularity': hence any worthwhile self-examination will underline these peculiarities, best summed up in the expression: 'We are different'.

The second is a kind of natural reflex, which is to lay the blame on the 'burden of history'. A more or less official stamp of approval invariably accompanies this reference, which may be insinuated in a variety of fashions, for instance, in the form of criticism of other countries ('They have a head start over us because of their time-honoured traditions') or, as back-handed praise ('They are ahead of us because they are not weighted down by the past, as we are'). This amounts to a self-justifying 'We're doing our best, but...'.

The third precautionary measure, and one which applies especially to the fields dealt with here, is to qualify a situation by invoking the rather negative attitude of the public at odds with the nation's industry. Invariably considered an important stance in analysing citizens' attitudes vis-à-vis technical advance, this position is expressed in more ideological terms: either we say how shocked we are at society's disregard for the benefits born of progress in technology, or we relish this underestimation. (Incidentally, how can we really expect civilians - men, women and

149

their institutions - to look with a kind eye towards industry when 'by definition' they find it bothersome, if not downright disturbing? Yet can we seriously picture a population as hostile to that very industry which sustains it. A paradox is already emerging: the same differences make up our distinctions. In other words, we are different in the same way.)

It is left to the reader to decide what expressly distinguishes the French example and to judge for him/herself the relevance of the aspects we have thought it our job to highlight. These tend in three directions:

• <u>Impacts on the Government decision-making process and ensuring political and social conflicts</u>. Here, one basic issue is how central government is to deal with technological and scientific choices, given that these are becoming increasingly complex and important in man's everyday life.

• <u>Impacts at the place of employment</u>. The main questions here arise from difficulties in adjusting the division of labour and job qualifications to progress in technology: this involves evaluating the social costs and benefits of consecutive changes in terms of company productivity and the worker himself, who is often faced with redundancy of his technical skills.

• <u>Impacts on the quality of life</u>. The introduction of new technologies is upsetting the classical balance in communications and information networks, and giving rise to cultural repercussions of incalculable consequence; the automobile is one example.

These three areas have been examined from various perspectives, including, for instance: the emergence of a 'new middle class' which affects citizens' attitudes towards environmental issues and technical innovation; the realignment of working relationships; and the shifting balance between leisure and working time, between government intervention and private initiative.

Problems related to the acceptance of technologies are no longer dealt with solely at a national level. They may be better understood through comparative analysis of the new relationships between the social management of time and the organisation of urban and interurban space, with a view towards the twenty-first century.

I Government to Citizen: Social Disputes and Technological Acceptance

From the point of view of the social acceptance of technologies, the situation in France, at first glance, appears ambiguous for several reasons:

° On the one hand, the shortage of technical training and the failure of traditional instruction to keep up with the development of industry are now regarded as major causes of the country's economic woes. The French government's Ninth Plan (1984-8) therefore emphasises that, 'a misunderstanding continues to exist between French society and French industry which is capable of stifling an attempt to renovate the production machine'.

• On the other hand, France is one developed country where the existence of a number of wide-ranging technological programmes has eliminated many obstacles in advance. The most outstanding example of this has been in the electronuclear field, where French foresight has enabled the nation to meet energy objectives set in 1974 in the aftermath of the fourfold increase in the price of petroleum. (Japan has done so as well, but to a lesser extent.) Similar examples are the development of the TGV (high speed train) and the telematic networks.

Historical factors are often cited in explaining this obvious paradox. Since the installation of political and administrative centralism by the Jacobins during the French Revolution, France has witnessed the growth of a bureaucracy that lacks the blessing of public consensus. This is used by the government to place controls on the citizenry by means of a state-run apparatus that is out of touch with the population.(1) The institutions of France's Fifth Republic established in 1958, and characterised by the strongest pre-eminence of the executive and the office of the President of the Republic in all Western democracies, have merely strengthened this tendency. Whenever government planners have managed to gain approval for long-range development programmes, French industry has been crowned with success, as in the case of the nuclear fuels project.(2) On the other hand, whenever innovation is linked to the free interplay of international market forces and the willingness of companies to take risks, French industry competes at a disadvantage in the struggle to achieve full exploitation of the country's excellent scientific potential. The facts of history go some way towards explaining this state of affairs, but they do not go to the real heart of the matter.

It was not until October 1981 that, for the first time in the history of the Fifth Republic, the energy and nuclear power policies of the French government were subject to debate and a vote in the National Assembly of France. At that time, one kilowatt-hour in three was already nuclear-generated; the power stations completed since then had raised this ratio to one in two by 1984.

Nevertheless, it would be rash to assume that the only lesson to be drawn from the French experience is that authoritarian and centralised procedures are effective instruments for directing a broad technological strategy. Moreover, France has experienced its share of public resistance to the construction of industrial plant, along with a growing awareness of technical progress as a fact of life. This has been the common lot of all Western democracies over the last 20 years. For this reason, the body of social science research carried out in France, and considered here, can provide a significant contribution to international deliberations about the acceptance and acceptability of modern technology. Our goal here is to analyse the nature of locally organised opposition to large energy products and industrial programmes. We shall also consider the broader sociological conclusions which may be drawn from the existence of such opposition. We will conclude with an examination of the current state of affairs in France, which has incalculable ramifications. The central question here is the extent to which the present economic crisis, coupled with the measures taken by government to stimulate industrial innovation, tends to create new conditions for the broad acceptance of new technology and rapid change.

During the 1970s and 80s a wide range of development programmes was set up, for example the Larzac Military Camp, the Rhine-Rhone Canal, the high-speed Paris-Lyon express, several local airports, roads, nuclear power stations, hydroelectric plants, dams, high-tension lines and industrial installations (i.e. Fos, Marckolshiem, Vezelay, etc.). Each has encountered local, or even national, opposition which has sometimes forced them to be modified, or even cancelled altogether.

Of course, the environment and wildlife have been topics of acrimonious debate for many years. The Société Imperiale d' Acclimation, which later became today's Société Nationale de Protection de la Nature, was first founded in 1857, and the first international environmental protection conference was held in Berne, Switzerland, in 1913. French railway construction in the nineteenth century and the development of hydroelectric power in the aftermath of the Second World War both met with staunch opposition at every major stage along the way. However, opposition has seldom gone beyond the local level. In most cases it has been confined to making claims known and gaining individual compensation. Reaction against industrialisation in general has involved either isolated cases of personal interest, or a conservative challenge to change and progress - in short, rearguard campaigns that are now seen by many as having had no real historical significance.

New Meaning in a New Context

Although social scientists and politicians have focused a great deal of attention on the protest movements of the last ten years, historical precedents are not to be found. Today's social conflicts about technologies are of a completely new kind. Occurring within a completely new set of social, cultural and political circumstances, they are charged with extra significance because of the social risks they imply. These far outweigh in importance any of the issues which are immediately at stake, and vary from one industrialised country to another.

Extensive Growth of Associations

Associations and special-interest groups have always existed in France, but they have never been so important in national life as they are today. Over 25 million people in France belong to at least one association, and over 25 thousand new associations are created every year. The purpose and aims of these associations vary (for instance, protection of wildlife, raising the standard of living, land and building conservation, consumer interests, neighbourhood and local development, safeguard of regional folk customs, sports and athletics, etc.). But the diversity of their plans of action should not conceal their common trait - this sudden increase in the number of associations and in their memberships translates a need on the part of large sections of the population to take an active part in community life in ways other than the usual membership of a political party or trade union.

Increasing International Awareness and Preoccupation about Danger to the Environment

As early as 1973, 84 per cent of the population of France felt, 'that the entire environment would be affected by pollution if it continued to grow at its current rate'. Another 67 per cent felt that technical advances were creating such an artificial world that, 'the generation to come was endangered'.(3) Opinion polls taken in France and in the United States have never given the slightest indication of a rejection of the ideal of industrialisation and technological progress.(4) Nonetheless, public opinion polls show overwhelming concern about the general basis of technical growth, and there is widespread distrust of the specialists and institutions associated with technological advances and applications. A 1978 survey of a cross-section of the French public showed that 69 per cent of them felt that, 'modern science has given powers to our researchers and scholars that can make them dangerous'.(5)

Systems Theory in the Field of Epistemology

One outgrowth of progress in biology, data processing and cybernetics has been systems theory, which is greatly modifying the approach of social science. Many are breaking with the eighteenth and nineteenth-century notions of mechanical causality and analytic reductionism in epistemology, in favour of this new transdisciplinary model of knowledge. Scientists and researchers alike are changing their attitudes towards their work and objectives along the lines of new postulates of the universe, man, nature and man-made relationships.

First Signs of Political and Administrative Recognition of Social Concern Regarding Technology

Latterly, political parties have become more aware of the public's concern with everything to do with nature and the environment. Nor are they indifferent to the sheer scope of the associated movement and the strength of the aspirations which it seeks to express. Hence, politicians of all political shades now tend to include regular references to environmental concerns in their public pronouncements, even though such references may often seem a trifle tardy and perfunctory. A case in point is a 1976 proposal placed before the French Legislature by two members of the RPR (Gaullist) Party, then in a majority, calling for the creation of an Office for Technology Evaluation within the legislature fashioned along the lines of the Office of Technology Assessment opened in 1982 by the newly elected Socialist government.(6) In a legal and institutional sense, the creation of a Ministry of the Environment and the subsequent Environmental Protection Law of 1976, which rendered environmental studies obligatory in the development of new industrial plants, was the first official acknowledgement of an entirely new set of socio-political problems. Moreover, it implied that these problems required a new array of institutions, leaders and guidelines for their resolution.

Emergence of Ideological Opposition to Industrialisation

Local disputes over the construction of nuclear power stations and other major facilities have become fertile ground for ideological protest movements. Such movements explicitly combine resistance to bureaucratic red-tape and rejections of 'quantitative economic growth impelled by technological dynamism'.(7)

The French ecology and anti-nuclear movements are akin in origin and objectives to similar groups in other European countries. Many of their leading activists emerged more or less directly from the student upheavals of May 1968. After a period of flirtation with the left and extreme left of the French political spectrum, which in many cases proved thoroughly disappointing, these militants joined the ecology movement. Though their ideology is still very diffused, it does contain a single theme which is solid and constant. This is an emphasis on 'rational and universal awareness' of relationships between men, as opposed to the Marxist ideal of 'class consciousness'.(8) It rejects existing political organisations because of their imprisonment by productivist ideology. Radical ecologists in France have managed to organise mass demonstrations on a huge scale, with occasional outbreaks of violence. These demonstrations have often been just as impressive as the ones mounted in West Germany - witness, for example, the protests against the Malville Fast Breeder Plant in 1977, against the Larzac Military Camp, against a nuclear reactor in Plogoff, Britanny, in 1980.

On the other hand French ecologists (unlike the German 'Greens') have not yet managed to build up a strong enough following to bring to bear real political and electoral weight. Their lack of impact stems directly from the effects of French political and institutional centralism, namely that central measures aimed at imposing government decision, especially on matters of nuclear energy have short-circuited the debate. While extremist parties were capable of rallying factions to their camp by direct mass action, and of increasing their following at a local level, they remained fundamentally weak through their failure to influence the decision-making process at its core.

The fact that French political life has been divided since 1973 into an electoral right and left has been a factor in politicising local opposition and the ecology movement as a whole. At the same time it has been difficult for the movement to make a structured, autonomous political impression of its own.(9) In the last presidential elections the ecologists never officially endorsed the Socialist candidate, although many of their leaders expressed their Socialist sympathies. These views were completely reversed in October 1981 when the Mitterand government gave the go-ahead to the nuclear programme, in direct contradiction to the Socialist Party's electoral manifesto. This decision alienated the ecologists from the new leftist majority.

Secondly, and specifically within the French context, the electoral success of ecology groups is no real yardstick for the true extent of public concern about technological issues. It is merely one expression, albeit the most spectacular one, of phenomena which are deeply rooted in French society. Apart from local debates, studies have shown that the ecology vote, along with general public concern about the environment, must not be confused with full-fledged anti-nuclear activism. The share of the vote taken by ecology candidates is not necessarily higher in districts where anti-nuclear activities are the strongest.(10)

There is a clear differentiation to be made between ecologists and anti-nuclear activists. Most such activists tend to be leftist oriented, whereas voters who support ecology candidates can sympathise with either the right or the left. Political analyses view ecology as a partial substitute for middle-party politics, which was virtually obliterated by the polarisation of votes between two candidates during the second round of presidential elections. A study carried out on sources of conflict within French society(11) shows that ecologist activists and voters share the same 'liberal-libertarian' moral ideals, such as legalised abortion, tolerance of homosexual lifestyles, abolition of the military draft. However, they are strongly at variance in ideological matters: the militants' radical progressiveness conflicts with the conservatism of voters, whilst their belief in the social role of government (in keeping with the Confederation Generale du Travail (CGT) trade unionists and the members of the Communist Party) is in direct contradiction to the pronounced 'modernistic' individualism of ordinary voters.

Sociological Analysis of Local and Regional Opposition to Industrial and Power Plant Installations

Much work has been done - among others by Nicolon et al.(12)(13), Touraine et al.(14), Moatti and Maitre(15) - based on field work, with a view to identifying the various social components that make up opposition movements to engineering projects. In this way conflicts arising from a simple coming together of social forces whose only common denominator was the fight against a given project, have been distinguished from others that have more profound structural significance within society. Of course, these conflicts have had different motive forces, depending on the kind of installation under attack, the interests at stake and the nature of the surrounding population. In general, however, the case studies reveal a number of similarities.

Rural Microsocieties

Under this heading may be placed the owners of small and medium-sized farms, politically passive for the most part and ready to place their confidence in the hands of local authorities in order to defend their interests and negotiate expropriations, compensatory payments and minor alterations to public works projects. If these microsocieties agree, often under the leadership of a local dignitary, to constitute a union, association or committee, they do so as much to offset the politico-ideological action of an active opposition as to strengthen the negotiating power of their chosen leaders.

Typical Local Dignitary-led Opposition
This type of opposition is initiated by local leaders who, sensing the mood of the local population, decide to seize the initiative from the opposition in order to control the resultant situation and direct it towards negotiation and compromise, while they themselves tighten their hold on local authority. In some cases dignitary-led opposition masks a behind-the-scenes battle for influence between groups of leaders (for example, leaders of the opposition party and the majority party). Both cases are typical, from a sociological point of view.

'Progressive' Farmer Opposition
This type of movement is made up of dynamic, often young, members, frustrated in their desire to extend their operations by the scarcity and expense of land, threatened in their present situation, or under the impression that agriculture is being sacrificed to urbanisation and industrialisation. This group is generally in favour of decentralisation in order to decrease the powers of the government bureaucracy and the technocracy of large state organisations. Some agree with reports criticising excessive use of agricultural chemicals by forming alliances with urban opposition nuclei on the issues of regionalism and 'environmental protection'.

Absence of Significant Worker Participation
This still holds true, even though some trade unions have come out in favour of particular issues; for instance, the Confederation Francaise Democratique du Travail (CFDT), the second largest French trade union, which has close ties to the Socialist Party, has contrived to play a decisive role in the government's electronuclear decisions without ever taking any official stance on the issue.

The Role of the 'New Middle classes'

Each dispute highlights the vital importance of a group of social categories, roughly approximating to that melting pot of workers which Wright Mills refers to as the 'middle classes'. These groups (especially teachers and executives in government and industry) furnish the leadership and basis on which the opposition relies. It is this class which, in every case, endows the opposition movement with its unifying theme: the defence of a way of life and protection of the environment.

Most French studies dwell on the aspect of 'territorial defence' as a motivating psycho-sociological factor in collective behaviour, leading people to reject the construction of some new plant. Of course, the environment and the defence of a particular way of life are sometimes a disguise for less praiseworthy aims (such as raising the price of expropriable land belonging to farmers, or defending the value of secondary residences). They can also be used simply to rally new recruits to the cause. But the fact that this theme always seems to be a legitimising central 'pretext' for an opposition movement is vitally important. The new 'middle classes' have a banner in the struggle against a broad array of technological decisions, or against the effects of those decisions on daily life and leisure. The weight of middle-class interest in these conflicts has also provided a basis for the various general sociological interpretations attributed to the opposition to certain technological decisions.

On Attempts at General Explanations

Some people have sought to explain the predominance of the middle class through an analysis of the process of participation. In order to express support or disapproval of a given technical project, an individual must enjoy a certain level of social standing. He must also be within a dynamically positive framework for social change (rising socially, or enjoying positive prospects in professional and private life). Similarly, declining social classes (small merchants, farmers) and the unskilled workforce (employees, manual workers) are likely to give only fitful support to movement leaders. The more positive their personal prospects, the more easily they are supposed to be involved in activist movements.

Others attribute ideas about the social control

of technology to change in 'cultural models' and value-systems (for instance, the rise of 'post-materialistic' values.(16) These views are sometimes related to the problem of the 'generation gap', and range from the rejection of patriarchal forms of authority to the search for new relationships with the Third World.

The more ambitious analyses attempt to explain the various aspects of protest in terms of 'social movements' including environmentalism, regional autonomy, women's rights, and consumer protection, etc. These analyses tend to agree that the movements in question are really a means of strengthening the socio-political importance of the new middle classes.

Another theory would have it that the goal of these movements coincides with the French bourgeoisie's search for new social alliances.(17),(18) According to this view, the intellectualised, urban lower middle class, an outgrowth of the 'capitalist modernisation' of the last 30 years, is taking over from the support classes which traditionally dominated this modernisation (peasant farmers, small businesses) and which are today in decline. Government decentralisation and the development of procedures for participation in decision-making has consolidated this alliance by offering as an outlet for this lower-middle-class group 'the possibility of subcontracting a position of dominance; hence a management partnership'. This line of analysis ends up, however, by representing every social conflict, aside from the working-class struggle in the factory, as a factor in the cybernetic self-regulation of the existing system; thus, unfortunately, ignoring the scope of the contradiction which surrounds technology and ways of life.

Some analysts prefer to see in the ecology movement of the last ten years, and more especially in the anti-nuclear groups, a foretaste of the social movement central to an emergent 'post-industrial society'. According to Touraine, 'the fundamental conflict in our society is no longer management versus workers in factories. It is the technocracy of information and decision-making versus the population which it dominates'.(19) We are witnessing the birth of a new social class, resistant to the polarization of ruling and working classes wherein we seek to contain it. It is a

160

class which is to some extent forgoing its own self-image and building its autonomy through movements of protest.'(20)

Such arguments are far from convincing. Any recourse to the vague concept of 'post-industrial society' is bound to be tautological. This is because, on the one hand, we are talking about protest movements made up of middle-class social elements, thereby justifying a transition to a 'post-industrial society', the exact economic rationale of which is never specified;on the other, we postulate a structural transformation of society which gives these movements meaning. The ambiguity of middle-class attitudes does not coincide with classic Marxist dogma at all. It is merely the product of empirical observations stemming from the study of protest movements. No analysis of the behaviour and habits of these social groups can avoid all consideration of their relationship to government and the machinery of government.

One cannot deny that, under the Fifth Republic, the social structure in France underwent 'an intense middle-class upheaval'.(21) Whereas the bulk of the working class remained relatively stable from 1954 to 1975, as did the management and professional classes, the traditional middle classes (small farmers, craftsmen and small merchants) saw their relative numbers within the working population diminish drastically from 31 per cent to 14 per cent. Over the same period of time, the group which included engineers rose from 0.4 per cent to 1.2 per cent of the population, and that which included technicians from 1 per cent to 3.5 per cent. Above all, growth in the public and para-public sectors has led to a 'boom' in the number of teachers, researchers and social and medical high-level staff who made up 6.5 per cent of the total working population in 1975 after having registered yearly annual increases of over 8 per cent since 1968.

The fact that these 'new wage-earning middle classes' were not integrated into the politico-institutional system, especially at a local level, is doubtless one of the major reasons behind the electoral advance of the French Socialist Party prior to 1981. To a lesser extent, this had fuelled their active participation in the anti-nuclear movement and in the fight against nuclear power station implementation projects.

Henry Lefebvre shows that these 'new middle classes' make up the 'social basis' and 'support' of the state apparatus. 'Emerging from a class group like this, with its support, the state is retaining favour in equal measure ... the state strengthens and feeds the middle classes, but this does not mean that they have political power.'(22)

This equivocal position vis-à-vis the state creates the essential ambiguity of the middle class, the choice between criticism and conservatism. (Cf. Les clivages idéologiques qui caractérisent les classes moyennes, Fondation Nationale des Sciences Politiques, 1983.) One section agrees with state rationalisation and administrative methods, along with political structuring of the social environment and society. The other rejects state tutelage and, with it, government's protective paternal role evidenced by excessive centralisation. In effect, one subsection of this group proves to be thoroughly conservative, dedicated to traditional forms of society (family and religion). Another strives to build alternative lifestyles that, without going as far as radical activism (counter-cultures, experimental communities).would give new content to the existing forms of social organisations, i.e. enrichment of family life by participation in associations, action directed towards changing the political and local leaderships, the building of a 'sympathy network' of social ties, thereby giving the meaning to neighbour relationships, and determination to escape the drawbacks of the industrial world without losing the material advantages it offers.

In disputes over the siting of large industrial plants, members of the middle class who end up assuming a critical attitude have tended to oscillate between categorical rejection of the working of the existing system and a claim for greater involvement in decision-making. In neither case is their line of thought carried through to completion.

Tentative Conclusions on the Social Acceptability of Technologies

Two general conclusions would seem to emerge from analyses of the strategic role of the new middle classes in these conflicts aroused by technology, both of which may have some operational value for policy-makers and government authorities:(i) the creation of a favourable setting for the social acceptance of technologies implies the careful adoption of measures to conserve the environment and improve the quality of daily life; (ii) acceptance depends fundamentally on the setting up of decision procedures which will meet aspirations for democratic political participation, especially on the part of the more highly educated elements of society. As Salamon points out, scientific and technical undertakings have become in themselves

topics of political debate ... This is the severe test for the notion of the people's will, as expressed in every democratic system via a majority vote. What exactly the general will might be, when it concerns esoteric questions with long range implications, is a problem that only specialists would presume to understand.(23)

The Current Crisis and Social Acceptance of Technologies

The present situation in France has two characteristics which may have changed the conditions for accepting new technologies, or which may at least have led to some progress in the ongoing debate. The first characteristic, shared by all the industrialised countries, stems from the economic crisis which the population now accepts as a continuing feature. The second characteristic is more specific to France and has to do with changes in the decision-process that have come about since the political shift in May 1981. But it needs to be emphasised that the major change here (the decentralisation reform set up by the law of 2 March 1982) brings the politico-administrative system in France more in line with developments in government territorial organisation in other Western European countries.

Changes in Public Opinion
A cursory look at recent opinion polls might give the impression that the economic crisis has had little influence on a trend that dates from 1965, namely, a continuing increase in public awareness of problems from the negative effects of technology. In 1981 95 per cent of the population still felt that environmental protection was important, and widespread uneasiness about the risks of pollution by chemicals, nuclear power and

petroleum remains general.

In fact, this remarkable unanimity hides a slow erosion, coupled with a broad social realignment since 1973.(24) Although uneasiness continues and is occasionally increased by accidental events (for example, the Seveso disaster), examples show that environmental problems have on the whole lost some standing in the hierarchy of public preoccupations. In 1973 a survey into public perception of threat ranked deterioration of wildlife in all its forms as first in order of priority, and fourth in order of probability. In 1982-3 the environment was only spontaneously stated as a serious problem in 2 per cent of the replies. The proportion of those who express dissatisfaction with their lifestyles (20 per cent) has decreased slowly since 1978. Complaints against pollution by individual industrial firms have likewise dropped since 1979. Parallel to this, questions bearing on ecological awareness have remained static or lost priority in public opinion since 1977. Preference is given to leisure and convenience over solar sources of energy, with the number of people opposed to nuclear energy and development diminishing.

Other issues such as fear of war, crime in the cities, and unemployment have widely outstripped problems of the environment as rallying points for indignation and anxiety. Nevertheless, they contribute to the fundamental urgency felt about protection against major risks (nuclear accidents, poisonous wastes, oil spills, floods) as expressed by an overwhelming majority of the population.(25)

The waning of interest in more general questions is offset by a heightened awareness of problems in daily life and the home/work environment, for instance, working conditions, quality of drinking water and related health considerations, overcrowding, lack of security, and unsightliness and noise in the cities. In a survey carried out in 1980 by the Société Francaise d'Enquetes par Sondages (SOFRES), traffic jams are mentioned as the principal inconvenience in daily living (20 per cent of those replying), followed by lack of social contact (11 per cent), long commuting journeys to work (12 per cent), noise, urban congestion and decay. Water pollution and estrangement from nature ranked eleventh (4 per cent). Another survey undertaken by the Centre de Recherche d'Etude et d'Observation des Conditions de Vie (CREDOC) in 1980 puts nuisances and inconveniences related to everyday living at the top, along with lack of prospects (60 per cent), shortage of green areas (53 per cent), architectural ugliness (48 per cent), noise (40 per cent) and traffic (35 per cent). Forty to 50 per cent of the wage earners questioned stated they were subject to daily nuisances at their places of employment. The crisis pervading government institutions has in a sense 'desocialised' the claims made upon them. Thus, priority is now being given to the search for individual solutions. People are trying to maintain their contact with nature in a variety of ways: cross-country biking, jogging, health foods, gardening (240,000 new gardens have been created yearly in the last ten years), pets, green plants, country walks; and there is an especially keen interest in detached houses and second homes. French people are focusing increasing attention on improving their lifestyles in every aspect, both working and non-working, urban and non-urban, natural and artificial.

Development of Decentralised Policy

The stimulation of industrial and technological innovation constitutes a major guideline in the medium-range strategy of the current Socialist government's approach to the economic crisis. The law of 15 July 1982 set aside 2.5 per cent of GNP for research spending until 1985. The National Colloquium on Research and Technology held in January 1982, was prepared by 32 regional boards and attracted nearly 25,000 participants. It constituted a first attempt at bringing together and mobilising the realm of research, along with its economic and social partners. Inevitably, this is viewed by the ecologists and anti-nuclear groups as a giant step towards narrow scientific productivism. Nevertheless, the policy announced by the administration enshrined a sense of preoccupation about social controls on technological advances. These preoccupations were concretely expressed by the following legislative and institutional innovations: creation of a parliamentary bureau for evaluating technological options; 'mobilisation programmes', with research priority given to the improvement of working conditions and the adaptation of transferred technology to local conditions in the Third World; an obligation to consult the 'employees' committee' led by union representatives before any new

view to providing greater openness and improved
access to information, during public enquiry
procedures prior to the construction of nuclear
industrial installations (July 1983). Above all,
whereas the defence of local democratic rights was
once a frequent topic for combined action by
protestors and a faction of the political elite in
the regions (especially some elected Socialists)
the decentralisation reform has now contrived to
relegate the more radical ecology groups to the
political sidelines.

The actual running of this administrative and
'Jacobin' centralism which has thus far dominated
the French political system, is in reality much
more complex than one might ordinarily suspect.
Granted, final decision-making remains the preserve
of Paris. Yet the authorities in Paris come to
decisions primarily under the impetus of local
authorities.

The imbalance of power in favour of Paris is not
necessarily what best characterizes territorial
hierarchy, and between the administration and
leading citizens. Influence is concentrated at
every level, not only at the highest level, in the
hands of only a few people. The absence of
participation is a consequence of this system
which, to work well, must remain in the
shadows.(26)

Economic and urban transfers over the last two
decades have multiplied the inconveniences of this
system, handed down as it is from the historical
traditions of rural France; and recourse to a
technocratic authoritarianism has offered the only
means of imposing coherence between objectives at
the local level and those at the national level
(the nuclear programme from 1974 to 1981 is just
one example of this type of 'resolution' of
contradictions).

One might hypothesise that the Socialist
government is working towards a model of 'corporative democracy', characterised by the reciprocal
play of influences among highly organised interest
groups. A model of this kind has never hitherto
been developed within the French socio-economic
framework, but for Wilensky(27) this corporative
democracy model can already be applied to the
Netherlands, Belgium, Sweden, Norway, Austria and,
to an increasing extent, West Germany. It is

distinguished by its 'special procedures, applied within a quasi governmental context and its object is to create, through negotiation, a general consensus on the principal questions raised by the modern political economy'.

Current reforms may be viewed as a change in the institutional machinery, aimed at encouraging widespread management of local affairs by the new middle classes, and their consequent rise to a new level of power and importance. It is obviously too early to say whether these changes will lead to a significant shift of power in matters of technology, (via co-operation, delegation or direct control), or merely constitute symbolic participation consisting of information and consultation) along the lines suggested by Arnstein.(28) But it should be pointed out that the setting up of mechanisms from 'dependent political participation' in itself (meaning that people are at least to be associated with the application of decisions if not with the taking of them) is in itself a major innovation in France. On the other hand, as Habermas has emphasised, this may be in keeping with the continuance of a technocratic pattern of rationality.(29)

Evolution in Terms of 'Social Movements'

Where anti-nuclear protests and environmental demands are diminishing, other initiatives aimed at promoting a 'technical culture' are on the increase. These are emanating from various social groupings: local communities, chambers of commerce and industry, teachers and researchers, unions and citizens' groups, networks of 'science boutiques', amateur computer clubs, industrial techniques centres, regional museums for science and technology, and so on.(30) Even with the women's rights movement, attention is now being directed towards women's relationship to technology in domestic areas and in the working environment.(31) We would make the following observation: in the 1970s the French socio-cultural field was dominated by anti-industrialist tendencies with roots in the past. Other tendencies were nourished by the unrest created by certain negative aspects of technical progress (for example, concerning the natural and human environment). A counter-current has nonetheless appeared, The spread of new technologies now seems, within the context of economic crisis, to offer a vehicle for positive changes in values and lifestyles.

Objectively, economic and social factors go hand in hand. This broadly explains the growth of the anti-nuclear and ecology pressure groups in the 1970s, and today it is working in favour of social control and technology. In our opinion, however, the political and ideological structure of the ecology and anti-nuclear movement in France will make it difficult, even impossible, for it to alter its position without a major upheaval. However, policy decision-makers would be mistaken if they drew the conclusion that the acceptance of technologies is no longer linked to strong social control over choices and broad public participation in the decision process: indeed, quite the opposite is true.

II Production

An Industrial Revolution?

New technologies are revolutionising businesses. Virtually everyone is agreed on this, although on reflection the fact may seem surprising. Is this due to a widespread misunderstanding concerning the problem and, ironically, to a refusal to examine thoroughly every aspect of the upheaval in the structure of society caused by technical progress? What exactly is a revolution in industry anyway? Many observers see it as a reversal of secular trends, leading, for example, to a growing division within the workforce, or to a permanent process of obsolescence affecting the qualifications of workers. In other words, a revolution in industry means a kind of deflection in the normal course of development, an event which unexpectedly jolts all known categories and standards of observation off course. This view can be narrowed even further: a revolution in production can perhaps be seen as a movement affecting the very forms of the activity itself, thereby throwing conventional job classifications, work and workers into disarray.

The basic references formerly used for future planning are thus radically altered. Followers of Saint-Simon's line of argument once considered that a crisis reflected a system which had not yet been properly organised, the components of which would not necessarily repeat themselves over time, nor yet lend themselves to a regular schematic presentation. The fact that today's descriptions of the nature of new industry are blurred may be because the true impact of recent forms of development has yet to be thoroughly clarified. On the other hand, it might well be that the appearance of these forms has already made traditional means of measurement redundant, and that the revolution is manifesting itself precisely by the apparent confusion and incoherence of the indices.

The New Technologies

What exactly is new technology? It is obviously more than something recently developed! It is possible for an instrument or method of management to be a novelty in only one branch of activity, or in one area where it has just been introduced, even though elsewhere its effects have long since become exhausted. A new instrument emerging as a result of the latest scientific developments may merely duplicate an already well-known technical model. In the steel industry digital direction had to wait for advances in electronics before it could achieve the same degree of autonomy as that provided for many years by simple Jacquard technology. Of course, the heterogeneity of these two processes still remained. Rather, as soon as we go beyond the general relationship of the operator to machine, this heterogeneity reappears in the functioning of design departments, as well as in work preparation and maintenance. It even extends to the place occupied by the business in the broader energy, information and services network. But this contrast must be described as that of two aspects of work, in which human activity and mechanical operations processed along different lines.

A distinction must therefore be made between the technical organism and those principles and considered actions which have led to its creation. This means that we should perceive a difference between the mechanism as it is described and characterised by the technicians who seek to perfect it, and the composite result of their work as seen by the social analyst - that is, the mechanism as the partner, rival, limiting factor, or even the sealed environment of human workers. This is not an easy distinction to make, and it is even less easy to uphold in terms of relativity and dynamic tension. Nevertheless, it is impossible for us to sort out the consequences of computerisation and electronics for the organisation of production without at the same time interposing consideration of technical entities which, though they would doubtless never have existed at all were it not for these inventions, nevertheless do have a separate character of their own.

In this field a number of attempts have latterly been made to identify the two clear aspects mentioned above. The result has been the creation of an analytical minefield - for example, many difficulties have arisen from a confusion of automation (in the sense of a tool which functions free of human operators during part of its cycle) with other recent discoveries that merely simplify the unravelling of complex operations. This has led to the erroneous conclusion that industrial history is drawing to a close with the dawn of automated devices, even though there is much evidence to show that this is a highly premature idea in terms of mankind's real progress. French folksongs have long told us about the mill that turns while the miller sleeps or makes free with the farm girls who bring him grain. Julien Sorel, in Le Rouge et le Noir, watches a mechanical water-saw which carries on by itself while Stendhal's hero reads the history of Napoleon. And even the simple snare used to catch animals in prehistoric times could be described as a perfect reflex machine.

Today's techniques are, in effect, no more than a form of co-operation between the community of men and the realm of machines, a specific combination of human and mechanical functions. Perhaps this is why we can best comprehend their more revolutionary and original aspects, particularly in respect of the qualifications they require, which is always a vital question in judging the effects of an industrial innovation. And maybe we can then go on to evaluate the probability that such new processes will be accepted without resistance by the employees who must work with them.

Work and Qualifications

The question of work qualifications is dominated by two competing formulae of interpretation, both of which have accumulated so much weight over history that the scattered information available on today's industrial conditions has trouble displacing them. These formulae thus tend to lay down in advance the results to be expected from technological innovation.

Take the team formed by a man and a machine. It goes without saying that the human operator should be better informed, advised and experienced, the more complex and powerful the machine he controls. Progress in tools has changed the qualifications needed of the worker. Doubtless, this trend could be blocked or held back. Alternatively, 'The meaning and content of development must be grasped in order to train employees for new jobs and strengthen their qualifications'.(32) Denial of this, in the long run, would be disastrous. The worker's responsibilities would increase without his being given the means to cope with it, which would lead to a corresponding increase in administrative double-checking and directives, and in turn to a strengthening of the hierarchy, all of which would ultimately be costly and inefficient.

With this obvious and well-tried approach, we may contrast the second interpretation, which draws a directly opposite conclusion. A given work station amounts to the pairing of a technical entity with a human being, and together they carry out a specific operation. Initially the worker occupied this place alone. Then the machine appeared as his helper, and little by little it took over his job, first his energy, then his skill, and now, today, his ability to adapt and calculate. The worker's helper has become his parasite by gradually taking away his job. He is reduced to being a reader of codes and orders and, finally, he is eliminated entirely. An operator on this interpretation no longer represents much more than a general resource, and an eventually replaceable one at that; he need be neither experienced nor knowledgeable, and he has no value, either in terms of ability or of exchange. He is used by a working organisation outside himself, until such time, that is, as he is replaced by a robot. 'Rationalization, automation and the mass-produced worker: such are the effects of the division of work and mechanization on the productive value of the labour force.'(33) The more complex the machine the more it dequalifies the human operator.

Each of these scenarios, taken separately, suits a given set of circumstances. What are they? When we establish the different conditions to which they apply, we discover the suppositions on which each is based and the respective limits of their effectiveness. For instance, it is clear that the first proposition supposes that the man/machine team reflects an adjustment stage, which in turn creates new and appropriate functions for itself as it goes along. It is ultimately a profitable partnership. In contrast, the second proposition defines this relationship as the manifestation of a fixed set of operations, a closed situation wherein one partner's loss is the other's gain. In other words, each formalisation supposes work re-organisation, some kind of intervention and realignment of specific job roles. Industry offers clear examples of both situations: in the one, man rides his machine; in the other, the machine exploits the man; and in any given situation one has to decide which proposition is relevant. The former applies really to the professional worker involved in design work, the latter to skilled workers subject to the kind of division of labour imposed by an assembly line.

It is easy to see the methodological means through which these partial interpretations have been insinuated into and strengthened within our attitudes. Every available study on the effects of technical advance on the factory floor has tended to neglect the factory itself, along with its characteristic method of labour; to neglect, that is, how the workshop structure is maintained throughout the observed changes, or how it gradually evolves. But both the interpretations we are considering here are evidently related to the area of work concerned. We cannot force business, engrossed as it is in a movement of limited modernisation, to fit into some broader concept of industrial history. After all, history does not necessarily help much in this process of 'job dequalification and requalification', although it can help to modify the means of work division and work assignments, and thus the direction, content and mechanisms which determine work qualification.

The Decline of Qualification

This is exactly what has been observed as a result of new techniques. The traditional machine tool paired a mechanical entity and a human being, both working together over the same length of time. The worker's know-how and experience were measured in terms of his ability to direct the tool more or less independently, checked according to how much he produced, and this was translated into a given work status. It is this one-to-one relationship which is now confirmed. All this points towards a gradual 'dissociation between machine work and human work'.(34)

New techniques which are now grouped together were formerly spread out in such a way that man had to direct, implement and co-ordinate them. Nowadays, actions are accommodated to their objective, follow one after another, and are interdependent with the result of the others. One of the new elements is the robot. Here again, one has a mechanical entity which completely occupies the work station vacated by man, replacing him on the assembly line and obeying the same division of labour. It is this very limitation, that is, the small number of functions accomplished, the limited timespan of intervention, and the adjustment and surveillance it requires, which qualify the robot as an artificial individual. This is how the robot appeals, even more than other new technologies, to the fantasies of our civilisation, which is fascinated by the hope, or fear, of work without the worker. However, the robot cannot become more complete and less demanding without getting lost in a new structure, a larger organisation with its various articulations arranged by different norms. It is here that men are called in to calculate, determine, and establish relationships between actions; these tasks are carried out by the use of successive instruments. Thus, the worker is in turn integrated into the system, caught in its toils, questioned by it and tied in via an intermediary.

'In this situation, automation is an essential but nonetheless subordinate function. It represents the capacity of a mechanism to work through a number of parameters that ensure its co-ordination within the whole. Of what, in this case, does human work consist? Enabling the machine to operate and produce, ensuring the smooth running of installations.'(35)

This pattern of events is not entirely new; certain branches of the chemical industry have long been operating along these lines. And, on reflection, the principal labour of farmers has always consisted in preparing the land for production. Their various operations are temporarily circumscribed and they fit into a larger framework of major cycles.

A business tends to become an integrated unit, in which the requirements of commercialisation, innovation and discovery are immediately translated into realignments and corrective action for the machinery of production. So strong is this momentum for integration that R. Galle and F. Vatin, in their study of the refinery, believed they were witnessing, 'the demise of the work process ... Fluidity at every level (technological, economic and social) tends to completely dissolve production into circulation.'(36) This opinion is obviously overstated, and indeed, somewhat absurd in this form (cf .P. Naville 37). Even so, the distinct characters of each of the different sectors of industry - work preparation, testing, maintenance and production - cannot be sustained. Distinct workers' groups are formed only when there are disagreements over areas of responsibility, the distribution of tasks, and communication gaps. These same problems crop up in the relations of business to its environment, because the degree of integration attained by the productive process is related to the integration of the process in the broader industrial network of which it is a part.

In the new business context, the links between qualification, know-how, responsibility, career and wages, are in the process of being dismantled. As long as this development continues within the confines of traditional employment categories, it will look like either an obvious re-evaluation, or an obvious devaluation. In fact, the deeper changes which are going on are affecting the evaluation process itself. The remunerative value of man's work and his status were based on the demands of tasks as assigned by an arbitrary division of labour. New procedures will enshrine instability in the distribution of assignments, favour one means of collaboration over others, and create the conditions for regulated mobility. In the traditional factory, orders from above were responsible for setting work objectives. Men exchanged products and information. They now activate the mechanism by which information and operation interact. Work thus reassembles consumption because nowadays consumption is more than just the use of products and services - it also means availing oneself of products on request, fashioned according to need, and obtained through a specific access to operational networks. In this situation worker and consumer knowledge depends on the same basic apprenticeship.

In the new industrial world, the structure of the working collective no longer reproduces that of technical individuals, whose evolution has anyway diminished and all but abolished individuality.

Team members are now simply organised according to a distribution of job assignments around machinery that is both flexible and modifiable. The former divisions between intellectual work and manual labour, for example, and even that which differentiates work preparation from execution have begun to blur and disappear. 'The changeover from traditional tooling to digital computer finishing can assign to the regulator the role of programming, in addition to creating programming specialities within the firms' ways and means section'.(38) Often, the prevalence of outmoded techniques (which are sometimes present even in the most up-to-date industries) along with a concern to avoid labour disputes, have enabled the old work classifications to remain despite the new situation.(39) Nonetheless, in general, the loosening of traditional work classifications marks the dawn of new freedom, along with new dangers. The prevalence of logical and scientific knowledge over that of experience and special competence, which used to be the guiding principles for worker autonomy and worker resistance, can today render workers defenceless and dependent upon an increasingly narrow work situation. On the other hand, it can also associate them with a more egalitarian and flexible working environment. If they are maintained, work-qualification indices will be unable to measure the increase of worker readiness and dynamism. The real movement of this type of skill will appear only through this instrument of measure as disqualification. But can we do away with these qualifications if we thereby lose, or risk losing, the reward for education and training, relative worker autonomy, guaranteed employment until retirement and the usual ties with the employing firm which job classifications endeavour to maintain? Although it is still too early to say, it would seem that the only stable organisation is a form of flexible job distribution which does not equate people with the tools they use, but makes use of their increasing detachment from those tools to develop new ways of deploying their availability and effort over more limited periods of time.(40)

The Acceptance of New Technologies
New technologies offer a variety of horizons to wage earners, and therefore stimulate lively debate. Young people do not react in the same way

as their elders; experienced workers able to adapt
to, and even find personal advancement in the
changes brought on by a new applied technology, do
not share the same attitudes as unskilled workers
without training. Factory-floor workers may not
agree with the head-office staff; employers in one
area may not see eye to eye with others in the same
industry. In these cases workers break into
different factions, and rivalry, even enmity,
appears.

In companies, a new technology's direction is
determined according to whether it increases job
perspectives or introduces unemployment risks.
Meanwhile, it is vital that employees as well as
employers talk over any planned innovations. In
this way compromises can be arranged, that is
mechanical modifications and institutional
transitions to ease the transition from one job
assignment to another. This is the specific
purpose underlying the 1982 Auroux Laws to open 'a
new space for economic and social democracy'.
There are, of course, limits to such discussions.
Company management personnel begin to be highly
unpredictable when it comes to consulting with
employees as company members and discussing their
own future in terms of their work value; this means
in effect that they are asked to set their own work
conditions and even decide on their own redundancy.

Each innovation tends to decrease the labour
time involved in the production of a given article,
and thereby causes changes in the workforce, even
job down-grading. Where this danger is reduced,
new technologies have a better chance of being
accepted and of being successfully applied. Worker
mobility cannot be increased unless we take into
account the needs of workers for stability and
their ever present need for training. The whole of
man's working existence must be organised in such a
way that he can cope with the tensions and movement
of industry, instead of always having to adjust at
the last minute and as an urgent resort. The
social status of workers is at stake in this issue:
effective employment and training can match one
another only in terms of the community as a whole.

III From Citizen to Consumer

The complex question of technology's acceptability
clearly requires a prior examination of how the
concept itself is genealogically structured. This

inevitably leads to the problem of the citizen vis-a-vis the consumer. But before going on to a broader topic of consumer acceptance of technology, we will consider the example of the microcomputer as a classic case. In conclusion, we will return to the differing attitudes of citizen and consumer in general discussion on the problems posed by deregulation.

Genealogy

The question of the acceptability of new technology is expressed in different terms, depending upon whether it concerns products of mass consumption or innovations within areas of secular government control. Whilst the development of large networks has kept pace with the development of commercial and industrial needs, mass consumption is a latecomer to the scheme. The development of the middle classes, higher earnings and the larger proportion of free time allocated to production workers have all contributed towards the opening of new markets. Today we face a dilemma which springs from a reduction in the earnings of the middle class involved in the working process, a lack of 'consumer' time available to those who work, and a shortage of money to spend for those who are without work.

In France the problem of innovation is basically the same whatever the issue under discussion: roads, telematic networks, railways, telephones or electricity. It is the control of technology by a centralised government whose power in every one of these spheres has done away with a feudal system only to replace it with a central power - royal, imperial, finally republican - which is handed down through regional leaders and local representatives of its ideas.

Social acceptance primarily involves acceptance by leaders of a technical-power system which will not upset the local social balance in terms of employment and regional consensus. In the event that the established equilibrium is threatened, the power pattern must strengthen the political polarisations which are to be re-defined once the innovation is applied. The telephone, for example, was first developed as a special tool for the new political classes and those in business. The democratisation of large technical systems has often been stimulated for technical reasons. (Charles Boileau remarked about the law of 26 May 1921 concerning hydroelectric plants on the River

Rhone that utilisation was vitally important: he stressed the need 'to incite the public to become consumers by every possible means, by educating and disciplining them to achieve the 6,000 hours of utilisation upon which the entire system's economy is based'.)

The heavy cost of new industrial construction, along with the increasing initial investment it entails, has led to acceptability being treated as a minor problem, even a subsidiary one, in relation to the general economics of the technical system under consideration. On the other hand, the danger of breakdown or sabotage is crucial when it comes to technical choices in cases where news or information is to be communicated, poison is to be transported, or explosives or nuclear energy are at issue. In these cases social acceptance is defined by the government, which exercises complete control over such strategic material in its role as guarantor of the security of the citizen. In the case of the transport and telephone systems, acceptability on the part of the citizenry is defined according to the relative risk involved: that is, government ensures a balance between good and bad commercial risks so that everyone can enjoy access to transportation and communication, even though installations in one region may be considerably more costly than in another. Heavy technology, high-risk technology and national technical systems are thus guaranteed to individuals in every region, however poor they may be. This is a problem for the citizen not for the consumer, since the relationship is characterised by universal equality of all, whenever, and wherever systems and services produced and commercialised by government are concerned.

In contrast, the problem of the consumer individualises the social relationship, differentiates it, divides it, and permits the free play of passing enthusiasms, fashions and consumer infatuation with certain products. And these responses do not grow out of the simple basic logic of primary needs, but derive rather from new trends in cultural goods. The citizen's problem is edging closer to that of the consumer, and the characteristics of innovation which structure networks, control and regulation by the state, are now being called into question. Indeed, where government management of large networks is concerned (part of the citizen's problem), innovation is somewhat rare, because it nearly always tends to upset subtle and complex balances. The problems of acceptance are seldom broached, and when they are, they tend always to be resolved through compromise technologies which preserve the macrobalance. The emergence of new technologies and their related products bring the public sector (systems traditionally run by government) and private sector (industrial production and distribution networks) into play, thereby modifying the rules of the game. The new dimension of Consumerism has outstripped Citizenship in the classic sense of the word, and the rules governing acquisition and ownership have been altered as a result.

Consumption

How can fads and enthusiasms about certain technical products be explained? How does one or other object come to fill a need it has itself created. Obviously, political, judicial and regulatory systems have no part in the determination of these choices. On the other hand, the media arouse crazes for new consumer items by launching or encapsulating symbols which the new consumers recognise as part of their badge of belonging.

Social science research has difficulty keeping track in this area: we all know how hard it is to analyse a body of images, and how diversé and scattered are the messages broadcast on television or used in advertising. Some research into consumerisation has focused on the dynamic of visual effects in relation to consumption, by showing how they follow the logic of economic classes, into which the media can penetrate. Other work has relied on the idea of object-systems and on the postulated appearance of a new object within such systems. Thus, the new technologies are giving rise to new forms of analysis, more oriented towards research into trends; this consists of piecing together evidence enabling researchers to identify new consumer behaviour, rather than working towards a general theory of the consumer or of his technologies. The imagination of the consumer is certainly one of the most promising directions in this form of analysis.

Impulses of the Imagination

Based on analyses made during microcomputer studies,(41) some criteria have been found for judging whether a given technical object will be acceptable and accepted by the user, whether it will 'make the grade' in approval. There is no doubt, for one thing, that the presence or absence of a social 'impulse' is a crucial determinant for acceptability. In the case of the microcomputer this impulse would seem to be present; indeed, this technical object symbolises the relationship between new technologies and data by virtue of the questions it raises.

Mastery of Knowledge: the discovery of the microcomputer - and of information technology as a whole - seems to have brought us to a turning point. It is now possible to imagine not only selective access to knowledge but also general and democratic access. This hope is based on the dream of an ideal data bank in which the whole corpus of human knowledge would be available at the push of a button. The ancient humanist myth of the encyclopedic man has returned in a different form. And on the margins of this dream is the simple fact that mastery of all this knowledge consists only in mastery of the machine which gives access to it.

Mastery of Action: (autonomy and independence) as opposed to the imprisonment of individuals and groups in sharply defined tasks, the microcomputer allows us to imagine a return to a more localised, personalised existence. A solution which seems far better than physical decentralisation, the microcomputer offers local participation in decision-making and subsequent action - hence a share in the harnessing of the future. Non-polluting and non-aggressive, the microcomputer is the absolute instrument of small-scale management. It is the last expenditure you need ever make, the embodiment of a dream.

Mastery of Relationships: (communication, clarity) Here we recognise all the themes that went into the development of the telephone: the instrument of 'universal' understanding, the instrument of the world citizen, over and above the limits of nation-states. Furthermore, the dawn of the microcomputer is seen as a fresh opportunity to do away with the ambiguities of communication. At last communication will simply become an exchange of formalised data!

The strength of these hopes and ambitions cannot be overstated. They stimulate production research, offering a broad industrial inventiveness. They create institutions with strong identities and bold aspirations; and on the way, they offer glimpses of a future bounded on the one hand by hell and on the other by paradise. The machine may yet exploit man, or man may tame the machine.

The Machine: Servant or Master

According to some, social segmentation will not long resist the advance of technology. Machines will eventually be insinuated into social relationships, which will be increasingly vulgarised. The individual will lose his identity, becoming more and more acclimatised to an increase of unreality: man will be left with no roots, and no past to fall back upon.

Machinery will multiply itself ad infinitum: machines for healing, thinking, knowing, learning, generating, managing, ordering, deciding, doing, travelling, living, playing, hearing, seeing, communicating, controlling, guarding and punishing. There will be an artificial solution for everything.

The notions of unreality and artificiality are very important in discussing new technologies, in that they reflect new divisions between man and his natural environment, man and his body, man and his neighbour. The atomisation of existence into functions taken in charge by machines and prosthesis would bring a definitive end to humanist culuture. Machines would merely simulate the functions they were supposed to carry out. Men would be alienated and dissipated by the resulant atmosphere of pretence, or unreality.

Thus, to be taken out of reality means to be cut off from ordinary existence: it means determination and loss of one's humanity, turning into a function, nothing more. It means being isolated from a rich, complex, natural and social reality for the sake of efficiency and profitability.(42) The variety of things we produce now sets the pattern for our work, because they make us operate within interconnected networks. These things would be replaced completely by consumer products. And these consumer products would make our lives colourless and standardised, because they would by definition, put an end to all our simple household rituals and practical skills, by eliminating the need for them.

Our endorsement of such classical reasoning stems from the fact that we need no longer conform to the old ways, and can create new ones of our own. But this new power clashes head-on with our social distinctions and differentiations, as well as with our culture and its various levels of attainment. A different argument would have it that man is like a jockey, riding and guiding his machine. This is the positive side of the coin; we are operating an intelligent tool which gives us a slightly better command of other tools. Thus, new working rituals are allowed to take shape.

Thus, in the case of the videotape recorder, an analysis by J.C. Baboulin, J.P. Gaudin and P. Mallein(43) shows that the purpose of this innovation is not to reverse a decline in television quality, so much as to make the flux of television 'portable' so to speak; hence, transportable over time. Today's viewer, overwhelmed as he is with images, now wants to control his film intake.

For those who defend this point of view, the innovations now appearing in the home environment presage a new form of collaborating between man and the machine. They are teaching people to adopt a new attitude towards objects in anticipation, perhaps, of new ways of living and working.

Specifically home-based, technologies such as the videotex, may provide adolescents, for instance, with a means of transition to more complex products - with wider possibilities for play, learning and programming. Behind these new technical objects lies the possibility of rebuilding a symbolic framework, which at present is defective. The two vital themes of this new beginning are education and communication.

Social Incentives

The accumulation of small functional objects seems to be some kind of response to the crisis which has overtaken large systems. Hence the microcomputer has a certain appeal as a means of escaping the economic crisis. Given an economic situation in which the keywords are 'reduction of working hours' and 'remote work', the microcomputer represents a pioneer tool in both the home and the office.

To the hope in renewed development held out by computer-related products, we add an article of faith: the microcomputer may be the key to a new level of productivity. The economics for work at home and work at the office now overlap, as opposed to forming successive entities. Hence, they compel us to reconsider the status of the worker himself.

The microcomputer is ushering in a new era - that of the double workday. The reduction in ordinary working hours created by the economic crisis has created the conditions for this in the following way: by forcing a 'work' crisis, with bonuses less and less evenly distributed; by forcing a 'family' crisis, in which the latter is threatened by the intrusion of the microcomputer. The constraints associated with time spent at home and living space none the less stand in the way of this so-called contamination of the home by professional instruments. Women are frequently the chief objectors, thus earning the label of anti-technologists. Research on the redistribution of leisure time and of modifications to the home setting brought on by these new instruments for processing data is still very meagre.

Today some observers have discerned the emergence of a double trend: on the one hand, professionalisation of the home area, where we are finding the same instruments as at work, and on the other, an increasingly relaxed atmosphere in the workplace. This double trend involves a minimum control of technological instruments and consequently the development of technical education. Training therefore constitutes one of the major conditions for the spread and adoption of new instruments. Particularly in the case of microcomputers, their potential has led to two distinct developments: computer education and computerised education. But will these developments be enough to support broader acceptance of the concept of a technical culture?

Among the social stimuli for acceptance, businesses directly involved in distribution hold a prominent position. Whether we take sides for or against the application of new technical apparatus, analyses of social factors and of ideas associated with technology, along with a cloud of words surrounding technical progress, are disproportionately prevalent. Doubtless, sociological analysis would be better served by the study of more concrete, practical indicators, such as wholesale and retail systems, maintenance networks, specifications and modes of application. There has been no investigation into commercial sociology as such, and certainly no study of the commercial sociology of new products derived from electronics. However, the cost and availability of instruments, commercialisation, the spectacular drop in the cost of components as well as in the

price of mini-calculators and microcomputers, have
all had a direct influence on demand. As indicated
in the 1978 Bureau d'Informations et de Prévisions
Economiques (BIPE) study on the mass market for
electronics, the drop in the price of mini-
computers and their availability at various
commercial outlets has placed them well on the gift
list because they outperform the ordinary gadget by
their highly functional applications. Nowadays,
objects like the credit card calculator-music box
can be included on this line of commercialisation.

The price reduction in certain technical
objects and their availability in the shops pose
questions about consumer habits and the commercial
networks which cater to those habits. Research
into the layout and business hours of retail points
is hopelessly inadequate, when it could be very
enlightening in the study of consumer demand and
motivation. Lastly, ease of choice and training in
the use of the tool have also played a part which
is sometimes difficult to evaluate.

The rivalry between microcomputers and other
systems, systems incompatibility, different
diskettes, software and base languages, all combine
to render consumer selection exceedingly hazardous
in terms of the evolution of the required services.
Nowadays, the cost of equipment is no longer the
only criterion for selection.

Experience of communication with users is thus
absolutely indispensable, from the marketing of the
product right through to self-instruction in
software by means of guides, and instructions
though sophisticated, are still not sophisticated
enough. Meanwhile, some manufacturers still
believe they can get by well enough without
instruction manuals. The information provided with
the products is all too frequently incomplete when
it is not frankly counter-instructive.

Taken as a whole, the commercial aspect of
products and services to consumers are nowadays
overshadowed by their ideological and cultural
ramifications.

Decontrol, Consumer and Citizen

The government is very much involved in the debate
over consumerism. Thus, by executive order, the
state acts as a springboard for industrial
development; it protects the citizen legally; and
it stimulates the creation of new products by the
use of incentives. This involvement also has an
effect on the industrial structure, since it can

hold down supply and control demand. Doubts about state regulation arise from the fact that consumers believe they will receive better service from private rather than public organisations, whether in terms of information, communication, transportation or energy. Consumers sense that these relationships, which once appeared against a background of strategic and military considerations, have now been reduced to a functional role, and no longer call for government control as emanating from the community.

In this instance decontrol merely acknowledges the fact of recently revealed shortcomings in the state machinery for responding to the aspirations of today's consumer. More troubling, however, are the doubts as to whether the government is really able to control science and technology. Uncertainty about this question could lead to a crisis of legitimacy. The rejection of technical decisions made by the government takes various forms: contempt of those political representatives in charge of mediation, and a lack of faith in the role of experts traditionally assigned to the political sphere; a growing autonomy and complexity of science in relation to citizens; and the absence of proper instruments for analysis, as brought to light by contradictory expert evaluations. In fact, public attitudes towards science and technology are less related to political, moral and ethical opinion than to the level of scientific information. In short, all this looks very much like a gigantic game of musical chairs. Whereas the management of large government-run systems treats citizens as consumers, new technological products a priori expects the consumer to act as a good citizen, in so far as the cultural dimension of his choice of product implies a choice of society. Thus, a report by Simon Nora and Alan Minc(44) has provided citizens and consumers with information about the political and social questions which are at stake in the computerisation of French society. The fact remains that only the government, through its regional institutions and technical resources, is in a position to cope with the many difficulties involved in these changes.

The powerlessness of consumers as such fuels their ambivalence towards government all the more: on the one hand there is a credibility gap, on the other, a search for a guarantor and protector. This is the new status of government. The public acceptance of new technologies hangs in the balance according both to whether a product is a commercial success (if it is bought, it is acceptable), and to the sense of responsibility felt by the citizen, who has a vague impression that his choices. And it is for this reason that he turns for guidance to the state, which in theory is the proper guardian of the public interest.

Notes

1. Crozier, M. in Huntington, Crozier and Watanaki, 'The Crisis of Democracy', Report on the Governability of Democracies to the Trilateral Commission, New York: University Press, 1975.

2. Fagnani, F. et Moatti, J.P., 'The Politics of French Nuclear Development', Journal of Policy Analysis and Management, vol. 3, No. 2, 264-275, 1984.

3. Le Progres Scientifique, 1974.

4. Boss, J.F. et Kapferer, J.N., Les Francais, la science et les media, La Documentation Francaise, 1978.

5. Ibid.

6. Moatti, J.P., 'Une democratisation des choix technologiques', Le Monde Diplomatique, septembre, 1982.

7. Salamon, J.J., Promethee empetre, Editions Futuribles, Pergamon Press, 1982.

8. Durand, M., La qualite de la vie. Mouvement ecologique, mouvement ourier, Paris Editions Mouton, 1977.

9. Fagnani, F. et Moatti, J.P., Environment Issues in France Social Claims and the Institutional Context, Conference ECPR 'Environmental Politics and Policies', Florence, 24-28 mars, 1980.

10. Boy, D., 'Les urnes en vert', revue Autrement, No. 29, fevrier, 1981.

11. Association pour l'etude des structures de l'opinion publique (AESOP.), 'Les structures de l'opinion publique en 1981 (le theme nucleaire

parmi d'autres)', tome 2, Paris, 1981.

12. Nicolon, A. et Fagnani, F., (sous la direction de), Nucleopolis, Presses Universitaires de Grenoble, 1979.

13. Nicolon, A. (sous la direction de), Analyse des mouvements de defense de l'environnement et du cadre de vie a partir d'oppositions locales a des projets d'equipment et d'amenagement, Ministere de l'Environnement, Paris, 1981.

14. Touraine, A. et al., La Prophetie anti-nucleaire, Paris: Editions du Seuil, 1980.

15. Moatti, J.P. et Maitre, P., Analyse des oppositions locales a des projets d'equipement: le cas des centrales nucleaires de Cattenam et de No gent sur Seine, Ministere de l'Environnement et du Cadre de Vie, 1981.

16. Inglehart, R., Changing Values and the Crisis of Environmentalism in Western Societies, Berlin: Wissenschaftszentrum, 1983.

17. Garnier, J.P. et Goldschmit, D., Le socialisme a visage urbain, Editions Ruptures, 1978.

18. Garnier, J.P. et Goldschmidt, D., 'Faux prophetes et bons apotres (a propos des nouveaux mouvements sociaux), Critique Communiste, janvier, 1979.

19. Touraine, A. et al. 'La Prophetie anti-nucleaire'.

20. Dagnaud, M. et Mehl, D., 'Des Contestataires comme il faut', revue Autrement, no. 29, fevrier, 1981.

21. Lipietz, A., 'Quelle base sociale pour le changement', Le Temps Modernes, no. 430, mai, 1982.

22. Lefebvre, H., 'De l'Etat', tome 3, Editions 10/18, Paris, 1978.

23. Salamon, J.J,. Piomethee empetre.

24. Theys, J., Evolution des problemes physiques et de la demande d'environnement, Colloque 'Les

politiques d'environnement face a la crise', Germes et Secretariat d'Etat a l'Environement et a la Qualite de la vie, Paris, 10-12 janvier, 1984.

25. Lagadec, P., Le risque technologique majeur: politique, risque et processus de developpement, Paris: Pergamon Press, Coll. Futuribles, 1981.

26. Crozier, M. et Thoening, J.C., Decentraliser les responsabilities. Pourquoi? Comment?, La Documentation Francaise, 1976.

27. Wilensky, H., 'Le corporatisme, le consentement populaire et la politique sociale' in 'Le'Etat protecteur en crise', OCDE, Paris, 1981.

28. Arnstein, 'A leader of citizen participation', The Journal of the American Institute of Planners, XXXV, July, 1969.

29. Habermas, J., La technique et la science comme ideologie, Paris: Editions Gallimard, 1973.

30. De Noblet, Manifeste pour le developpement de la Culture Technique, Paris: Centre de Recherche sur la Culture Technique, 1982.

31. 'Femmes et Techniques', revue Penelope, Universite de Paris VII, no.9 octobre, 1983.

32. Lucas, Y., Codes et machines, Paris: PUF, 1974.

33. Coriat, B., L'atelier et le chronometre, Paris: Bourgois, 1978.

34. Gaule, A. et Granstedt, I., Les incidences de l'informatique sur les conditions de travail et d'emploi, Grenoble: IREP, 1971.

35. Daniellou, F., L'impact des technologies nouvelles sur le travail en postes dans l'industrie automobile, Paris: CNAM, 1982.

36. Galle, R. et Vatin, F., Le modele de fluidite. Etude economique et sociale d'une raffinerie de petrole, Bandol, 1980.

37. Naville, P., La maitrise du salariat Paris: ANTHROPOS, 1984.

France

38.   Merchiers,   J.,   'L'informatisation   des
activites d'etudes', CEREQ, Paris, 1984.

39.   Bertrand,   O.,   L'automatisation de l'usinage
et le   developpement de   la   commande   numerique,
Paris: CEREQ, 1984.

40.   Naville, P., 'La maitrise du Salariat.

41.   Chopplet, M., Gabillard, E. et Hermant, C.,
Le   micro-ordinateur,   technologie   et   societe.
Itineraire   vers   le   futur,   1983,   EUR   8746   FR
Commision de Communautes Europeennes.

42.   Chesnaux, Jean, De la Modernite, Paris:
Editions Maspero, 1983.

43.   Baboulin, J.C., Gaudin, J.P., et Mallein, P.,
INA,   Le   magnetoscope   au   quotidien,   Abuier-
Montaigne, Paris; 1982.

44.   Nora, S et Minc, A, L'informatisation de la
societe, La Documentation Francaise, 1978.

Chapter Seven

PUBLIC ACCEPTANCE OF NEW TECHNOLOGIES IN AUSTRALIA

Peter Stubbs

Introduction

The analysis of public attitudes towards new technologies in Australia is hamstrung, as it is elsewhere, by serious inadequacies in the data. This is true not only as regards the range of new technologies for which we have any measures of public opinion but also longitudinally: there are hardly any objective means to assess the extent of changes in public attitudes through time. There may be some improvement in the near future, following the recent launch of the Australian Values Study Survey. The survey has recently published information about public attitudes to technology expressed in 1983 and it is intended that it will be repeated periodically, thus enabling the construction of a profile of attitudes over time.

Historically, one would expect a predisposition towards technology among Australians. Just as economic conditions in the United States, especially the expanding frontier and a relative shortage of labour, are held to have contributed to the growth of American inventiveness, so too in Australia there are celebrated cases in the standard histories of the inventor-technologist as hero, spurred on by the prospects of Australia's two most successful industries, agriculture and mining.(2) Shaw (1962, pp. 155-6) mentions several agricultural innovations: the Ridley stripper of the 1840s speeded reaping and economised expensive labour; the stump-jump plough coped with difficult local conditions; and H.V. McKay's combine harvester of 1884 brought manifest advantages. In 1860 James Harrison, a Scot living in Geelong, pioneered shipboard refrigeration to facilitate meat exports. Necessity was also fecund in the mineral industry, although Australians were equally eager to apply foreign technology. The flotation process of ore treatment was born at Broken Hill, and Robert Sticht perfected pyrite smelting at Mount Lyell in Tasmania. Of the late Victorian era, Geoffrey Blainey has written 'In innovation and adaptation no mining country matched Australia in this triumphant era of world metallurgy'.(3)

There are also successes in the twentieth century, frequently connected with those same two industries: CSIRO's development of myxomatosis for rabbit control; the sugar cane harvester and the atomic absorption spectrophotometer are exemplars. CSIRO - the Commonwealth Scientific and Industrial Research Organisation - dominates civilian research expenditure and stands high in public esteem.

Geography has also emphasised to Australians that technology has brought many advantages in communications. The steamship, the telegraph and the airliner have all reduced the psychological distance between Australia and the countries from which her settlers came. Internally, the outback shrank successively with the emergence of river steamers, the telegraph, railways, automobiles and aircraft, including the world-famous flying doctor service. In the future, satellite communications will further improve links at home and abroad.

Current Public Attitudes Towards Technology

The Australian Values Study conducted its first national survey in August 1983. It was based upon international standards and included questions relating to technological change, and is the widest and most recent investigation of public attitudes, with a nationwide survey population of over 11,000, though the questions treat technology in the broadest sense without addressing questions about specific technologies. A repeat is planned for 1985.

The picture that emerges from the Study is that Australians are, on balance, favourably disposed towards science, technology and change, but also that a significant proportion are worried by the pace of that change, and are cautious about new things and new ideas. Asked about general attitudes to change, no less than 58.7 per cent agreed that 'everything is changing too fast', with only 38.6 per cent disagreeing. However, it is not

191

clear that technology is the prime culprit for this misgiving, so that we cannot compare it with the 53 per cent of Americans in 1979, noted by Dorothy Nelkin in chapter 3, who felt that science and technology were 'changing life too fast', since barely one-third of respondents felt that Australia would be a better place to live in if there were a return to the standards of their grandparents; and almost half felt that scientific advance would help mankind in the long run, compared with less than a quarter who felt that it would not; 27.7 per cent felt that there would be pros and cons arising from scientific advance. Attitudes to innovation, in things or ideas, were varied, as the following responses show in Table 7.1.

Table 7.1: Attitudes to New Things and New Ideas

	%
Very much attracted	7.5
Attracted on the whole	26.0
Attitude varies ('it depends')	31.5
Cautious on the whole	31.4
Very cautious	3.6
	100.0

Source: Australian Values Study, 1983.

It may be, however, that the qualifications or misgivings noted above were not exclusively concerned with technological change: they could encompass changes in social attitudes. As a force for good, new technology was seen as worthy of more emphasis by 58.4 per cent of respondents, against 14.5 per cent who thought more emphasis a bad thing, the balance of 26.5 per cent being neutral. A slightly more specific question elicited a more positive response. Asked how they thought technological developments would affect Australia, the Survey population generally foresaw benefits as in Table 7.2.

Australia

Table 7.2: Expected Outcome of Technological Developments for Australia

	%
Many more benefits	20.8
More benefits	33.5
Benefit and disadvantages	31.5
More disadvantages	7.6
Many more disadvantages	1.6
Not accounted for	5.0
	100.0

Source: Australian Values Study, 1983.

Table 7.3: Elements Judged Important for Solving Social Questions

Elements	Cited by percentage of respondents
Education system	48.2
Economic development	47.6
New technology	38.5
Changes in human nature	30.5
Church (religious organisations)	26.5
Charitable organisations	22.7
Dedicated, visionary individuals	21.9
Parliament	18.4
Law courts	18.1
Labour Party	16.5
Time - most problems solve themselves or go away	14.4
Government departments	11.6
Trade unions	10.8
Industrial organisations	8.6
Liberal Party	6.9
Australian Democrats	3.8
National Party	3.3

Source: Australian Values Study, 1983.

On this scale barely one person in eleven is unequivocally pessimistic about technological change, and this apparent optimism is reflected in the high importance accorded to new technology as a solution to social problems, shown in Table 7.3. It is interesting to note that in this context, new technology is seen as a more potent solution than changes in human nature, and than the influence of all political parties added together; and vastly more important than the trade unions.

The survey enables some stratification according to the age, sex, education level, political affiliation and domicile of respondents. Liberal Party (right wing) supporters were very slightly more supportive of technology than were Australian Labour Party supporters, though both exceed the survey average and held views more strongly than other groups, which had higher responses in the neutral category. Supporters of the centre Democrat Party were the most fearful group, though still expressing majority support for technological change; a higher proportion of them - almost half - felt that technology would change their jobs. Men displayed more optimism about technology than women, although more men expected it to have an impact on their jobs: women were less convinced of the benefits of technological developments. Young people expressed much more confidence in technology as a solution to social problems, with the under-30s attaching more than twice as much importance to it as did the over-70s, and with a continuous diminution in confidence from teenagers onwards.

People with tertiary education expressed the most anxiety about the impact of technology on their job, but whether this reflects their intelligence or the fact that their 'human capital' has the most to lose if technology leaves it outmoded, is an open question. Twice as high a proportion - nearly 23 per cent - of tertiary graduates expressed this concern than people with only a primary level of education, notwithstanding the high probability that the latter would include a higher representation of the conservative older age groups. Since the question addressed 'job or main activity' it is, of course, possible that pensioners felt less threatened than employees.

Geographically, Western Australians displayed the most favourable attitude among the states towards the development of new technology, possibly reflecting the economic buoyancy which the state has enjoyed in recent years. However, the tendency for rural respondents to be more concerned about disadvantages than were metropolitan city dwellers is most pronounced in Western Australia, where 74 per cent of people in Perth favoured more emphasis on developing technology, compared with only 48 per cent outside Perth.

Useful though the Survey is, it can do no more than lay a foundation for further work. It will be particularly helpful to explore changing attitudes. So far as it goes, it suggests majority support for technological change with some significant qualification: there is clearly concern among many people that technology brings problems and that the pace of change is too fast for comfort, though there is scant support for turning the clock back. The major limitations of the Survey (which, it must be stressed, was not intended as a specific study of technology but rather of broader attitudes) are that it treated stated rather than revealed preferences, and that it did not explore attitudes to specific technologies, or among a wider variety of interest groups.

The point about revealed preference will be familiar to anyone with a nodding acquaintance with economics. People may say one thing but do another, may lament the passing of the corner shop while actively deserting it for the supermarket. This is a problem one faces in all attitudinal surveys; but we need to observe actual behaviour in relation to technology. And even then it will be necessary to ask how much real freedom of choice the public or individual subgroups have accepting or declining new technology, for if the choice is constrained it could explain some part of the ambivalence between stated preference and observed behaviour. However, that ambivalence may be born of unrealistic expectations by consumers who want the wide choice and low prices of the hypermarket and the cosy familiarity of the corner shop.

The point about technologies and interest groups is that technological change almost axiomatically redistributes income and wealth, and is therefore likely to create interest groups. Innovators, and those employed by them directly or indirectly, will gain on balance from changing technology and seek the promotion of it and of circumstances conducive to it: vested interests in superseded technology will tend to oppose its replacement, or extract some measure of compensation from innovations and/or governments. We need then to examine the evolution of these different groups and, if possible, some instances of technical change to examine their conflicts of interest.

Interest Groups

There are no simple categories which lead us to easy solutions of the distributional dilemma which changing technology poses, and one person's selection may be as good as another's; the ones offered here reflect my own predispositions as an economist, and emphasise the distributional problem.

Consumers
Consumers are the biggest and most significant group affected by technological change. Though they are often singularly ill-organised, the massive power of their expenditure should not be underestimated. While many consumers may be ignorant of the economic costs of tariff protection or industrial inefficiency, they know a bargain when they see one and that is enough to establish their power en masse.

The consumption of durable consumer goods is encouraged by high levels of disposable income and, inasmuch as consumer durables enable the substitution of expensive labour services by domestic capital equipment, by high wage-levels also.(4) By international standards, the average Australian consumer enjoys a high real income and is relatively sophisticated. Ownership of consumer durables is high (see Table 7.4); overseas travel and the prevalence of overseas media, such as British and American cinema and TV, have promoted an awareness of innovations in consumer products, while high income-levels enable their absorption into consumers' budget plans fairly quickly. However, the introduction of innovation often lags behind that in, say, the United States or Japan. Few new products are first introduced in Australia, and the limited local market and minor incompatibilities of design or regulation may delay their Australian debut, broadly consistent with the international product cycle paradigm propounded by Vernon. Yet paradoxically, this may well result in a very receptive consuming public. Not only is there the international demonstration effect already mentioned, but the customer-pioneering has already been done abroad. The worst bugs in the new products will have been sorted out, so that on their Australian launch, new products generally meet an informed, wealthy and confident local market.

Table 7.4: Household Ownership of Consumer Durables

Percentage of households owning:	Austr- alia	Austria	Bel- gium	Can- ada	France	West Ger- many	Italy	Japan	Nether- lands	Spain	Sweden	Switz- land	UK	USA
Refrigerator (1980)	99	81	91	99	95	95	92	99	98	83	95	94	92	100
Freezer (1980)	43	44	41	--	27	49	29	--	46	12	68	--	36	--
Washing machine (1982)	95	85	83	88	82	90	90	98	88	80	70	58	86	95
Dishwasher (1980)	14	12	13	--	16	22	17	--	12	10	23	22	5	--
Microwave oven (1980)	4	2	1	--	3	4	.2	--	2	1	2	--	2	--
Colour TV (1980)	86	54	57	--	56	75	38	--	76	36	80	63	74	--
(1984)	95												86	
Video recorder (1982)	14	4	4	--	6	13	1	--	7	3	15	9	25	--
(1984)	36												35	
Telephones per 1,000 persons (1980)	489	401	369	686	459	464	337	460	509	315	796	727	477	788
Cars per 1,000 persons (1980)	407	299	318	428	357	377	310	203	304	202	347	354	276	537

Sources: Australian data for 1980: National Energy Survey; Car data: MVMA, World Motor Vehicle Data 1982; Telephone Data; UN Statistical Yearbook, 1981; other data: Euromonitor Publications, various.

Transport costs and high costs of local manufacture and assembly will tend to delay diffusion, but the evidence shows that, despite later introduction, the diffusion of colour television and video recorders has been faster and gone further in Australia than in Britain. Specifically Australian characteristics could be influential: a hot climate encouraged the early use of refrigerators; high levels of home ownership are likely to boost the demand for consumer durables; the high degree of urbanisation (in fact, the highest in the world) would encourage emulation by consumers; while low-density urban housing and the city/rural population distribution, with limited public transport, provide a stimulus to car and telephone ownership.

The Australian consumer thus appears to be enthusiastic in the acceptance of the new technologies: some would even say too enthusiastic, linking the proliferation of fast and convenience foods to levels of obesity in Australian children which are above the American and well above the British levels.

Producers
There are some analogies between consumers and producers. Overseas travel is a prime source of information about product and process innovations(5), and imitative lags in innovating(6) give local producers the subconscious impression that they must keep running eternally just to stay abreast or within sight of the advance of technology overseas - an impression gained first from Europe and North America, and which has since been compounded by the rise of Japan to technological eminence.

It may be possible to draw a distinction between those sectors of the economy which are largely subject to the direct disciplines of the world market, and those which are insulated by the protections of tariff, quota or transport costs and cater largely for the domestic market. Mining and agriculture are still strongly dependent upon export markets (as they were so heavily in the nineteenth century), and in general both have applied new technology in a necessary search for competitive efficiency. The government has encouraged agricultural research through CSIRO, and the dissemination of results to farmers with manifest success though there remain pockets of inefficiency, and the agricultural lobby is no less astute at seeking government support than any other.

The growth of that part of the manufacturing sector devoted largely to import replacement has drawn heavily upon overseas technology, as one would expect since innovations are preponderantly foreign. The high level of multinational corporate investment in Australia, induced in part by trade barriers erected by government, has proved a steady and long-observed source of production technology (7), though it begs the usual questions whether the net effect has been to stimulate domestic R & D and technological output, or to condemn local industry to technological serfdom. Parry has shown the chemical industry to take overseas technology with little modification, despite the fact that the local scale was sometimes inappropriately small. The costs and risks of developing local alternatives would hardly appeal to parent companies, though some expenditure on modification might be justifiable: however, in the recent context of reducing and levelling tariff protection, it is perhaps not surprising that there has been a reluctance to initiate expensive R & D. Naturally, this is an emotive issue, and one that has become topical again in recent years.

Australia's small domestic market and her distance from the prime industrial markets of the world are chronic and severe disabilities in the struggle to develop competitive manufacturing industries. Moreover, the natural protection of distance, augmented by protectionist government policies soon after federation and explicitly after the Second World War, led to a plethora of small, inefficient local manufacturing industries and singularly little manufacturing specialisation on Swedish, Swiss or Dutch lines. While the present government recognises the need to rationalise and restructure industry, it is a difficult task. Entrenched manufacturing groups threatened by increasing international competition tend to become defensive, and innovation and technology will create schisms among producer interests.

Vested interests among local producers sometimes oppose the introduction of new technology which might supplant their own product. This opposition may be manifest through lobbying, seeking regulatory restraint of tariffs or quotas against rivals. If they seek protection for local production, of know-how and/or innovations, the impact of this (certainly in the short run) is to raise the cost and price of new technology and therefore to delay its diffusion; the application of a bounty on local production is a possible exception.

The nature of industrial competition, with its emphasis on the Schumpeterian dynamic of innovation, poses serious problems for a country like Australia which has little comparative advantage in industrial innovation, and much of whose industry is multinationally owned and quite rationally has the status of a technological client. From the corporate standpoint of private benefit, it may be more lucrative to devote rewards and talent to finance, marketing and lobbying, rather than to technology, for which imperfect, if tolerable, overseas substitutes can be bought at relatively little more than marginal cost. The gradual move to a less protected economy in the 1970s and has brought tangible benefits for consumers, as the performance of the whitegoods industry (refrigerators, cookers etc.) exemplifies, but the innovation performance of Australian industry, which has failed to maintain its expenditure on R & D, has been disappointing to proponents of an increased contribution from Australian technology. The government therefore committed itself to incentive schemes, and to additional widespread tax deductability for indigenous R & D in the election campaign of November 1984. The opportunity costs of this policy go largely unremarked, but it is seen as a desirable counterweight to a severe natural disadvantage, and one that is necessary to assure firms of the benefits of technology, and to allow the fullest realisation of social benefits which might exceed private returns to innovators.

The corporate sector has been assiduous in seeking government support of the sort just described. It may therefore be thought strongly pro-technology, but if government support policies are financed by general corporate taxation, those profitable multinational subsidiaries with access to low-cost parental technology may consider themselves to be losing on balance; so too would profitable firms in non-scientific sectors of the economy. Hence one would not expect government support for local <u>production</u> of technology, as against consumption of technology irrespective of its origins, to be welcomed unanimously within the corporate sector.

The general judgement on the corporate sector is that there is widespread acknowledgement of the need to apply technology as a vital instrument for international competitiveness, but serious shortcomings in its application. Good engineers are scarce by international standards, and limited markets inhibit movement along the learning curve, which is necessary if local producers are to compete internationally. In turn this inhibits entrepreneurship. Thus, while industrialists hope for sectoral and national benefit from technology, and in surveys by management consultants or in responses to the National Technology Strategy discussion draft a high percentage of them emphasise the importance of technology, in private some of them are apprehensive about the outcome of international technological competition.

Labour
Apart from capital, represented by the corporate sector, various labour groups display sectional interests. Some are very specifically supportive of technology, such as the Australian Industrial Research Group (AIRG), a group of senior industrial research managers; employers' associations such as the Business Council of Australia; and professional scientists and technologists in various employments such as academics and CSIRO scientists, some of whom sit on committees such as the Australian Academy of Sciences. Their public visibility and esteem are almost certainly disproportionately large compared with their numbers when they speak in a representative capacity for their respective or collective disciplines.

Attitudes among labour unions vary. The position of the Australian Council of Trade Unions (ACTU) has changed little since it was reported to the committee of Inquiry into Technological Change in 1980.(9)

Union responses to technological change generally have been to seek prior consultation over its introduction by employers to ease any problems of labour-displacement which would result, and to seek a share of the benefits for their members. Sometimes this has led to major conflict with employers, as in the Telecom dispute over the displacement of telecommunications engineers, in various wharfside disputes, and in problems over the introduction of advanced newspaper technology. While there are occasional calls for technological change to be halted or postponed,(10) senior unionists generally concede the ultimate need for change, and there is often more realism in the bargaining process with employers than public rhetoric and media reports imply.

This realism is not surprising in the fact of the economic power of technology. Where union power has secured high economic rents or returns for members there is a very strong incentive for employers to subsistitute capital for labour: high wages as well as high profits might be seen as a signal for Schumpeterian innovation. While a resolute and solid union may hold new technology at bay, it does so at the risk of damaging its own industry, and in the end it tends to compromise. It is difficult to access the performance of unions in dealing with problems raised by technology unless they have an explicit maximand. Some have been unable to prevent a severe decline in numbers employed, as on the wharfside, but may have succeeded, as there, in securing very high rates of pay and benefits for their remaining members. It has been suggested that the decline of employment in insurance compared with banking reflects the stronger union structure in the latter, but this almost certainly is an over-simplification.

What is beyond doubt is that certain subgroups in the labour force are especially vulnerable to the effects of technological change. McCann, Fullfrabe and Godfrey-Smith (11) mention women, older workers, and immigrant workers, to which one can add young people and the handicapped (12).

As Selby-Smith (13) has shown, women in the labour force are concentrated in a relatively narrow range of occupations, and may be more susceptible to job loss through technological change. They are generally less well qualified educationally, and are penalised also by being housebound during child-rearing, and being geographically restricted in looking for work if their husbands are tied to their jobs. These difficulties may help explain why women in the National Values Survey expressed misgiving more widely than men about the disadvantages of technological change.

Older workers are at risk if technological change erodes the value of their skills, which they find difficult or impossible to replace. As early retirement is extended to younger age groups, problems in seeking re-employment will increase, although social attitudes toward retirement may become more approving as it becomes more common.

Migrants are often perceived as vulnerable to unemployment due to technological change, especially if their English is limited. They tend to be concentrated in manufacturing and construction work, with their womenfolk in process work and labouring jobs. However, it is difficult to say what the net impact of technology is: they may be at high risk of displacement, but more willing to accept other relatively unskilled jobs.

Young people are also at risk if there is, as now, widespread technological change and high rates of unemployment. Where businesses introduce labour-saving technology and are trying to minimise redundancies by relying upon wastage, recruitment suffers, and this bears heaviest upon the young and inexperienced. While the young have the capability to benefit from training and the federal government has established various instruments such as the Trade Training Programme, the Skills Training Programme, the National Training Council, Work Experience and Training for Young Persons, as well as the present Kirby Committee of Inquiry to review labour-market programmes, there is no doubt that as a whole they are disadvantaged. Ford (14) has argued that Australia has tended to neglect the skills required for modern high-technology processes, has emphasised the traditional academically oriented skills, and has become increasingly reliant on 'technological packages', imported, installed and even maintained from overseas. If this is so, it poses very serious problems for the young in job markets, but as we saw in the National Values Study, it does not seem to have coloured the attitudes of the young against new technology.

Some commentators have taken the view that labour has consistently 'lost' to capital in Australia as a result of technological change, and that innovation and its associated problems 'have hardly been considered at all, in their own right, within the arbitration system in the years since 1945'.(15) However, analyses of this sort usually view the issue of distribution from the standpoint of existing, often unionised, interests, and neglect the fact that new skills and occupations are being included by the same innovation, and that these are a substantial benefit to labour. There is no inexorable downward trend evident in the ratio of wages and salaries to gross domestic product.

The Legal System

The legal system as an intendedly neutral arm of the state both affects and is affected by technological change. As such it can obviously have a major impact on the public acceptance of new technology, and the rules established by parliament and the legal system are crucial to the distribution of costs and benefits arising out of changing technology. The issues are vast and many of them are presently under consideration(16) with the government seeking advice from bodies such as the Law Reform Commission and the National Health Technology Advisory Panel.

The full compass of interaction between technology, the law and society is obviously beyond the scope of a brief survey like this one: indeed, it is a subject which has been little explored so far but which will grow in importance in the future, just as biomedical ethics has emerged as a subject in its own right in the past decade. Apart from the unprecedented issues raised by biomedical engineering, there are thorny problems to be addressed about personal privacy in an age where electronic data are mushrooming, about intellectual-property rights, about compensation to the victims of technology (raising the issue of class actions in the United States style, where groups of individuals can seek legal redress against corporations), and many others.

The net effect of the system is thus a matter for conjecture at present, and any full assessment of its impact would probably have to extend to include the effect of the taxation system on incentives and social stability. A highly permissive system, which some industrialists allege exists in many less developed countries, and further allege offers such countries major industrial advantages, could be explosive in a highly educated and vocal society, where there are already widespread fears about technology running too far ahead of the comprehension and legal rights of the common man. In any case, product regulation is weaker than in the United States. On the other hand, it could be argued that elements in the legal superstructure, such as the long tradition of government involvement in industrial arbitration and wage-setting, have built rigidities into the ecomonic system which are of doubtful advantage in the process of technological changes where flexibility is a vital element in international competitive success.

Recently there has been a professed Accord between the Labour government, industry and the unions, and in the latter part of 1984 organised labour gained some significant concessions relating to technology. In July three major private banks agreed to consult the bank employees' union (ABEU) about their plans for technological change, and two of them gave written assurances that no employee would suffer loss of salary from the technological changes, and offered appropriate retraining to redeployed workers. Soon afterwards the Arbitration Commission supported the ACTU's test case for consultative mechanisms about technological change, implying measures of industry-wide consultation and compensation. Some private employers, principally large companies, had already adopted similar measures, but the Commission's ruling can be seen as an attempt to win union acceptance and buy industrial peace over the issue of technology. The Commission's deputy president Mr Justice Ludeke noted in the banking case 'Experience in other industries has shown that if the employer is too reticent about the outcome of the introduction of technological innovations, industrial relations consequences are simply not worth contemplating'.

Environmental and Other Groups

As in North America and Western Europe, several groups have grown which seek to protect the environment or disadvantaged minorities from developments which they judge to be undesirable, and which can bring them into conflict over technological change. By 1982 there were 11 conservation-lobby groups, active in various ways. The older-established ones tended to raise funds for educational and research purposes, like the National Parks and Wildlife Foundation (NSW), but others such as Friends of the Earth and the Tasmanian Wilderness Association campaigned much more actively. Among issues which they addressed were the mining, use and exporting of uranium, and the preservation of wilderness and forests in Tasmania, Western Australia, Gippsland and Queensland. There has also been strong and effective pressure in favour of aboriginal land rights.

Objections to Australian participation in the Vietnam War had raised the visibility and effectiveness of public protest, and while there was no real parallel to the involvement of United

States scientists in caballistic policy-making, there was strong objection to pork-barrel politics involving the sacrifice of Tasmanian wilderness to dams for hydroelectric power. The focus of opposition was not always technological, however, but two particular issues did excite controversy: one was nuclear power and the other was a more diffuse unease about technological unemployment which emerged at the end of the 1970s, and probably contributed to the setting up of the Myers Committee of Inquiry into Technological Change in Australia in 1979.

Political parties, especially when faced with the realities of office, have supported technology as an instrument for economic growth. This has tended to leave the peripheral groups somewhat stranded and driven them to adopt public protest, as in 1984 when nuclear protestors camped outside the Labour Party conference in Canberra. In the 1984 election, the fledgling Nuclear Disarmament Party gained about 7 per cent of the vote, and a seat in the Senate for Western Australia. But the extent of protest movements is limited: there is no real equivalent of the Greens in West Germany, or the Greenham Common Women or CND in Britain, or various protests in the USA. No Office of Technology Assessment has been established, despite suggestions, and the issue of technology seems to have been received altogether on a lower key in Australia.

The Public Service

If governments and the public service are generally favourable to technology, it cannot be assumed that all is harmony within the corridors of power. Hearsay has it that the two most recent Prime Ministers have concluded personally that an added emphasis on technology is essential to Australia's continued prosperity: Mr Fraser is said to have reached this conclusion during a period of reflection in a protracted convalescence from a back injury, Mr Hawke following an overseas visit which opened his eyes to the technological strides being made in Pacific Basin countries. But the means to promote technology nationally give rise to varying opinions in the public service, which often survive changes in administration. A wide range of policies and instruments has been deployed over the past two decades to encourage local technology, but with disappointing effects on the whole. This has contributed to divergences of opinion within the public service.

Several departments have been strong advocates for technology, but with subtle yet important differences. The Department of Science and Technology under the Ministry of Barry Jones, a very experienced propagandist for science and for the importance of positive policies to harness it for social goals, has advocated much increased local production of technology, invoking social benefit to justify higher subsidies. The same themes have been followed by CSIRO and ASTEC (the Australian Science and Technology Council). In other departments, however, such as the Treasury and the Department of Finance, there is considerable scepticism about the efficacy of high degrees of subsidy. The Bureau of Industry Economics (BIE), within the Department of Industry, Technology and Commerce, has pointed to the benefits of technology and the dangers of trying to restrain it in the interests of groups who would suffer immediate, local disadvantage; but the Bureau takes a much more sceptical view of the case for lavish support of the local <u>production</u> of technology rather than the application of it, irrespective of its national origin. The newly formed Economic Policy Advisory Committee secretariat takes a similar view. In December 1984, following an election promise to increase the level of tax deductibility for industrial R & D to 150 per cent, responsibility for technology was shifted from the Science portfolio to the Department of Industry and Commerce. It was thought more appropriate that industry and technology policies should be framed together, possibly as part of the corporatist approach of the Accord, which also involves the Department of Employment and Industrial Relations.

On the surface the public service is united in favour of technology but subject to internal dissension about where the responsibility for administering the technology should lie, and what degree of subsidy should be given to local production of technology. In recent years many important Reports of Committees have been published: the Crawford Report on Structural Adjustment (1979)(17), the Birch Report on CSIRO (1978),(18) the Williams Report on Education, Training and Employment (1979)(19), the Myers Report on Technological Change in Australia (1980)(20) and the Kirby Report on Productivity and Innovation Programs (1981) (20), all of them

involving, in greater or lesser degree, science and
technology. In all of these, public servants have
provided a vital secretariat and in various ways
the reports have raised the 'visibility' of
technology in debate, and increased the national
perception of its importance. In this sense, the
impact of public servants has been pro-technology
and the qualifications expressed within reports
have often been ineffectual. For example, many of
the recommendations of the Report of the (Myers)
Committee of Inquiry into Technological Change
(1980) involved safeguards for labour, and met
little acceptance from the Liberal government of
the day, though a recent test-case ruling on
consultation of employees is belated acceptance of
the Committee's first recommendation.

While there have been some regulatory measures
concerning technology, as we shall see in the cases
considered next, the thrust of government and
public service involvement has been to laud and
strengthen technology. By the mid-1980s there
seems to be a wider measure of public acceptance of
the imperative of technology in a competitive
world. The worst fears of the late 1970s and early
1980s, often rather luridly and negatively
publicised(22), have been overshadowed by the
realisation that Australian living standards are
being overhauled by the more successful Pacific
Basin countries such as Japan and Singapore, that
the growth of these countries represents both a
challenge and an opportunity for Australia, and
that since Australia cannot dictate the
international pace of technology she must remain
receptive to it. However, the recent subordination
of technology to the Industry portfolio could
signal that the honeymoon with science is souring,
that too much has been claimed and that deus ex
machina is faulty.

State governments have also become active in
the encouragement of technology, vying with each
other to attract or germinate high-tech companies
in newly established science parks. Initiatives
have been taken to reduce wasteful counterbidding
by rival state governments, but it is questionable
whether the intentions of these initiatives will be
realised.

The Media
The media play a unique and vital role in public
awareness and debate concerning technology. Shades
of opinion expressed within the media range from

black to white, and their net effect must be disparate.

Certain minority interest groups, such as the environmental and anti-nuclear protestors mentioned earlier, achieve a steady presence in the media. Occasionally, astute journalists or television producers succeed in boosting the ratings and their own standing by making a national issue of technology and stimulating public debate: in 1979 the BBC TV film, Now the Chips are Down, had a signal impact in Australia as well as in Britain. Against the low warning voices and the occasional dramatic splash, however, must be ranged the steady flow of technologically persuasive messages, through everything from personal computer magazines to the television series 'The Inventors' and the radio programme 'The Science Show'. The strength of that flow is its pervasiveness. In a radio review broadcast the present writer had once to forbear, reluctantly, revealing to the author of a book critical of microprocessors that the book had been printed using computer-typesetting.

In the absence of objective evidence, the effect of the media on public attitudes is a matter of judgement, yet it would need a bold observer to conclude that it was at all negative. Though consumerist organisations have often been less than enthusiastic for some innovations, such as computer-coding of supermarket goods and computer threats to privacy, the very act of testing and reviewing new goods in journals such as Choice tends to legitimise them for many consumers.

Special Issues in Public Acceptance

Nuclear Power and Mining

In 1979 a Morgan-Gallup poll(23) sampled 1,077 people, and found 52 per cent in favour of developing nuclear power and 35 per cent against. Women were much less in favour than men (44 as against 66 per cent), and so were ALP (the Australian Labour Party) voters when compared with Liberals: and whereas 49 per cent of Liberal voters said they would not object to a nuclear plant near their homes, only 31 per cent of ALP voters felt the same. In an international survey (24), Australia emerged as the second most favourably disposed of 14 nations towards the continued development of nuclear power, second only to South Korea, though Australians were only ninth in their willingness to have one sited close to home. But the nuclear power question is essentially an academic one now, as there is no likelihood of plants being constructed given the country's vast reserves of low-cost fossil fuels. The nuclear power lobby has failed, and at the time of writing was preoccupied with the modest pursuit of a cyclotron for the local production of radio isotopes.

A majority polled in 1981 was opposed to Australian possession of nuclear weapons, and there was widespread opposition to the entry of American or British warships into Australian ports if there were any chance of their carrying such weapons. A poll in 1982 showed respondents almost evenly split on the weapons issue, though less concerned about nuclear-powered ships. A more pertinent issue for most Australians is whether to mine and export uranium ores. Polls in 1983 and 1984 (25) showed roughly two-thirds in favour of mining and exporting uranium for peaceful purposes, but an increase of 6 percentage points in the proportions opposing it. Opposition was stronger among the young, and lowest among Liberal voters. Even most supporters of mining professed to be worried about it, especially those with tertiary education.

A report by ASTEC was commissioned, at the direct request of the prime minister (26), and produced under the chairmanship of Professor Ralph Slatyer. It examined Australia's nulear safeguards, in the context of bilateral and multilateral agreements and her prospective influence to advance the cause of nuclear non-proliferation. The recommendation of the report, that mining and exports should continue, in the ostensible hope that Australia could sit as an active party at non-proliferation conference tables, was seen by many ALP supporters as a betrayal of their pledge to abandon the industry.

Nuclear issues will clearly remain contentious, especially as New Zealand's new Labour government has taken a harder line on US naval nuclear vessels entering its ports. Australian attitudes hardened against nuclear weapons on several counts, notably the French tests at Mururoa atoll, during the deliberations of the Australian Royal Commission in 1984-5, which revealed that the British nuclear tests in Australia during the 1950s had released much more radioactivity than commonly acknowledged, and during the controversy in early

1985 whether Australia should offer facilities for United States aircraft to monitor MX missile tests. Public disquiet could conceivably reverse current government policy on mining, though for the moment economics and realpolitik rule.

Medical Technology
Medicine has a high professional standing in Australia. Despite widespread feelings that many doctors have become singularly greedy, they still carry the greatest public esteem (see Table 7.5).

Table 7.5: Public Judgement of High Ethical Standards in Selected Professions

Percentage judged high in:	1976	1981	1984
Doctor	62	63	64
Bank manager	66	61	61
Schoolteacher	56	55	55
University lecturer	47	51	51
Federal MP	21	16	17
Union leader	9	8	6
Car salesman	4	3	3

Note: n = 2,228
Source: Morgan-Gallup poll, The Bulletin, 22 May 1984, p. 32.

The intellectual standing of the profession is enhanced by the fact that each of Australia's three Nobel laureates have emerged within medical fields, and Australia has done pioneering work in the early detection of rubella, and thalidomide deformities, and recently has broken new ground in in vitro fertilisation. New medical technology, such as the contraceptive pill and CAT scanning, has diffused quickly, and Australian public opinion, where measured, has usually appeared more receptive than is British public opinion to test tube babies and surrogate motherhood. However, recent polls suggest a slight hardening of attitudes as public discussion has increased.

Table 7.6: Attitudes Towards In-vitro Fertilisation

	For otherwise infertile couples			Frozen eggs for later implant	Frozen for experiments
	1981	1983	1984	1984	1984
Percentage who:					
Approved	77	74	73	45	32
Disapproved	11	13	11	39	54
Are undecided	12	13	16	16	14

Source: Morgan-Gallup poll, The Bulletin, 3 July 1984, p. 23.

The inference from Table 7.6 is not only that a limited element of doubt is growing, but also that as one moves from the personal context of the childless couple to the clinical neutrality of the laboratory, public sympathy wanes. Some public misgiving is also evident about the morality of using animals indiscriminately for laboratory tests and research (27).

While 59 per cent of respondents approved compared with 33 per cent against, approval depended strongly upon the purpose of the experiment: 77 per cent approved for medical research, 65 per cent for veterinary research, but only 14 per cent for testing cosmetics or household cleaners. The advance of medical technology will pose much more direct and difficult ethical questions in the future, and these are likely to promote increasing public controversy.

If the birth control pill represents the most sophisticated technological approach to contraception, there has been a reduction in public acceptance of technology there. National data are inadequate for finely detailed analysis, but the work of the Australian National University Department of Demography suggests some reaction against the pill 20 years after its introduction, despite the fact that side effects have failed to affect mortality statistics. The intra-uterine device (IUD) has also lost popularity, mostly to sterilisation but to some extent to more traditional means such as the diaphragm or sheath.

Despite the evidence noted elsewhere in this chapter that more men are committed to technology than are women, the 1979 Canberra Population Survey showed (female) tubal ligations to be more than twice as common as (male) vasectomies, and for the ratio to be widening further, even though vasectomy is surgically simpler. Conceivably, the use of the pill may have resulted in responsibility for contraception being viewed as the province of the female.

Electronic Technology

This encompasses a vast range of issues: what follows here is the merest mention. Public reception of electronic capital goods used by consumers, like that of consumer durables, has been good. Automatic bank telling machines and the use of computerised payout by the Medibank health insurance scheme have been well received. Though unions have been defensive, the growth in demand for banking services has cushioned job loss, and the banking union agreement noted earlier should smooth the path of progress.

Public acceptance of information technology is difficult to predict with any accuracy, given the uncertainties about costs, diffusion and the nature of coverage at the household level. The Minister for Science, Barry Jones, envisaged it as a key technology for Australia in the National Technology Strategy discussion draft(28), and there have been official reports on Telecom services, cable TV and videotex. The satellite issue, in the finance of which Telecom will retain a major stake, is potentially very far reaching; but here policy formulation seems likely to precede public awareness and participation.

The author's own impression is that the Australian consumer will accept the new information technology as widely as they have other consumer innovations. Australia, a little surprisingly, is the world's most urbanised country, and, sunshine notwithstanding, her people watch more TV than do their British cousins. If cable TV, entertainment led, were to be the herald of the household information technology revolution, Australia could be well placed, though American experience suggests that it may take considerable time as interactive services have made very little headway there, and limited Australian trials have not been encouraging. However, a strong commitment to the AUSSAT satellite-broadcasting services(29) led a recent authoritative report to prefer satellite TV, by which rural users might be cross-subsidised by urban users, though it will probably deny the prospect of interactive services to the latter. Whatever the pros and cons of such a decision, it has to be taken in advance of any wide public appreciation of the issues and choices because of the long lead times and heavy investment required.

Regulation and Road Safety

Australia has lagged in the enforcement of product-safety standards. The Australia Federation of Consumer Organizations noted in 1984 that only 11 product-safety standards were mandatory, the remainder being voluntary. A regulated vehicle-recall code has been advocated by the Australian Transport Advisory Council, and though there has been some progress on pollution levels, a tougher vehicle-emissions strategy has been mooted.

In one area, however, Australia has been in the forefront of mandatory measures. In seat-belt legislation she has been many years in advance of Britain or the United States. The State of Victoria in 1970 became the first legislature in the world to make the use of seat belts compulsory, and their national adoption is estimated to have saved between 5,000 and 10,000 lives over a decade and prevented about 100,000 injuries.

Table 7.7: Road-vehicle Related Fatalities

	Fatalities per 100m vehicle km travelled 1980	% change 1968-80	Fatalities per 1m population 1981	% change 1970-81
Australia	3.0	-45.5	224	-26.3
Austria	-	-	253	-25.4
Belgium	5.3	-56.9	225	-26.2
Canada	2.7	-42.6	223	-6.3
France	4.5	-56.7	246	-23.8
W. Germany	4.2	-45.5	189	-40.2
Italy	4.1	-58.6	151	-26.0
Japan	3.0	-76.9	96	-54.1
Netherlands	3.3	-59.8	127	-48.0
Spain	7.5	-	170	+4.9
Sweden	-	-	94	-42.3
UK	2.3	-39.5	109	-22.1
USA	2.2	-33.3	214	-16.7

Source: A. Altshuler et al., <u>The Future of the</u>
<u>Automobile</u>, London: Allen & Unwin, 1984, Tables 3.2
and 3.3.

Pressure for the installation and mandatory
use of seat belts did not come from the general
public. Although as can be seen from Table 7.7
Australian accident rates have been measurably
higher than British and, particularly, American
ones, it was institutional rather than public
initiative that promoted the change.   In the
development of the Snowy Mountains hydroelectric
scheme, cars went off mountain roads and met with
serious accidents. Engineers set up a dynamic test
rig, and this, plus the work of Professor Joubert
at Melbourne University, led first to a static test
standard, and later to a dynamic standard.
Victoria required floor anchorages in the mid-
1960s: a 1967 joint select committee, which called
Joubert as its first witness, recommended
compulsory wearing of belts, and this was enacted
in 1970.
Public acceptance has been high. In 1970 an
estimated 20 per cent of drivers used belts, but by
1981 the compliance rates in New South Wales were
84 per cent for drivers, 79 per cent for front
passengers and 26 per cent for rear.   Recently
random breath testing for drivers has been
introduced.   Clearly,   this has considerable
potential to lower accident rates, and this
author's casual impression in Canberra is that it
is proving effective and is widely seen as a
justifiable regulation.

Some Observations

The four cases briefly examined above show the
public acceptance of technology to be high and
unequivocal where it is simple and unlikely to have
hidden and belated side effects; hence the lack of
controversy about seat belts, despite the fact that
they were imposed by law.   But where there is
uncertainty and vexed moral issues, as in nuclear
technology, public doubt is widespread and real.
Similarly, in medicine, despite admiration for its
status and achievements, there is mistrust of
technology which can bring very asymmetric risk,
which might maim or kill a few, yet benefit many
thousands.   Ironically, road use involves a similar
profile of risk but provokes a different public

215

reaction: a woman may face much greater likelihood of death or injury on the road than from using the contraceptive pill, but the latter involves a technological mistrust. However, the individual has a measure of choice about contraception. There may be less choice in the revolution of the electronic office, which at the time of writing is exciting public controversy because of the unusually high incident of tenosynovitis, an irreversible condition of the wrist joints associated with long exposure to modern electronic keyboards. Further work on accurate diagnosis is required, as some doctors are puzzled by international differences in its incidence and wonder whether for some operatives the problem is more imaginary than real. The response is likely to be some restriction on hours of operation rather than a reversal of the technology. In the world of cable and satellite TV there is little or no public choice: full public awareness will come only with usage.

Conclusions

Australians emerge from this study as generally favourably disposed towards new technologies, whether for use in consumption or production. However, there can emerge opposition from interest groups where the threat of immediate, tangible and specific loss to the group exceeds any notional and generalised long-run gain from technological change: in most cases, the group would not conceive of its opposition to a specific microeconomic technological change seriously impeding the general flow of benefit from technological change at a macroeconomic level, which they would assume to continue. However, these individual defences of sectional self-interest can sum to a degree of rigidity which does impede growth and change at the national level. Mancur Olson has suggested for some of the advanced economies that the rigidities consequent upon the collective action of interest groups are a crucial explanation of poor economic performance, and that the insulation of Australia through high tariff protection has created an entrenched interest group and a serious misallocation of resources(30).

Hence, the acceptance and speed of diffusion of new technology may be different among different sectors. It may well be that, in principle, incremental innovations are more readily acceptable than radical ones, since the latter involve much more obvious change and attract a more focused opposition. However, we require a further categorisation of innovations in relation to public acceptance; the following classification is offered as a tentative beginning:

(1) There are systemic innovations which, once made, effectively pre-empt any immediate alternatives for public acceptance. They are usually introduced, supported and regulated by governments or major instrumentalities. Examples would be a road system, a major airport, the fluoridation of metropolitan water supplies or the establishment of a satellite communications system. Many of them involve natural monopolies in the economist's sense, and would require massive public involvement to deflect their formation or reverse them.

(2) There are dominant innovations, which succeed by supplanting inferior products or techniques at the microeconomic level. Innovations using computers and microprocessors exemplify this category, and may be applied to a very wide range of established products and innovations. The superiority of a dominant innovation is such that, sooner or later, failure to apply it can put the user at a critical disadvantage. As such, dominant innovations are more likely to dictate to employers, though it is possible to see the telephone and the car as enlarging the job market for an individual so as to make it a dominant innovation for them.

(3) Some technological changes are optional, or consumer sovereign innovations, which leave each individual with an open choice whether to use them or not. Most leisure goods would fall into this category.

(4) Essentially a subset of (3) above, which is appended for perspective and completeness, is the failed innovation. Apart from the high proportion of research and development that does not lead to innovation, many production innovations launched fail to survive (31). Part of the success of technology arises from the fact that, in effect, both firms and markets apply to it stringent screening tests.

Opposition to new technology also falls into a number of categories. Individual attitudes to change may depend in part upon personal psychology - in the jargon of economists, whether a person is risk-averse or not: for example, an optimist may feel that technology will soon provide a solution to the disposal of radioactive wastes, but sceptics would differ. Such categories might be as follows(32).

(a) There may be genuine uncertainty about the technology, as distinct from risk in Frank Knight's sense of measurable uncertainty. One cannot assign probabilities to unique single events, so that individual perceptions and hunches can divide opinions sharply, as in the case of radioactive waste mentioned above. Life and health-threatening hazards are seen as a particular danger, with risks arising from many quarters, ranging from the prospect of nuclear holocaust to disasters like thalidomide, or industrial tragedies like Bhopal. Since health hazards can strike anyone and anywhere and defy any immediate remedy, they have become the new devils of our society.

(b) Different groups may have different social time-preference rates, reflecting their attitudes towards the welfare of future generations and the sacrifices that present consumers should make for consumers in the future.

(c) There are loosely defined fears that technology marks a loss of innocence which, like the opening of Pandora's box will usher uncontrollable evils upon mankind. Technological change, particularly in the sphere of biological engineering, poses immense ethical questions and the prospect of severe and far-reaching social change. There are also vague fears that technological change will split society more or less sharply between the beneficiaries and controllers of new technologies, whose lives will be enriched and fulfilling, and the displaced on the other hand, condemned at best to menial jobs.

(d) Technological change raises questions about positional competition, as Hirsch called it (33), which suggests that if consumer satisfaction rests not upon absolute levels of consumption but in significant measure upon levels relative to other consumers, then expectations will not be fully realised through material growth. Increased positional competition may involve a deterioration in the environment, evident ultimately even to those who would not consider themselves environmentalists. Certain consumers, presently

well placed in the positional stakes, may protect their advantage by objecting to the technology which enables rising consumers to encroach upon them: owners of quiet beaches, for example, would decry the noisy water-scooter.
(e) Most obviously, technological change will meet opposition if it threatens the monopoly rents of existing, vested interests, be they labour unions, professions or owners of capital.

Governments at state and federal level support technological change and have usually devoted much more money and effort to its implementation rather than to the mitigation of its social upsets. The current Labour government is inclined to a corporatist approach, with more intervention than its Liberal predecessors, who were psychologically more inclined to leave technology alone in the belief, still widely espoused, that in the long run the market mechanism would take care of problems of adjustment and redistribution. However, Galbraith's depiction of a technostructure(34) would seem to be generally apt, in that industry and government (of whatever likely political complexion) have a common interest in the encouragement and employment of technology, although they may display very considerable inefficiencies and misunderstandings in this.
Public opinion, however, has made its voice heard, independently of sectional interests of unions or capital. The quality of life matters increasingly to the public once their basic consumption needs are met, and some evidence of a slowly rising tide is evident in Australia. The young, while favouring technology, are critical of its undesirable side effects, and public awareness of these effects has grown in recent years, promoting suspicion and a more questioning attitude than before. But as we have seen from the four cases briefly surveyed earlier, the power of public awareness to influence choices is imperfect; and this fact alone will guarantee continued debate into the future, as will the ubiquity of technological change.

Notes

1. I owe thanks to many people who helped me in the preparation of this paper. In particular I should like to thank Anne Gorman of Social Impacts for data from Australian Values Study, and the editor of The Bulletin for permission to cite from its Morgan-Gallup polls. At ANU, Professor Max Corden of the Research School of Pacific Studies most generously provided a room, telephone, and access to library and secretarial facilities which were invaluable to me; Professor Tom Parry was instrumental in this good fortune; and Dr David Lucas of the ANU Demography Department was most helpful. Many public servants were extremely helpful to me, of whom I should like to single out Mr Richard Joseph of the then Department of Science and Technology, a former pupil whose unstinting help has repaid many-fold any debt he might have owed his teacher. I should also like to thank Dr Bruce Middleton of ASTEC and Mr John Bamford of the Department of Transport; and Ms Patricia Boyce of the Department of the Prime Minister and Cabinet, and John Nightingale of the University of New England for comments on an earlier draft, The usual disclaimer applies: the views expressed are attributable to the author alone.

2. Shaw, A.G.L., The Story of Australia, London: Faber, 1962, pp. 155-6.

3. Blainey, Geoffrey, The Rush that Never Ended, Melbourne University Press, 1964, p. 255.

4. Vernon, R., 'International Investment and International Trade in the Product Cycle, Quarterly Journal of Economics, vol. 80, 1966, 190-207.

5. Stubbs, Peter, Innovation and Research: A Study in Australian Industry, Melbourne: Cheshire, 1969.

6. International Technical Services Ltd, A Study of the Rate of Diffusion of New Technology within Australian Industry, Canberra: AGPS, 1972.

7. Brash, D.T. (1966), American Investment in Australian Industry, Canberra: ANU Press, 1966: Parry, T.G. and Watson, J.F., 'Technology Flows and Foreign Investment in the Australian Manufacturing

Sector', <u>Australian Economic Papers</u>, vol. 18, 1979, 103-18.

8. Parry, T.G., 'Plant Size, Capacity Utilisation and Economic Efficiency: Foreign Investment in the Australian Chemical Industry', <u>Economic Record</u>, vol. 50, 1974, 218-44.

9. CITCA, <u>Report</u> of the (Myers) Committee of Inquiry into Technological Change in Australia, Camberra: AGPS 1980, vol. 4.

10. Stubbs, Peter, <u>Technology and Australia's Future</u>, Melbourne: AIDA, 1980, p. 117.

11. McCann, P., Fullgrabe, K. and Godfrey-Smith, W., <u>Social Implications of Technological Change</u>, Canberra: Dept. of Science, 1984, pp. 28-9.

12. ASTEC, 1984 <u>Technology and Handicapped People</u>, Canberra: AGPS.

13. Selby-Smith, J., 'Developments in Micro Electronic Technology and Their Impact on Women in Paid Employment', <u>Australian Quarterly</u>, Summer, 1983, 415-31.

14. Ford, G.W., <u>'Human Resource Development in Australia and the Balance of Skills,</u> paper presented to 52nd ANZAAS, Macquarie University, Sydney, 1982.

15. Fisher, Chris, <u>Innovation and Australian Industrial Relations,</u> Canberra: Croom Helm Australia, 1983, p. 213.

16. McCann, P. et al., 1984, <u>Social Implications,</u> pp. 60-9.

17. Study Group on Structural Adjustment, <u>(Crawford) Report</u>, Canberra: AGPS, 1979.

18. Committee of Independent Inquiry into CSIRO. <u>(Birch) Report</u>, Canberra: AGPS, 1978.

19. Committee of Inquiry into Education, Training and Employment, <u>(Williams) Report</u>, Canberra: AGPS, 1979.

20. CITCA, 1980, <u>Report</u>.

21. Department of Science and Technology, National Technology Strategy Discussion Document, Canberra: Department of Science and Technology, 1984.

22. Reinecke, Ian, The Micro Invaders: How the New World of Technology Works, Ringwood: Penguin Books Australia, 1982.

23. The Bulletin, Sydney, 30 October 1979, 25.

24. The Bulletin, Sydney, 18 December 1979, 46.

25. The Bulletin, Sydney, 29 May 1984, 34.

26. ASTEC, Australia's Role in the Nuclear Fuel Cycle, Canberra: AGPS, 1984/6.

27. The Bulletin, Sydney, 28 February 1984, 59.

28. Department of Science and Technology, National Technology Strategy Discussion Document, Canberra: Department of Science and Technology, 1984.

29. Australian Broadcasting Tribunal, Cable and Subscription Television Services for Australia, Canberra: AGPS, 1982.

30. Olson, M., The Rise and Decline of Nations: Economic Growth, Stagflation and Social Rigidities, New Haven: Yale University Press, 1982, p. 132-5.

31. Midgley, D. F., Innovation and New Product Marketing, London: Croom Helm, 1977, p. 164.

32. The categories draw upon suggestions made by Ronald Dore at the Technical Change Centre conference on public acceptance of new technologies, in London in January 1985.

33. Hirsch, F., The Social Limits to Growth, London Routledge & Kegan Paul, 1977.

34. Galbraith, J. K., The New Industrial state, Boston: Houghton Mifflin, 1967.

Chapter Eight

PUBLIC ACCEPTANCE OF NEW TECHNOLOGIES IN ITALY

Gabriele Calvi, Andrea Colombino, Piero Fazio and
Giuseppe Zampaglione

Foreword

This chapter takes up several aspects of the
problem of technological innovation and focuses, in
particular, on the cultural and social aspects
which have most characterised the attitudes of the
Italian public, and on various surveys conducted in
Italy on this topic.

These surveys reveal that Italians have,
today, a more positive and optimistic outlook on
technological innovation than they did in the past.
In particular, they disclose that: firms now
possess a greater capacity to create and
incorporate technological innovation in both
processes and products; the political awareness of
the parties, trade unions and other social groups
has grown in regard to innovation; the image of
technological innovation as an instrument able to
help the country out of its economic crisis has
been bolstered. These changes have taken place
despite the persistence of a situation in which
public policy towards innovations is rather uneven
and unco-ordinated, and in which information in the
mass media is still quite confused and incomplete.

In this scenario, formulation of a strategy to
improve the level of acceptance of new technologies
can be furthered by a deeper understanding of the
meaning of the concept of 'acceptance'. In Italy
the debate on this theme is still marred by
equivocations about, and misunderstandings of, the
terms of the question and about their relative
conceptual universes.

The term 'acceptance' has an inordinate
semantic breadth. We can endow the term with a
generic meaning, the preference for a product or
project revealed by an opinion poll, or we can have

it refer to the concrete adoption of new technologies. This second meaning may, in turn, operate at two different levels: at the level of the individual who applies or uses technologically new products; and at the societal level when the decision is much larger and more complex, and concerns a far greater number of people. In the latter case, the decision to be taken does not so much concern innovation in itself but, rather, the 'social representation' which has formed around it as a result of communications influenced by ideological, political and economic debate.

A small industrialist adopts new technologies according to the first type of mechanism, and only after the true benefit of the new method has been demonstrated to him. The consumer may be influenced by an embryonic social representation of the new product (for example, publicity for a liquid-crystal watch); but his decision is an autonomous one, and he adopts the product only after he has seen, touched and tested it. In certain cases the consumer adopts products unaware that they incorporate new technologies (for example, a driver may be uninformed about all the new technologies incorporated in his car).

Totally different is the case of technologies requiring for their adoption a consensus of institutional, political and technical bodies, and of public opinion. Nuclear power plants are one example of this. If we analyse the decision-making mechanism activated in these cases, we note that, frequently, inadequate attention has been given to the fact that the decision hinges more on a 'representation' than on reality, in other words, on a symbol constructed by cultural groups and which relates to cultural and existential spheres.

The Italy of the 1980s is starting out along a new path and strongly feels the need to work out a cultural strategy aimed at innovation and socio-economic development. Acceptance of the new technologies will depend on how this strategy is formulated. The social representations of these innovations must be proposed and communicated by the media, in all their aspects and with all their implications, direct and indirect.

Information is precisely the field in which interesting prospects are emerging for the development of a policy to achieve broader public acceptance of the new technologies. In Italy information is still scarce both at the popular level and at the level of transferring scientific and technical know-how to industrial operators. This observation applies to both the quality and the quantity of information, and also to comparison of these levels with those achieved in other countries. Possible developments in the field of information must be identified immediately. This must be accomplished on the basis of events in other industrialised countries (growing interest of the mass media, identification of the themes of public debate, increasing influence of the 'green' movements, greater attention to the problems of information by decision-making bodies). It must also be accomplished on the basis of considerations concerning the structure of the Italian decision-making process. This structure possesses in embryo a conspicuous potential for very broad participation in decision-making. But this potential can be effectively activated only if the wide gap between the decision-making bodies and those interested and affected by the decisions is reduced in terms of both know-how and knowledge. If it is desirable that the social groups directly concerned with the processes of technological change should be able to have a voice in decision-making, then it is essential that they should be as well informed as possible.

Closing this gap is necessary and indispensable if the technological innovation process is not only to be realised, but to be realised adequately to meet the needs of the country's economic, social and cultural development. This endeavour must be set in motion in the awareness that the processes of technological change are primarily processes of social change. They may affect, in other words, particular social groups, both big and small, who may anticipate, depending on the case, costs or benefits from technological change. The groups in question will oppose or favour the process in accordance with their identification of costs and benefits.

These observations are highly pertinent to the changes that will take place in labour, in both quantitative and qualitative terms. On this point we must specify that, on the one hand, adequate information needs to be supplied on the structural changes (especially in terms of employment and the new professions), and on the other, that satisfactory answers need to be given to the growing demand for redefinition of the professions.

(Reference should be made here to the work conducted by the Higher Institute of Sociology of the University of Milan and by ENEA, the Italian Commission for Nuclear and Alternative Energy Sources (ENEA).

If, as noted above, information must expand to close the gap between decision-makers and those interested and affected by the decisions, formulation of this information must, then become a moment of theoretical probing of the aspects of acceptance. The possible fall-out which public-acceptance initiatives may have on the different institutions (governmental, trade union or political) must not be underestimated, particularly in a country like Italy, in which there is no cultural tradition in this field and in which the formulation and exploration of certain facets of technological innovation is still experiencing delay.

This chapter is a partial contribution to the study of the problems connected with acceptance of the new technologies and is largely based on already existing studies. It focuses in particular on the attitudes of Italians to nuclear energy and on the opinions of certain social groups (researchers, workers, etc.) about the introduction of new technologies. The level of acceptance in Italy is also compared to that in other European countries. It would be worthwhile to follow up this work by exploring a number of equally interesting problems: the way Italian scientific and technical institutions perform research; the public perception of the employment cost/benefit balance connected to the introduction of new technologies; the impact of environmental considerations on decision-making processes; the phenomenon of environmental protection movements; the role of marketing, publicity, information and entertainment techniques (cinema, television, major exhibits) in the processes of introduction and acceptance of the new technologies.

The Need to Overcome Dated Stereotypes

There is a widespread conviction that science and technology are substantially removed from the basic values, life styles and culture of the Italian people. We will attempt below to verify whether this long-standing opinion corresponds to current group attitudes and behaviour, and the extent to which it can be attributed to the 'persistence of images' in the mass media and institutions, of a social and cultural situation which is today obsolete. In the 'genetic code' of national character, there exists a historical-cultural heritage that casts a remote, but not wholly interrupted, echo of suspicion on technological development:

• the delays and limits on industrial development in Italy reveal a series of explicit 'inherent defects': 'scarce capital (lack of original accumulation), scarce raw materials, absence of a big market (political division of the country)...Inherent flaws which accompany its growth, keeping its economy behind that of the other leading industrialised nations.'(1)

• the fact that industrialisation takes place 'without a vigorous ideological spur', which in other countries was generated by different currents: laissez-faire economics in England, Saint-Simonian doctrine in France, nationalism in Germany and Marxism in Russia.(2)

For this reason, the transformations in Italian society and production structures have not allowed a culture imbued with technical and scientific values to consolidate. The specificity of the Italian situation, very clear in the early years of industrialisation, is marked by a substantial ambivalence

'which allows it to be... modern without abandoning the mythical gifts of backwardness and its inherent wisdom and nobility; which does not exclude the advent of industry, while retaining the virtues of a system of 'human' values tied to the farming economy'.

The offshoots of this type of ideological scheme take root in the conviction that 'native intelligence is superior to method, improvisation superior to preparation, the heart, zeal and the "Garibaldi" approach superior to organisation'.(3)
With the Fascist regime, this consolatory ideology of the nineteenth century become rhetoric.

'In Europe, we Italians represent a vital element of opposition to the triumphant spirit of the northern nations: we have a very ancient civilization, based on all the physical, material and mechanical values, to defend. Anglo-Saxon modernity is not for us: to assimilate it would lead us to irreparable decadence.'(4)

In the period between early industrial development and Fascism, to which the previous citations refer, and our own day, institutional, economic and social events (the birth of the Republic, the economic boom and the modernisation of may backward areas) have created a new cultural substratum. So, from the end of the Second World War to our own day, the values of development and efficiency have found the occasions and the instruments to be amply channelled to all social interstices. Today no social elements and political sections express substantially negative attitudes towards scientific and technological progress and economic development. The older culture does transmit more or less clear echoes in certain forms of group behaviour and in certain institutional choices and trends. But the status quo is not explicitly ideologised, and there is no assumption that the national character is contrary to the scientific technological development of social and economic structures. Certain delays can be perceived at the political and institutional level. Scientific research spending has for many years constituted only a trifling share of gross domestic product, and the threshold of 1 per cent was crossed only in the 1980s. The first law providing incentives for industrial research was approved only in the late 1960s. The flow of funds to industry from research began to assume significant proportions only in the early 1980s, but it has still not exceeded 2 per cent of the national budget.

In the last decade, although cloaked in still widespread stereotypes in the mass media and public opinion, and despite scanty innovation policies, the Italian situation has changed perceptibly. The capacity of industry to create and to incorporate innovation has grown. In the meantime, all the social institutions, from the trade unions to the political parties, have expressed an increasingly firm interest and openess towards themes and problems of technological innovation. And there have been concurrent trends in public opinion. Italian society is today far more mature in evaluating the importance of scientific and technological development, and can turn a deaf ear to myths and preconceived notions.

Italy

Surveys on the Attitudes of Italians Toward Science and Technology

Les attitudes du public européen face au développement scientifique et technique (Brussels, 1979) provides significant insights into societal attitudes towards scientific and technological innovation. This survey has a twofold interest:

it was conducted during the late 1970s, at a time when public opinion was highly critical of the economic situation. The apocalyptic tones used in official documents and the mass media have considerably softened in subsequent years, and in the past decade the Italian production system has also undergone profound transformations. However, industry has tended both to reorganise its production processes and to innovate products.(5) The climate as perceived by the public, was one of widespread crisis and, presumably, the responses were affected by this perception. The need to qualify and to innovate Italian production structures has led, as we will see, to a recognition of the crucial importance of the role of science and scientific policy in the framework of strategic choices for the economy and for society;...

... the survey was conducted on a European scale, allowing international comparisons to be made. Since other nations possess a longer scientific tradition, a more consolidated technological development, and a more profound socialisation of modern culture in the population, it might be assumed that public opinion in those countries would be better able to evaluate the importance of a commitment to science and technology and the consequences of this for the economy and society. In this regard the results offer a few surprises. We must first emphasise the substantially optimistic Italian view of the role and of the effects of scientific and technological development. The idea that science is one of the principal factors in improving the quality of life is explicitly accepted. This attitude is, on the average, stronger in Italy (76 per cent) than in the other countries of Europe (74 per cent). Only Great Britain expressed a more optimistic outlook (79 per cent).

We can recognise in Italians, therefore, a more genuine awareness than in the past of the positive role of science and technology, an awareness which excludes unequivocal attitudes. In a book widely read two years ago - II secondo pianeta- Colombo and Turani stressed the extensive change which had occured in the attitudes of international public opinion:

'from a phase in which it expected technology to completely solve all the problems of mankind, it has entered a phase of growing suspicion. Suspicion which is not confined to public opinion, usually ill-informed on these matters, but which extends to those in power and, in many cases, to those who deal directly with technology, in other words, scientists and heads of industry'.(6)

This picture contains particular nuances in the attitudes registered in Italy. We can assert that there is no sign of a strong culture of opposition to scientific development. Proof of this is the limited appeal of ecology movements, at least compared to the following they have gained in other industrialised countries. The main reason for this lies in the want of a broad socialisation, in Italy, of scientific optimism during the years in which the messianic expectations for science and technology were particularly fervent in other social and economic contexts. Italy lacked this culture of optimism, so the disenchantment of those with failed expectations is today imperceptible in our country.

The prevailing attitude towards technological and scientific development is one of confidence tempered by awareness of its critical implications and of the contradictions inherent in the application of research findings. The European Community survey reveals that the highest consensus on the ambivalence between the 'positiveness of scientific know-how and the problems connected to its applications' is to be found precisely in Itlay (73 per cent, compared to an EC average of 69 per cent).

It is clear that the positive attitude chiefly refers to scientific development, which is seen as promoting cultural and economic advancement. The attitude toward technological development is more complex and contradictory. This does not imply that the expectations of Italian public opinion tend to encourage prudence in development options and in the applications of research findings. The reigning conviction is that new scientific and technological developments will allow the negative consequences of their application to be controlled. It is not technological development in itself which must be controlled, but rather the capacity of those in charge of directing, choosing and managing it. In this regard we must point out that:

. among EC countries, Italy registers the highest consensus on the statement that 'policy-makers do not assign sufficient importance to scientific achievements';

. at the same time, Italians disagree most with the statement: 'scientific options and discoveries are placed at the service of the public interest'.

Italians are highly critical of the country's institutions for their role in the field of national scientific policy. It is interesting to note that this opinion matches the one expressed by research bodies:

'We can undoubtedly affirm that Italy's secondary role, in terms of overall funds allocated to scientific and technological research, represents a coherent quantification of weakness and casualness of political commitment and of the overall knowledge gained in this sector.'

Although there have been scientific and technological developments of great significance,

'the institutional capacity to receive and to channel these inputs into operative strategies and commitments coherent with social and economic needs has been wanting'.(7)

The weakness of political bodies in the scientific field not only determines the amount of funds allocated to research, and the significance of the choices the relative programmes assume in terms of the public interest, but it also helps define the level of diffusion and consolidation of a modern culture in a population.

It is no coincidence, therefore, that in the EC survey Italians claim to be least prepared to be able to speak in a considered way about science. If we attempt to overcome the all-encompassing (and necessarily generic) dimension of the questions, the analysis becomes even more significant in regard to the Italian attitude toward technological and scientific innovation.

In the first group of technical and scientific sectors, Italian acceptance of research and experimentation is the highest among all EC countries. This applies to the development of nuclear power stations (Italy 53 per cent, compared to an EC average of 44 per cent), genetic experimentation (Italy 49 per cent, EC 33 per cent), organ transplants (Italy 90 per cent, EC 82 per cent) and centralised data banks on individuals (Italy 47 per cent, EC 22 per cent). The underlying motives of such positive attitudes reveal a very pragmatic outlook on the problems in the different fields of health, national bureaucracy and energy. A good many commonplaces regarding national culture and its alleged attachment to traditional values, and to dogmas inculcated by the dominant institutions, are left by the wayside. The 'common sense' of the Italian public results in acceptance of the applications of scientific know-how in areas bearing upon the quality of life. Its 'sense of reality' prevents it from seeing the new as a demon.

In some scientific areas the Italian consensus for pressing on with experimentation is quite pronounced, and always superior to the EC average. This applies to spending for new energy sources, where only the Danes expressed a higher consensus. It also applies to the development of satellites for controlling and studying atmospheric phenomena (only the Dutch and the Danes voiced wider agreement on these). Highly positive attitudes emerged towards the endeavours of national research in all the advanced technology sectors.

A second group of experimental sectors concerns synthetic materials and artificial nutrition. On the former, the Italian response, like the EC average, is not negative, although some reluctance towards new materials exists which may be due to an undiminished taste for high-quality natural fabrics such as cotton, wool and silk, regardless of their higher prices vis-à-vis the newer synthetic materials. On artificial nutrition, the Italian response is rather different from the EC average. This is due, perhaps, to the widespread Italian desire to interpret food not only in terms of satisfying a primary need, but also as the celebration of a rite and of a tradition.

Artificial nutrition would leave very little room for this custom, at least not until experimentation had reached a very advanced stage. For this reason, it might be better to pursue a strategy of second best rather than expect to play an innovative role, with the burdens deriving from such a choice.

Another survey conducted by DOXA several years after the EC poll(18) revealed an even more favourable attitude toward scientific and technological development and their social consequences. By the early 1980s many industries had undergone reorganisation. The large industries had considerably reduced personnel, and the use of the Cassa Integrazione Guadagni had spread significantly. (The Cassa Integrazione Guadagni is a public indemnity temporarily to assist workers who have been laid off by their own plants, or in a given sector in an attempt to streamline operations.) In the context of this profound transformation of production structures, and in the massive application of new, labour-saving technologies, there was still no particular attitude of rejection towards technological innovation. On the contrary, the overriding attitude was one of positive opinions on the present situation and on future trends.

The question of whether, in the transformations which have taken place in the past 25-30 years, the discoveries and applications of science have played an important role received a clear answer: 60.9 per cent of those interviewed recognised that the impact of science and technology had been very important, and 32.0 per cent regarded it as quite important. This opinion was virtually uniform. Women attached less importance to scientific and techological development than did men (59.0 per cent compared with 62.4 per cent); the lower social class less importance than the higher (53.0 per cent compared with 70.6 per cent); the population of the south less importance than that of the northeast (56.5 per cent compared with 68.9 per cent); and high school graduates less importance than university graduates (64.4 per cent compared with 75.0 per cent). In any case, none of the groups interviewed saw the impact of science and technology on the economy and on society as of little importance. Lower-class subjects (5.2 per cent), southerners and the 21-24- year age group (4.0 per cent) and women (3.9 per cent) most frequently took this view.

The general consensus on the dimensions of the impact of science and technology in social terms corresponds to the Italian public's substantially favourable view of the quality of life. Of those interviewed by DOXA, 56.1 per cent recognised that change resulting from science and technology was positive, 26.3 per cent felt it was negative, and 17.6 per cent were uncertain. In this case, however, the divergences in opinions expressed by the various groups regarding the impact of science and technology were more marked:

- men were more optimistic than women (61.6 per cent compared with 49.4 per cent);
- the upper class was more optimistic than the lower class (66.0 per cent compared with 49.5 per cent);
- the inhabitants of the northeast and of the islands of Sardinia and Sicily were more optimistic than those of the south (61.7 per cent compared with 52.9 per cent);
- university graduates were more optimistic than high school graduates (65.4 per cent compared with 60.4 per cent);
- the public in towns with less than 10,000 inhabitants were more optimistic than the public in the big cities (59.3 per cent compared with 52.9 per cent);
- students were more optimistic than non-professionals (62.2 per cent compared with 46.6 per cent);
- those with a high socio-economic status were more optimistic than those with a medium-low status (61.4 per cent compared with 52.3 per cent);
- those with a high cultural level were more optimistic than those with a medium-low cultural level.

Two elements stand out from the overall findings of this survey. A mostly positive attitude is normally expressed by those who recognise the great impact of scientific and technological development in social and economic terms. At the same time, the categories and groups most affected by economic development are more likely to express positive opinions, while those excluded from it (non-professionals) or those who have been less affected by post war development (southerners and women) tend to be less favourable. (In the south there has been disillusionment with development policies, especially after the substantial failure of big industrialisation projects, which often ended as 'cathedrals in the desert', without generating diffused economic development.) The fact that affirmative replies are greater in the northeastern regions than in the northwestern (61.7 per cent compared with 57.1 per cent) and in the small towns compared to the big cities (59.3 per cent compared with 52.9 per cent), underlines the significance of the quality of development which has taken place in the attitudes expressed by the sample. In the more developed areas and in those more affected by congestion, the opinions expressed are more articulated and qualified. In any case, in all groups a positive attitude towards development which assimilates the new technologies is clearly overriding. To repeat, the DOXA sample was taken in the early 1980s, at a time when the effects of industrial reorganisation on employment were clearly being felt. This did not result in rejection of the new technologies, although there was an awareness and fear of their consequences.

In this context, the results of an opinion poll taken by the Atlantic Institute of International Affairs in October 1982 are also interesting. The poll reveals that unemployment is one of the greatest fears among Italians, at 61 per cent only 1 per cent behind criminal violence. We can conclude that Italians are convinced that technological development is substantially positive, even though it may have adverse social consequences.

The most significant information emerging from the DOXA survey concerns opinions on science. Almost 80 per cent (79.9 per cent) of the sample recognised that science is one of the most crucial factors in improving the quality of life, while only 3.2 per cent believed it created more disadvantages than advantages, and only 2.6 per cent believed it dangerous. Consequently, we can affirm that a clearly prevalent attitude among Italians is an expectation that society will act to incorporate a higher level of scientific know-how and new technologies. The conditions are emerging for a progressive shift in the axis of our attention 'from a scenario based on references to the past to one looking toward the construction of the future' - a development which Aurelio Peccei has defined as essential in building tomorrow's world.(9)

We must also stress the strong similarity of attitudes toward science of the different social groups and of the population of the different areas. Differences in culture and in social status undoubtedly affect the formation of a positive attitude, but the divergences in the attitudes expressed are quite limited, and in no case is there outright opposition:

- men are more favourable than women (81.4 per cent compared with 77.9 per cent);
- the upper and middle classes are clearly more favourable than the lower (84.3 per cent compared with 73.5 per cent);
- university graduates are more attached to the values of science than are the other groups (86.7 per cent compared with 83.6 per cent);
- residents of the chief towns of a region are more apt to recognise the positive impact of scientific development (81.1 per cent) than are the residents of smaller (79.5 per cent) and medium-size towns (79.0 per cent);
- individuals with a higher socio-economic status are more favourably disposed (86.1 per cent) than are all others; individuals with the lowest socio-economic status are the least favourably disposed (69.9 per cent);
- there is also a definite correlation between cultural level and recognition of the importance of science, though the highest value is actually expressed by individuals with a medium-high cultural level (85.8 per cent), as against those with a high cultural level (83.5 per cent). Those at the lowest cultural level express the least attachment to scientific values (71.8 per cent);
- the northeastern regions are the most favourable (85.4 per cent), while the southern ones are the least favourable (78.7 per cent).

The groups which see greater drawbacks than benefits in scientific development are young southerners (5.6 per cent) and lower-class youth (4.8 per cent) while those most likely to consider science dangerous are inhabitants of northeastern Italy. As we can see, these percentages are insignificant and, therefore, do not reflect a culture oriented towards tradition. The acceptance of science as a positive factor in development and cultural and social growth does not reflect an uncritical and unique viewpoint on all fields of technology. We can recognise a capacity to express priorities and to suggest curbs, and these, in certain cases appear to be a criticism of political choices in this field.

Public opinion holds that the disciplines and technologies to which priority should be attributed are:

Italy

1. medical and pharmaceutical research   58.8%
2. research to exploit agricultural
   resources   57.6%
3. pollution reduction and control   44.2%
4. prevention and cure of drug addiction   43.9%

Public opinion holds that the sectors in which scientific investment should be limited are:

1. arms and national defence   54.4%
2. space exploration   35.0%
3. meteorology and climate control   27.3%
4. rapid transit   19.5%

The 1981 DOXA survey furnishes additional information on the attitudes of Italians towards science and technology. Those interviewed were asked how much attention they devoted to technical and scientific information reported in newspapers, in periodicals and on television. Well over 70 per cent of the sample declared they devoted special attention to it, with 'peaks' of interest among college graduates (89.8 per cent), the upper class (83.5 per cent), students (80.6 per cent) and residents of the chief regional towns (78.9 per cent).

The fact that the inhabitants of the northwestern regions devote less attention to scientific and technological information supplied by the media than do the inhabitants of the other regions, and that the same applies to individuals with higher socio-economic status leads us to presume that inhabitants of the more developed areas and those with a higher status have other opportunities and channels for receiving such information.

In Italy considerable attention is, then, devoted to issues of science and technology. The growing circulation of several popular science magazines in recent years offers testimony of this.

The reasons behind this increased consumption of scientific information can evidently be traced to the widespread awareness that science and technology are essential components of modern culture, and the belief that there is felt to be inadequate preparation in them. We should recall here the charges of inadequate education expressed by the Italian sample in the EC survey on the attitudes of the European public toward scientific and technological development. A second reason can be identified in the immediacy of the effects of

Italy

the new technologies. Having to live in everyday
life with new technologies, instruments and
equipment, obliges the public to acquire greater
knowledge and awareness of their contents and
implications.

Surveys on energy

Gli Italiani e il problema dell'energia: due
indagini demoscopiche - 1980 e 1983(10) is a
sectoral study on energy which allows us to grasp
the attitudes and opinions of the public in a very
specific and important field. The major interest
in this study lies in the fact that it was first
conducted in 1980 and then repeated in 1983. This
permits us to make a diachronic analysis, and thus
perceive the change in attitudes of the population
with reference to energy problems in general, and
to nuclear energy in particular.
   Our first conclusion concerns the image of
nuclear energy. It is a high-risk energy source.
This can be recognised in the number of people who
see it as a grave danger to the population; who
stress the problem of waste and its disposal; who
point to the risk to workers in the industry; and
who recognise the considerable investment nuclear
power entails. However, Table 8.1 shows that the
trends in nuclear power's image improved
considerably between 1980 and 1983.
   The factors concerned may be divided into
'items of interest' and 'items of risk'. The first
group are those on which consensus reflects a
positive attitude of those interviewed (in the
Table these are items: (h), (i), (m), and (n)). The
second group are those which reveal a hazard or, at
least, a significant problem in the opinion of
those interviewed (in the Table these are items:
(a), (b), (c), (d), (e), (f), (g), and (l)). It is
worth noting that the percentages associated with
all the items of risk diminish in the period 1980-
3, while the percentages associated with only one
of the items of interest (i) falls. It is also
worth stressing the different hierarchy of items in
1980 and 1983:

	1980	1983

1980   1983

1. It has dangerous wastes   1. It has dangerous
                                 wastes

2. Involves serious danger   2. It is the energy of

238

Italy

for the population
3. Hazardous for workers

4. Plants are too costly
5. It is the energy
   of tomorrow

tomorrow
3. Involves serious danger
   for the population
4. Plants are too costly
5. Hazardous for workers

Table 8.1:   Answers to Questions on the Image of Nuclear Energy*

		1980 %	1983 %
(a)	Plants are too costly	32.9	32.8
(b)	It is very expensive	28.0	21.5
(c)	It has dangerous wastes	42.8	41.3
(d)	It involves excessive danger in transport of raw materials	19.9	16.9
(e)	Involves serious danger for the population	40.7	37.8
(f)	Hazardous for workers	37.6	32.5
(g)	Pollutes the environment	30.2	27.0
(h)	In Italy there is little of it	13.0	15.2
(i)	Few countries possess it	17.3	13.8
(l)	I would never adopt it for Italy	24.2	16.7
(m)	I would adopt it for Italy	13.0	16.0
(n)	It is the energy of tomorrow	13.2	40.8
(o)	No answer	20.7	19.8

*   More than one answer could be given.

Source: ENEA   and   EURISKO,   Gli Italiani e il problema dell'energia: due indagini demoscopiche (1980 e 1983) Notiziario dell 'ENEA, Rome, February 1984.

Clearly, there is a substantially critical opinion toward nuclear energy, but judgements have become open (item 5 of 1980 becomes item 2 in 1983) and less preconceived (items 2 and 3 of 1980 drop to 3 and 5 in 1983). We must stress, in fact, that between 1980 and 1983 there was a great change in opinion over the stand those interviewed would take were there to be a referendum on the installation of nuclear power stations in Italy (Table 8.2). The total percentage of those in favour rose from 32.5 per cent to 41.1 per cent, while the percentage of those against fell from 40.7 per cent to 33.1 per cent.

Italy

Table 8.2: Answers to the Question: 'If a referendum were to be held on the installation of nuclear power plants in Italy, would you be for or against such plants?'

	Total		Education							
			University degree		High school		Junior High school		Elementary school	
	1980	1983	1980	1983	1980	1983	1980	1983	1980	1983
For	32.5	41.1	44.3	40.6	34.4	44.2	34.4	46.4	26.5	35.7
Undecided	26.8	25.8	16.4	19.5	20.8	21.3	22.1	22.9	37.1	32.8
Against	40.7	33.1	39.3	39.3	44.7	34.5	43.5	30.7	36.4	31.5
Total	100.0	100.0	100.0	100.0	100.0	100.0	100.0	100.0	100.0	100.0

	Total		Status					
			Medium-high high		Medium		Medium-low low	
	1980	1983	1980	1983	1980	1983	1980	1983
For	32.5	41.1	42.2	49.7	32.9	41.8	28.0	36.5
Undecided	26.8	25.8	18.7	20.2	22.7	23.7	37.3	32.0
Against	40.7	33.1	39.1	30.1	44.4	30.1	34.7	31.5
Total	100.0	100.0	100.0	100.0	100.0	100.0	100.0	100.0

Source: ENEA and EURISKO, Gli Italiani e il problema, dell'engergia: due indagini, demoscopiche (1980 e 1983), Notiziario dell'ENEA, Rome, February 1984.

We should point out that:

the lowest levels of acceptance of nuclear energy were expressed by individuals with a low educational level and low social status:

• the highest levels of acceptance were expressed by individuals with a medium-low educational level and medium-high social status;
• the only group which registered a drop in consent between 1980 and 1983 was the one composed of university graduates.

But the most interesting aspect of the study lies in the attitudes expressed over the possibility of installing a nuclear power plant in the town in which the respondents resided, so that in this case the issue has a direct bearing on their living environment rather than being an opinion on the general question. Table 8.3 lists the highly differentiated opinions expressed. Those favourable to installation formed the largest component (22.5 per cent), while those opposed were the next largest group (20.8 per cent), followed by the undecided (20.6 per cent), then by those who were opposed but were prepared to accept installation with reservations. The range of opinions was extremely fluid. But in any case, those totally opposed to installation formed a distinct minority (one-fifth of the entire sample), while some 60 per cent were prepared to accept installation, although not all on the same grounds.

The socio-cultural and political profiles of these different opinion groups can be outlined as follows:

Favourable: this group is predominantly composed of men, individuals with a high socio-economic status, individuals on the centre-right of the political spectrum, industrialists, professional people, white-collar employees and tradesmen. The prevailing conviction of this group is that nuclear power is inevitable for further socio-economic development. This group's fear of nuclear power is limited, and it has an optimistic outlook on Italy's future.

Favourable but with reservations: this group is chiefly composed of individuals with a medium-high socio-economic status, young people, individuals on the centre or centre-left of the political spectrum. Many of these people have the opinion that nuclear power can be used temporarily, until a less hazardous technology can be made ready. The group believes nuclear hazards will persist, not in connection with the technology itself, but rather with its use in Italy; that is, there is a lack of

Italy

confidence in Italian capabilities to operate nuclear plants safely.

Table 8.3: Answers to the Question 'If the decision were made to build a nuclear plant in your home town or in towns nearby, what position would you adopt?'

	Total	Education			
		University degree	High school	Junior high school	Elementary school
Favourable	22.5	25.5	25.2	23.8	18.7
I would accept, but I am not completely favourable	16.4	17.2	17.5	19.2	13.5
I would be undecided	20.6	13.0	13.9	19.7	28.9
I would be against, but I would accept	19.7	14.5	21.4	20.5	19.3
I would be against and I would oppose it	20.8	29.8	22.1	16.8	19.6
Total	100.0	100.0	100.0	100.0	100.0

	Total	Status		
		Medium-high/ high	Medium	Medium-low/ low
Favourable	22.5	30.9	22.0	20.2
I would accept, but I am not completely favourable	16.4	19.8	17.5	12.7
I would be undecided	20.6	11.7	17.6	29.9
I would be against, but I would accept	19.7	17.1	20.9	18.5
I would be against and I would oppose it	20.8	20.6	21.9	18.7
Total	100.0	100.0	100.0	100.0

Italy

Source: ENEA and EURISKO, Gli Italiani e il problema dell'energia: due inolagini demoscopiche (1980 e 1983), Notizario dell'ENEA, Rome February 1984.

Undecided: this group is predominantly composed of women, individuals with a low socio-economic status, the elderly, individuals with an elementary or junior high school education, housewives and pensioners.

Opposed but with reservations: this group, which would go along with a majority decision to install nuclear power plants, is mainly composed of individuals of the socio-economic middle class, individuals at the centre-left of the political spectrum, young people, individuals with a medium-high education level, students and the unemployed. The preponderant conviction of this group is that 'there are no guarantees of safe operations' and that 'with a bit of inventiveness we will be able to do without nuclear power plants'.

Opposed: this group is largely composed of people between the ages of 36 and 45, at the highest educational level (university and high school graduates), on the left of the political spectrum, teachers and white collar workers. This group does not equate progress with nuclear power, and feels that Italy is too densely populated and lacks suitable areas for the installation of nuclear power plants. Almost 50 per cent of this group would be willing to accept a lower standard of living rather than resort to nuclear power.

Some observations can be made on the typology above. The socio-cultural profiles of those favourable and those opposed are somewhat similar. Those in favour may be wealthier, those against are better educated, but these are not crucial differences. The really significant feature is the gap between these two groups and the undecided, mainly individuals with lower educational level and social status. The ENEA and EURISKO survey has shown that information on energy is lacking in Italy. None the less, as shown in the above tables, between 1980 and 1983 it appears that opinions on nuclear energy became less characterised by emotional reactions (see Table 8.1 and the comparison 1980-3 for items (b),(e),(f),(g),(h), and (i)). This might be due to

243

an increased and improved diffusion of safety, health, environmental and socio-economic information on this subject at all levels of government (central as well as local) and throughout the public sectors.

The Attitudes of Certain Social Groups Towards Science and Technology

Several polls and studies allow us to identify or infer the attitudes of certain social groups to scientific development and technological innovation. Through these analyses we can grasp eventual differences in attitudes and behaviour in a social situation which, although homogeneous in certain respects, is marked by significant fragmentation.

The Trade Unions

The attitudes of trade unions are essential in analysis of the positions expressed on scientific and technological development, since the application of any type of innovation in the production system sets accelerated changes in the organisation of industrial systems, and as well, in a number of cases, human labour is replaced by automated instruments and systems. Until the mid-1970s, Italian trade unions were strongly opposed to any management initiative which involved autonomous control over the human factor in the plant. In a period of from five to seven years a system of trade union rights was evolved to protect workers from the adverse effects of technological and organisational innovation. These rights introduced a certain rigidity into the economic system. This rigidity then led to an impasse in strategic and operational choices.

Today, trade unions no longer believe that a regulatory system is sufficient to protect workers and control plant working conditions. The peculiarities of the technologies now being, or soon to be, applied (from robots to flexible manufacturing systems) do not permit any such constraint.

FIAT provides us with an interesting example of innovation. It installed

'a plant to assemble mechanical sub systems at fixed work points, served by a moving system on automatic trolleys; all the movements of the components, a mix of great complexity, are managed by a computer. This system is extremely vulnerable to 'variations' which, if not rapidly brought back to normal, provoke a chain reaction. To restore normal operations, workers must intervene actively and swiftly, and the nature of this intervention cannot be prescribed by norms. In this regard, the situation can be considered rather typical of current automation trends.'(11)

The application of new technologies requires a flexible and adaptable plant system, and this is incompatible with traditional trade union positions. The organisation of labour becomes a necessary variable of technological choices.(12)

'Plant automation makes it impossible to negotiate (in the sense of defining) rhythms, work loads, numbers of workers, terms of using technologies and forms of labour relations, since the design has already largely determined and set the possibilities in regard to defining the content and forms of human labour.'(13)

Today, the Italian trade unions are reformulating significantly their positions. They are aware of the objective need for and ineluctability of technological innovation in the plant, and of the inadequacy of traditional methods of control to cope with it. However, it is not easy for trade unions to define a policy allowing a greater acceptance of new technologies without adversely affecting the current levels of employment. The trade unions recognise that they cannot continue their struggle against every single application and impact of the new technologies. The bone of contention is now the very conception and design of the single technology. In other words, they have realised that in order to be effective they must acquire knowledge and new understanding in fields far removed from their present sphere of influence and interest. It is imperative then for Italian trade unions to begin forging adequate research instruments in order to formulate and build up a culture of innovation.

Today employers have a firm position on the issue of the application of the new technologies. They argue that the organisation of labour can no longer be negotiated in industrial relations, since it is up to technology to determine it mechanically.

The Workers

The position of the trade unions is both the cause and the effect of new attitudes recognisable in workers. There are no surveys which allow us to visualise completely the attitudes of workers towards scientific and technological development, but there are some indications of their positions.

One such indication can be drawn from a sample of 1,496 industrial workers taken to clarify their positions on important questions regarding Italian institutions, the economy and society.(14) Asked, 'Why are industries in crisis?', their answers frequently suggest an implicit attitude towards technological innovation:

· 57.6 per cent affirm that there are too many workers (61.6 per cent of trade union members believe this);
· 35.8 per cent believe this a consequence of old and obsolete plants and production methods (31.2 per cent of trade union members answered this way);
· 22.1 per cent state it is a consequence of the lack of necessary investment (20.4 per cent of trade union members are of this opinion).

These answers reveal an attitude which interprets technological development as a factor encouraging progress for the economic system. The rationalisation of production methods and higher investment are necessary for plants that must grow and compete in the market even when this means reducing employment.

A second indication of workers' attitudes can be gleaned from research on office automation at FIAT Auto. This process is now underway and is divided into several phases.

'A first, current phase involves the introduction of machines which stand alone. A second phase, which begins in 1984, will involve linking up one or more offices in local networks, leading to complete integration with the central FIAT Auto system and the diffusion of additional technologies (electronic postal systems, languages for more sophisticated processes, etc.).'

We should point out that in the period 1980-3 company personnel were reduced by 25.4 per cent and white collar employees by 21.9 per cent. A

profound transformation process in internal
organisation and in the redefinition of the
functions and number of workers has been set in
motion. And this process is extremely risky for
the personnel involved. Even so, management
spokesmen can still state that 'this is the first
time that the introduction of a new technology at
FIAT has not provoked resistance and tensions from
below'.(15)

A third research project relevant in this
context was conducted by the Higher Institute of
Sociology of the University of Milan on a sample of
406 workers from a group of firms in industry and
services in Lombardy, all of these firms already
being involved in technological innovation.(16)
The most significant features revealed by the
research data concern: the methods, quantity and
quality of change within the firm as perceived by
the workers involved in it; the forecasts,
scenarios and expectations of those interviewed on
the transformations taking place inside and outside
the firm as result of the inroads of technological
innovation; the measures needing to be enacted to
limit the negative consequences of the introduction
of advanced technologies in the firm; criticism and
suggestions on trade union action as a whole.

The firms in which the interviews were
conducted all rely on the new technologies to some
degree. Just over half the workers (54 per cent)
felt they were directly and personally involved in
their firms' technological innovation process.

A series of questions was put in order to
measure both the subjective perception of the
workers who felt they were involved in innovation,
and also the aspects this involvement took. For
these workers, the most significant improvement in
their work concerned an increased quantity of
information about the production process (76 per
cent felt there had been an increase; only 1 per
cent felt there had been a decrease) and an
increase in professional qualification (63 per
cent).

The questions can be subdivided into three
groups referring, respectively, to the working
environment, the quality of the individual's work,
and the worker's control over his own work. The
answers reveal:

(a) in the perception of workers, the working
environment does not seem to have changed
significantly;

(b) particularly significant improvements have occurred in the manner in which workers perform their duties: to perform their specific tasks they clearly feel they need both to be professionally up to date and to augment both their technical capabilities;
(c) the changes in their control over the production process have been less marked, but are still conspicuous.

In conclusion, the perception of workers directly involved in their firms' technological innovation seems especially to concern a change in their qualitative relationship with their work, as they become aware that the changes occurring affect their professional role and point to a more responsible and stronger job in terms of the market.
Among those workers who declare they are already involved in the technological innovation process, a sizeable proportion hold that, until now, the changes in their duties have not been substantial, but practically all those interviewed feel that the innovation process holds important consequences for the future. In particular:

• 67 per cent of the workers interviewed are concerned about keeping their jobs and believe there will be cutbacks in the short and medium term. The opinions of blue- and white-collar workers are in marked contrast on this subject, however: only 55 per cent of white-collar workers feel jobs in their sector will be curtailed, compared to 80 per cent of blue-collar workers;
• the majority believes the working environment (72 per cent) and the quality of work (75 per cent) will improve, and that these changes will benefit the majority of workers (64 per cent of the sample expressed this view);
• only 25 per cent are of the opinion that many of their current tasks will be able to be performed in their own home in the years to come, while the majority (60 per cent) believes that many of these will disappear in the very near future;
• over two-thirds of the sample (70 per cent) believe that the firms' decision-making process will move to a higher echelon and be concentrated in top management;
• more generally, individual liberties are not felt to be threatened by technological innovation (only 35 per cent feel that innovation will

diminish them), but 60 per cent feel that control over the citizen's private life will increase conspicuously;

* in general, the workers interviewed feel that collective and individual consumption will become more standardised (55 per cent) and the mass media more uniform (62 per cent).

According to over 95 per cent of the sample, the worker most favoured by technological change has the following profile: young, male, skilled, with a high educational level and a background in technical and scientific studies.

Half of the sample feel that those willing to perform repetitive tasks will be favoured by technological innovation; the other half feel that those prepared to perform 'variable' tasks will benefit. This division conceals a disparity in the opinions of white-and blue-collar workers. The latter feel they enjoy an advantage in variable work, the former in repetitive work. There is an identical split regarding intercompany mobility: blue-collar workers believe continuation in the same firm will be more renumerative; office employees feel they have more to gain from frequent changes. Over 50 per cent of the sample believe that there will be a generalised reduction in jobs and in working hours. This may imply that the possibility of reducing hours to forestall waning demand is not seen as a realistic and feasible alternative.

The expectations of the sample regarding work prospects are also of interest:

60 per cent of those who feel involved in their firms' technological innovation process (and, in general, the young workers) believe their work prospects have been greatly improved; only 20 per cent of those respondents who do not feel involved in innovation believe this.

In short, those who have experienced a change feel they have or can acquire professional qualifications which will be easily 'spendable' in future demand in their own firms or in firms in other sectors, and especially in firms in the same sector as that in which they are currently employed. A significant majority calls for professional training within the firm (64 per cent) and an improved educational background (53 per cent) as among the most important measures to blunt the eventual adverse effects of innovation.

With regard to the trade unions, the workers interviewed feel that, up to now, they have been principally concerned with defending workers who lose their jobs (64 per cent) and with requests for wage increases for workers directly involved in innovation. The majority holds that, in the future, the trade unions must chiefly solicit measures to create new jobs (52 per cent), to improve the quality of work (42 per cent) and to inform workers of the changes taking place (30 per cent). Requests for wage increases seem to be of scant importance (only 10 per cent of the sample mention this). Without distinctions according to sex, age and position, the workers feel that trade union objectives should be to offer proposals and stimuli.

One of the clearest problems to emerge from the entire survey concerns 'the will to know'. Particular importance is attached by the respondents to the need to know what is taking place in the firm, to keep abreast of the technological changes, and to study in depth the consequent qualitative changes. Some of the workers who have already experienced change are now experiencing this process ('information on the production process is increasing') and feel this must become a priority objective in the near future.

Industrial Researchers

Additional insight into social attitudes towards science and technology comes from a survey of industrial researchers.(17) This study reveals the image that the protagonists of scientific and technological research have of their jobs and their role. The attitudes of these subjects are of great importance in the broad attitudes of the general public towards science and technology, since they deeply condition group awareness, albeit through the mediation of other political and social institutions. The principal findings of this study, at least as they concern this chapter, relate to opinions about science as a factor in social progress, the relationship between scientific and political institutions, and the researcher's role in the social system and his career prospects.

The sample gives a highly positive opinion about the role of science as a factor of social progress, and 89.6 per cent expressed this view, with minimum values in the metallurgical (83.4 per

cent) and various manufacturing (83.3 per cent)
sectors and maximum values in the pharmaceutical
(94.5 per cent) and food and textile (94.2 per
cent) sectors. This favourable attitude is tied to
the conviction that the impact of science and of
its application is important: the majority of
those interviewed feel scientific research is very
important as a factor of cultural innovation and
social change, and 86.0 per cent express this
opinion; there is less agreement on the statement
that the contribution of research is vital to the
country's economic and social development. Only
67.3 per cent feel that this is the case, and the
divergences from one sector to another are
considerable. The maximum value registered by
respondents in the R & D departments of mechanical
industries (79.3 per cent) and the minimum value in
some specialised research centres (53.9 per cent).
This latter value can perhaps be attributed to
researchers' distance from the production process
itself and from the moment when experimental
findings are actually applied.

We must stress that a positive opinion on
science and its applications is stronger among
younger (94.0 per cent) than among older subjects
(84.2 per cent). This evidently reflects, to some
extent, the greater enthusiasm of the up-and-coming
cadres and the progressive disenchantment of the
senior ones. This is also significant information
if we bear in mind the average age of the sample
(37.9 years). Male researchers and scientists tend
to be more favourable towards science than do
women (90.2 per cent compared with 89.6 per cent),
but the disparity is too small to allow us to infer
significant differences in attitudes between the
sexes.

The conviction that the positive role of
scientific and technological research needs to be
recognised corresponds to the conviction that the
public's confidence in it should grow. Some 90 per
cent of the sample express this opinion.

The favourable attitude of researchers toward
scientific and technological research corresponds
to the conviction that research should increasingly
meet the needs of society. There are three
different positions on this question: the majority
(53.2 per cent) believes that the planning of major
research objectives is a positive step, since it
acts to guide the choice of scientific operators; a
second group (30.1 per cent) is not only in favour
of defining major objectives, but also of defining

scientific priorities in political forums; a small minority (4.5 per cent) denies the importance of scientific planning and opts for a free definition of the fields of commitment by researchers.

The overriding attitude is, then, to anchor sectoral choices to objectives of general interest which must be defined in political terms. We should point out that less than 40 per cent of the sample hold the view that the work of scientists is more important than the work of politicians. They do not feel, however, that action should go beyond the definition of the major issues, in order not to fetter the autonomy and vocation of the individual researcher, or the freedom which must, perforce, prevail in this sector.

Recognition of the importance of scientific and technological research corresponds to recognition of the responsibilities of the researcher because of the ramifications of science for society. Of the sample, 63.3 per cent admit they feel this responsibility. And it must weigh considerably on them, since they believe the consequences of science and technology on society are likewise considerable. But despite this responsibility, they do not feel they enjoy a sufficient role. The researcher's social position is felt to be a superior one by one 26.7 per cent of the sample (the minimum, 5.6 per cent, is registered by respondents in the iron and steel sector). Only 34.0 per cent maintain that researchers enjoy significant social prestige. Only 20.9 per cent believe that the income of researchers is higher than the national average.

For this reason, the prevailing position seems to be one of frustration at the social role allocated to scientific operators. Those interviewed feel they are shut into their 'small professional garden', although not without receiving a certain gratification (66.8 per cent are satisfied with their work). Yet they do not feel their merits and responsibilities are recognised socially. And the career prospects of researchers are confined to this 'small garden'. As many as 74 per cent expect their professional growth to be exhausted in research. The sample can be broken down as follows:

36 per cent expect to continue their research activity in industry;
12 per cent expect to shift their research activity to the university or to other research

bodies.

•26 per cent expect to remain in research but with managerial responsibilities.

Only the remaining 26 per cent expect to assume responsibilities and roles in other professional and institutional spheres.

This problem is a significant one in terms of the social acceptance of the new technologies, since limited functional and institutional mobility obstructs the full socialisation of the values and culture of the researchers. This problem is also significant in terms of socio-economic development. The countries with highly innovative economies can boast a sizable number of researchers willing to assume management roles.

The Role of the Mass Media

In the last ten years the mass media have focused growing attention on scientific and technological information. Numerous dailies publish articles and supplements on science and technology; numerous popular periodicals dealing with these subjects have seen their circulation expand; radio and television have developed specific programmes on these themes.

In itself, this phenomenon has positive aspects, although the qualitative results are still not comparable to those achieved in other countries. For example, we might mention a survey conducted by the Committee of Italian Scientific Associations (COASSI) on a sample of 1,042 articles devoted to scientific and technological issues published in dailies and periodicals in the first six months of 1983.(18) This survey found that scientific information in Italy is still qualitatively inadequate. Constructing a 'quality index' based on the reliability of the facts reported in the articles, the survey brought to light that only 28 per cent of the articles analysed warranted a mark of ' good ' or ' excellent ', while 39 per cent could be termed ' satisfactory' and the rest ' unsatisfactory '. This survey also pointed out that among the different scientific disciplines dealt with in dailies and periodicals, the ones which received the greatest attention were the ones able to 'make news ' (astronomy was covered in 13 per cent of the articles, medicine in 11 per cent). Other disciplines (such as chemistry or mathematics, 2.5 per cent each) were neglected because they were less striking and 'newsworthy'.

In broader terms, the role of the Italian mass media in the field of scientific news can be outlined as follows:

• several of the leading dailies (Corriere della Sera, La Stampa, Il Messaggero, Il Tempo and Paese Sera) publish a weekly science supplement. The objective is to take up questions of specific concern or else burning issues (environmental protection, health, seismic phenomena, physics research, computers). There are now three scientific monthlies with a wide circulation - Le Scienze (707,000), Scienza e Vita Nuova (637,00) and Scienza 2000 (239,000).(19) It should be emphasised that their circulation has been expanding only in the past few years;
• RAI, the public broadcasting corporation has, for several years, been transmitting a highly sophisticated science programme, 'Quark', conducted by the jounalist Piero Angela. Initial interest in this programme has ballooned into an enormous success with audiences and the transmission has been rescheduled to prime time. There is even a second, afternoon edition for a younger audience;
• all the leading weeklies and monthlies give steady coverage to the most significant scientific events, again in regular columns.

Several observations can be made from this brief sketch:

• in Italy there is a demand for information on scientific and technological problems; although it is growing, this demand is still below the level reached in other industrialised countries;
• this demand has been met principally by traditional channels of information (newspapers, periodicals, television) and by certain specialised channels tied to market trends. For example, computer journals have been increasing in number, especially in relation to the diffusion of personal computers;
• although the qualitative level of information is quite high in certain periodicals, it is, on average, still unsatisfactory;
• scientific and research institutions have been

unable satisfactorily to meet the need of developing a scientific culture; the latter is a fundamental element in achieving broader acceptance of the new technologies.

Every serious strategy to spread scientific and technical information must comprehend from its inception that the mass media cannot supplant other institutions in the formation of a basic scientific culture and can, therefore, play only a complementary role. The commitment of these other institutions is irreplaceable in this field and needs to act to guarantee:

• the direct involvement of scientific centres, especially the university, in spreading information; this can be achieved by greater openness to the outside world and through the preparation of informative material on the different activities performed;
• massive utilisation of schools to diffuse scientific and technological culture to teachers and students through didactic experimentation; for example, investment in small laboratories in elementary and junior high schools is still limited, and there is a paucity of audiovisual facilities which would allow science films to be shown.

Radio and television are still vastly underutilised, although a programme like 'Quark' shows that a greater commitment to scientific information can become an outstanding success with the public. Considering the scant communication between universities and schools, and accepting that the public's interest in scientific and technical problems is high, investments of the type made for the BBC's Open University programmes are certainly justified.

An important role in the diffusion of scientific and technological culture can also be played by the political parties, the trade unions and manufacturers' and employers' associations. These institutions could improve their capacity to involve their own members and other institutions through public debates, conferences and seminars devoted to scientific and technolgical problems and their socio-economic consequences.

The steps in this outline are but a series of suggested actions which could be taken in the field of information, but they need to become part of a comparative strategy corresponding to the expectations of a public that is increasingly more mature and attentive to the new technologies.

Conclusions

The data assembled in this paper are partial and cannot be taken as an unequivocal index of social attitudes toward science and technology. The scenario which does emerge holds, however, an undoubted importance, since it permits us to debunk the image of a national culture strongly oriented towards traditional values and behaviour.

Very untraditional attitudes emerge, both in general surveys and in cross-sections surveying the trade unions, workers and industrial researchers. These surveys reveal, therefore, wider scope for the political sector to act on behalf of scientific development and technological innovation than is implied by the current figures on national R & D spending. If this fails to happen, it will probably be because of the resistance of certain obstacles which are hard to identify, given our present state of knowledge. In this regard, we could suggest a hypothesis requiring empirical confirmation. The series of data analysed in this chapter lacks a probe of the culture and attitudes of the political world towards scientific and technological research. It may be that the obstacles we refer to are to be found precisely in those institutions, where, if we are still to adhere to those traditional stereotyped images of society, policy-makers choose options which are far removed from the new awareness and culture of the public.

Notes

1. Pietro Grifone, Il capitale finanziario in Italia, Turin: Einaudi, 1971, p.5.

2. Alexander Gershenkron, II problema storico dell'arretratezza economica, Turin: Einuadi, 1965, pp. 84-5.

3. Giulio Bollati, L'Italiano, Turin: Einaudi, 1983, pp. 114-5.

4. Curzio Malaparte, Italia barbara, Longanesi, Milan: La Voce, 1928, p. 36.

Italy

5. CENSIS, Dal sommerso al post-industriale, Milan: Franco Angeli, 1983.

6. Umberto Colombo e Giuseppe Turani, II secondo pianeta, Milan: Arnoldo Mondadori Editore, 1982.

7. CENSIS, Criteri per la definizione di una politica e di una strategia nel campo della ricerca scientifica e tecnologica in Italia, Roma, 1981, pp. 7-8.

8. DOXA, Indagine sui valori degli italiani, 1981.

9. Aurelio Peccei, 'II mondo di domani' in AA.VV., Verso il duemila, Bari: Laterza, 1984.

10. ENEA e EURISKO, Gli italiani e il problema dell'energia: due molagini demoscofiche (1980 e 1983), Notiziario dell'ENEA, Rome, February 1984.

11. Giuseppe Della Rocca e Matteo Rollier, Azione sindacale e mutamenti dei processi produttivi, Progetto, 1984, 201.

12. CESTEC, Robotica, Milan: May 1983, p. 97.

13. Luciano Rouvery, Cohtrattazione e tecnologia, Progetto, no. 7, 1982, 24.

14. Giulio Urbani e Maria Weber, Cosa pensano gli operai, Milan: Franco Angeli, 1984.

15. Beppe Croce (ed.), Automazione d'ufficio alla Fiat Auto, in LITO Newsletter, no.1, January 1984.

16. Instituto Superiore di Sociologia, Universita di Milano, Gli operai e l'innovazione tecnologica, to be published.

17. Paolo Bisogno (ed), Il ricercatore nell'industria italiana, Milan: Franco Angeli, 1984.

18. Proceedings of the meeting on 'Divulgazione Scientificae Didattica delle Scienze', Florence, November 1984, Mariano Bianca, Mario Rigutti and Maria Antonio Sontaniello (eds.), to be published in Quaderni del Centro di Documentazione, by Provincia di Firenze.

257

Italy

19. <u>Mediabook</u> no. 6, March 1984, 70.

Bibliography

G. Balcet, U. Colombo, G. Lanzavecchia and G.B. Zorzoli, <u>La speranza tecnologica</u>, Milan: Etas Libri, 1980.

L. Bannon, U. Barry and O. Holst, <u>Information Technology: Impact on the way of Life</u>, Fast Program, 1982.

P. Barron and R. Curnow, <u>The Future with Micro-electronics</u>, Milton Keynes: Open University Press, 1979.

Paolo Bisogno (ed.), <u>II ricercatore nell'industria italiana</u>, Milan: Franco Angeli, 1984.

Paolo Bisogno, <u>Prometeo</u>, Milan: Arnoldo Mondadori Editore, 1982.

Giulio Bollati, <u>L'Italiano</u>, Turin: Einaudi, 1983.

H. Brooks, J. Adams, U. Colombo, M. Crorier, C.K. Aysen, T. Kristensen, S. Okita (e altri), <u>Scienza Sviluppo e Societá</u>, Milan: Franco Angeli, 1971.

Giuseppe Carqvita, <u>Vecchie e nuove professioni: i limiti dell distruzione creatrice</u>, Lito Newsletter no.4, April 1984.

CENSIS, <u>Dal sommerso al post-industriale</u>, Milan: Franco Angeli, 1983.

CENSIS, <u>Criteri per la definizione di una politica e di una strategia nel campo della ricerca scientifica e tecnologica in Italia</u>, Rome, 1981.

CESTEC, <u>Robotica</u>, Milan, May 1983.

Umberto Colombo, <u>Avanzamento tecnologico e competitivita: perche non dobbiamo perdere la speranza</u>, in inea d'orizzonte: scritti in onore di Vittorio, Valletta, Fondazione Angelli, June 1983.

Umberto Colombo e Giuseppe Turani, <u>II secondo Pianeta</u>, Milan: Arnolodo Mondadori Editore, 1982.

CNR, <u>I progetti finalizzati del CNR, esperienze e prospettive</u>, 1981.

Italy

CNR, Linee di intervento per l'innovazione tecnologica, Rome, 1982.

CNR, Rapporto sullo stato della ricerca scientifica e tecnologica, various years.

Comunità Europea, Les Attitudes du public Européen face au dévelopment scientifique et technique, Bruxelles, 1979.

Confindustria, Orizzonte 90. Incontro sul futuro, Rome: SIPI, 1984.

Beppe Croce (ed.), Automazione d'ufficio alla FIAT Auto Lito Newsletter, no. 1, January 1984.

Guiseppe Della Rocca, Relazione industriali e incentivi, Te Progetto, No 27, May - June 1985.

DOXA, Indagine sui valori degli italiani, Milan, 1981.

ENEA e EURISKO, Gli italiani e il problema dell'energia: due inadagini demoscopiche (1980 e 1983)', in Notiziario-ENEA, Rome, February 1984.

R. Ferrata, Progresso tecnologico e occupazione, Milan: Giuffrè, 1983.

FAST, Le trasformazioni occupazionali indotte dalla rivoluzione tecnologica, Milan: 1981.

C. Freeman, H. Clark and L. Soëte, Unemployment and Technical Innovation - A Study of Long Waves and Economic Development, London: Frances Pinter, 1982.

G. Friedrichs and A. Schaff, Microelectronics and Society: For Better or for Worse, Oxford: Pergamon Press, 1982.

Alexander Gershenkron, II problema storico dell'arretratezza economica, Turin: Einaudi, 1965.

Pietro Grifone, II capitale finanziario in Italia, Turin: Einaudi, 1971.

IIASA-IRPET, Long Waves, Depression and Innovation: Implications for National and Regional Economic Policy, Florence, 1983.

Italy

Istituto di Studi sulla ricerca e documentazione scientifica, La formazione e la mobilità dei ricercatori: una indagine sulla situazione in Europa, Quaderni, No. 13, Rome, 1984.

Giuseppe Lanzavecchia, II cammino dell'innovazione: dalla societa meccanicistica alla società cibernetica, Milan, 1984.

La situazione sociale del Paese, Quindicinale di note e commenti de CENSIS, no. 16, 1983.

Mediabook, no. 6, March 1984.

National Research Council, Outlook for Science and Technology, San Francisco: W.H. Freeman & Company, 1982.

OECD, Information Activities, Electronics and Telecommunication Technologies; Impacts on Employment, Growth and Trade, vol. 1, Paris, 1980.

Aurelio Peccei, 'II mondo di domani', in AA.VV. Verso il duemila, Bari: Laterza, 1984.

Luciano Rouvery, Contrattazione e tecnologia, Progetto, no. 7, 1982.

H.A. Simon, Technology and Environment, Management of Science, June 1968.

Alvin Toffler, The Third Wave, New York: William Morrow, 1980.

Trasformazione technologiche e processi decisionali, Rapporti CESOS, No. 13, December 1983.

Giuliano Urbani e Maria Weber, Cosa pensano gli operai, Milan: Franco Angeli, 1984.

260

Chapter Nine

PUBLIC ACCEPTANCE OF NEW TECHNOLOGIES: THE
CANADIAN EXPERIENCE
Michael Gurstein and Arthur J. Cordell

Introduction

Canada, except in certain limited fields, has not
played a major role as an initiator of new
technology; nor have there been, on the surface at
least, significant features characterising the
dissemination of new technologies in Canada. In
technology, as in other areas, Canada might be said
to have graduated from being in a largely dependent
relationship with the dominant power in one era, to
being in a largely dependent relationship with the
dominant power in the next - from the 'British
connection' to the 'American connection'. On the
other hand, reflecting on the subject of the public
acceptance of new technologies in the Canadian
context may have the effect of making visible the
broader relationships between technology, history
and economy, which are the ultimate framework
within which the public acceptance of new
technologies takes place.

For Canada and the Canadian public the
acceptance of new technologies is not a simple
matter of presentation, followed by acceptance or
rejection in the marketplace. Rather, new
technology is necessarily mediated through a
variety of factors whose affect is to determine,
if, or how, new technologies are accepted. Among
those factors are: the unique geographical
circumstances, including the size of the Canadian
land mass and the thin dispersal of its population;
Canada's position as an advanced and developed
economy based largely on resource extraction yet
with a substantial majority of its secondary
industry and a significant proportion of its
overall economy foreign owned; the dilemmas
presented by looking to incorporate two distinct

261

language and cultural groups within a single state; and finally the particular circumstances of sharing a continent and a language with the world's most vital culture, which latter is in turn supported by the most powerful concentration of communications media of the twentieth century.

What is interesting about Canada in the context of the public acceptance of new technologies is not how Canada differs in terms of acceptance, but rather how an apparent similarity in patterns of technology acceptance as between Canada and other developed countries reflects some deep particularities about Canada's historical and political circumstances.

The Unique Conditions Governing the Public Acceptance of New Technologies in the Canadian Context

Economic Conditions

The Canadian economy is characterised by the significance of the foreign-owned component. Foreign control of Canadian corporations reached a peak of 35 per cent in the early 1970s. Since then, as a result of mainly government and private acquisitions, foreign control has declined. As of 1981 the share of assets of foreign-controlled corporations in the non-financial industries stood at 25.5 per cent.

For individual sectors the extent of foreign control is quite dramatic. Some examples of foreign control are: transport equipment (primarily the automotive industry) 70 per cent; petroleum and coal products 61 per cent; electrical producers 54 per cent; chemical and chemical products 75 per cent; machinery 48 per cent; rubber products 90 per cent. Canadians have done better in the resource sector, with metal mining being 28 per cent foreign controlled, paper and allied industries 29 per cent foreign controlled, and all other mining activities 34 per cent foreign controlled. Two sectors particularly critical in the Canadian context (as we shall see below) have remained, in large part, Canadian owned; as of 1981 only 13 per cent of communications media were controlled by foreigners, while transportation carriers were only 6 per cent foreign controlled.

From the perspective of the public acceptance of new technologies, the issue in this instance is not public acceptance but rather public availability. The effect of Canada's particular structure of ownership has been to decrease innovation both in producer goods and in consumer goods. Foreign-controlled corporations do less research and development in Canada than do Canadian corporations. Thus, in 1979 foreign-controlled firms spent nearly a third less on R & D (as a per cent of sales) than did Canadian-controlled firms. Furthermore, the R & D that is done by branch plants of foreign firms is often at the transfer of technology developed abroad into the Canadian market.(2).

In this context the opinion of the average Canadian on the capability of Canadian industry to innovate is shown by the answer to the following recent polling question: 'Which kind of company do you think is more advanced in its use of new technology, a Canadian-owned company or a foreign-owned company operating in Canada?' While 21 per cent said Canadian, an overwhelming 62 per cent found foreign-owned firms more advanced in the use of technology (13 per cent said they use the same, while 3 per cent had no opinion).(3)

A relative lack of competition in the small domestic market has meant that companies have been less impelled to innovate in the goods offered and Canada frequently lags behind others, notably the USA, in the availablity of certain types of new technology items. In fact, of course, the peculiarities of Canadian industrial development have meant that whole areas of industrial activity have hardly developed locally at all such as, for example, the machine tool industry and the computer industry. Because of this, the argument has been made that Canada is making the transition to a postindustrial or information society without ever having been a fully industrial one.

One by-product of Canada's development path has been to foreclose options. For example, Canadian efforts in robotics have been severely constrained by the lack of a strong machine tool industry, since robotics has proven to be a natural follow-on of developments in machine-tools, linked with imaginative microelectronics applications.

The areas in which indigenous Canadian innovation and technology development have been commensurate with Canada's industrialised partners and competitors have been where industries are

direct responses to the particular needs of
Canadian geography or climate.(4) Canadian
technologies in communications and transportation
are internationally competitive. In these areas
government policies over time have been such as to
ensure indigenous ownership and control and, as
well, to maximise employment and investment
opportunities for Canadians. The highly successful
'Skidoo' snow tracked vehicle developed by Armand
Bombardier of Quebec was designed to remove the
isolation that so many experience in the peripheral
areas of Canada during the long, cold winters. The
company , Bombardier, has since gone on to become a
major force in the railcar and subway car industry.
The development of the bush aircraft - the Beaver
and Twin Otter - by de Havilland for use in the
Canadian North, has by now become legendary
worldwide. These aircraft act as the packhorses
and main-transportation links for many northern
communities. Here again, it should be noted that
the currently successful STOL (short take-off and
landing) aircraft such as DASH-7 and DASH-8 are
evolutionary developments of the Beaver and Twin
Otter.

Geographical Conditions
Canadian geography has shaped the public acceptance
of new technologies in ways similar to those in
which it has conditioned all other aspects of
Canadian life. The size of the country has meant
that there has been a dependence on technologies of
transportation and communication for all aspects of
economic and social life up to and including the
very existence of the country. (5) The northern
climate has meant that there has been a continuing
concern for shelter and for energy production; and
the limitations on arable land and its proximity
to the US border have meant that there is an
accelerating influence of America-based electronic
communications. These conditions have, to a
considerable degree, shaped how public (and
corporate) acceptance and technological innovation
has taken place. While for Canada geography is not
quite destiny, it is not possible to examine any
element of public life without at some time
determining the effect of geographical conditions.
      Canada as a cold country has had to take its
energy needs very seriously. Major gas and oil-
exploration programmes are augmented by an
intensive effort in nuclear energy. Atomic Energy
of Canada Ltd. (AECL), a Crown (government-owned)

Corporation, was created in 1952 to take over the operation of nuclear activities in Canada from the National Research Council. Over 90 per cent of research and development in nuclear energy in Canada is conducted by AECL. Commercial electric power generation from a nuclear reactor (the CANDU) began in 1962. While Ontario is most heavily dependent on nuclear energy for electrical power (about 25 per cent of electrical energy comes from nuclear), small reactors in both Quebec and New Brunswick indicate nuclear may soon play a role in meeting the energy demands of these provinces as well.

The development and commercialisation of nuclear energy in Canada has been marked by a rigorous public debate on cost and safety, but it appears that Canadians continue to have faith in the country's scientists and engineers to solve energy problems. As an example of the Canadian public's overall faith in science and technology, in a 1983 survey, 82 per cent were of the opinion that they could depend on science and technology to provide a long-term solution to the energy problem, (12 per cent disagreed, while 6 per cent didn't know).(6) This faith may also have contributed to the relatively quiet acceptance of nuclear power generation in Canada, as compared for example to Europe or even the United States.

The dependence in Canada on transportation and telecommunications technologies to bridge distance and tie together dispersed populations has meant that an atmosphere supportive of technology acceptance is built into the culture. In fact standards of living would be unsustainable in most parts of the country without commensurate modern technology. The consequence is an openness towards new technologies which may be surprising to those from less dispersed or less climatically hostile countries.

In more recent times, the accelerating importance of advertising and market information, especially via television - Canada forms a subregion for many US media markets - has meant that information about technology innovations (and about reactions to those innovations) have been available simultaneously in the US (the source of many of these innovations) and Canada (for the most part an industrially underdeveloped consumer of technologies). While being forced to confront the problems of distance and northern climate, Canada has thus been free to share and absorb the

technological developments of its great neighbour
to the south, a significant portion of whose
population live in regions with a similar climate.
This latter circumstance has been both an advantage
and a hindrance in that we Canadians have had
brothers in adversity in the United States in many
of our areas of possible singularity, and thus have
had to share the opportunities which these
adversities present. However, except for the areas
outlined earlier, Canada has, for the most part,
had little need to develop unique technology
solutions to its problems.

Political Conditions

Canada, formally independent in 1867, was
relatively slow in developing certain of the
institutions normally associated with advanced
industrial nations. It was probably the First
World War and the associated decline of the
imperial power of Great Britain which provided the
spur to the development in Canada of domestic,
intellectual and cultural bodies. The consequence
of the imperial link from the perspective of
technology and technology acceptance, was that
virtually all technology requirements in Canada
which could be imported, were imported. This
occurred even in many cases where the
importation of European, and especially British,
methods and devices made little or no sense in the
climate and geophysical conditions of the country.
In addition, there was little incentive or
encouragement for the development in Canada of the
informal institutions from which technology
springs, such as world-class universities,
professional journals, research institutions and so
on. Further, there was in Canada little of the
self-confidence in one's abilities and achievements which comes, for example, from being at the
centre of a world-spanning empire, and out of which
world-class accomplishment appears to spring.(7)

After the First World War, Canada began to
develop many of these foundation stones to
innovation, but in an environment which inevitably
invited comparison if not competition with the
already developed and world-class institutions in
the United States. And these, of course, were able
to draw on a population base some ten times greater
than the Canadian.

The consequence of this has been, that for
almost the entire length of its history, Canada has

found itself to be primarily a consumer of technology developed elsewhere. From the perspective of public acceptance of new technologies then, acceptance has been a matter of coming to terms with a position of colonial dependence, first on the British Empire and latterly on the American one. Resistance to new technology has, to a considerable degree, taken the form of resistance to a perception of foreign technology as being equivalent to economic and political subordination - on the part for example, of the Metis (half-breeds) of Western Canada who resisted the 'new technology' of the professional land survey as representing an attempt by the forces of the Empire to seize their land; on the part of the progress-resisting French Canadians who identified new technology with domination by Anglo-Saxons; and most recently, by romantic anti-technologists such as George Grant, who identify new technologies with Americanisation and incorporation into the 'liberal' and 'progressive' empire of the United States.(8)

Technology in Canada's History

<u>Introduction</u>
For Canada, as for other developed countries, the story of the country's history can in part be seen as an account of the application of technologies to specific situations. However, perhaps more than for other countries, Canadian history can be seen as a chronicle of the continuing interaction between men and the singular geography which Canada presented, and in turn the technologies which they adapted, or in some cases devised, to assist them in responding to and mastering that geography. From the days of the fur trade, whose technology was largely adopted from native Indians, through the waves of logging, mineral extraction and farming, the exigencies of extracting a livelihood and sustaining life amidst vast distances and harsh climate has provided the central impetus for new technology acceptance.

Moving from a period when technology was emphasised for economic purposes, Canada then entered into a long period during which technology was applied to the solution of the political problems of a country scattered across a continent. The Canadian National Railway system was built explicitly as a means to ensure the participation of the western regions in the Canadian Confederation. In a later era the development of the transcontinental highway system, the Trans-Canada Telephone System, the Canadian Broadcasting System, Air Canada, and the development and use of communications satellites - all represented the application of technology solutions to the problems of ensuring participation, co-operation and communication for a regionally, economically, and culturally dispersed population. Canada quite directly responded to the problems of nation-building through the use of the transportation and communication technologies of the times.

More recently, it has been in the area of 'social' technologies that Canada has begun to look to resolve the problems which it must currently address - pressure for independence from the Francophone population in Quebec; the requirement socially and culturally incorporate remote regions and aboriginal populations in step with the extension of resource extraction industries into previously inaccessible parts of the country; the need to assimilate a massive and highly diverse post-Second World War immigration; and the shift in the international economy away from the need for relatively high-cost raw resources and the consequent unemployment and social dislocation which has resulted. New technologies have also presented new challenges to Canadians with the massive penetration by US television signals, with the increasing requirement for sophisticated consumer products, and with the decline in the significance of transportation costs as a factor in production.

Canada has responded to these challenges in a variety of ways - the development of direct broadcast satellite systems to bring Canadian television content and telecommunications to all parts of the country; the development of bi-lingual national institutions; the development of means for linking dozens of ethnic communities into a cultural mosaic; and a variety of group and community initiatives in creating local employment opportunities for otherwise unemployable populations. The public acceptance of social technologies has been remarkably open and, regardless of the government of the day, there appears to have been an acceptance of social innovation and new social technologies to the point where they have become embodied in changed institutional behaviour and in the creation of new and potent social institutions.

Canada and Two Languages/Two Cultures

Inevitably in Canada there is the question of language and culture. Canada consists of two founding linguistic and cultural groups - the French and the English - each looking back to a separate and substantial history, and each looking to maintain both continuity with and distance from that history. In the case of the Anglophones, the continuity has meant a continuing suspicion of the other English-speaking metropole, the USA. In the case of the Francophones, the continuity is represented by continuing ties with France, and increasingly with 'la Francophonie', but always with a certain ambiguity, since the separation from France occurred before France was transformed by the French Revolution, and the re-establishment of contact has taken place only in the period since the First World War. In short, French Canada has had to find continuity in a context where such continuity has almost been totally lost in the 'Mother' country. There are several other consequences of Canadian bi-lingualism related to the public acceptance of new technologies:

• the existence of two languages means that there is always a time lag in the general dissemination of technologies across the nation. Often the lag is simply because of the time required for translation of documentation from one language to another:

• the in-built bifurcation in national cultural, intellectual and scientific institutions as between the two language groups (although it has declined markedly in recent years) has tended to fragment energies, talents and resources to the detriment of all;

• the development of English as the universal language of science in general and computing in particular has presented a variety of problems both to national bilingual instructions in science and technology and to those concerned with these areas in Quebec. Being within the US orbit for many of these developments has meant that technology development, translation, and the development of marketing and training materials in French has become the responsibility of Francophone Quebecois, and thus has presented opportunities as well as costs as Quebec attempts to carve out a role for itself as the translator of US high technology to the French-speaking world.

All of these consequences, moreover, add a time lag as well as a financial and innovative drag to the dissemination of technology in general and to public acceptance in particular.

The Role of the Canadian State

The state in Canada has been inordinately active in all areas of technology development and technology implementation. To a very significant degree new technology development in Canada has been mediated through the state and reflects this circumstance. Thus, for example, the relatively low degree of foreign ownership and the associated pattern of extensive technical innovation in both the transportation and the communications sectors reflects the early and continuing involvement of the state in these areas. In other sectors, such as agriculture, the Canadian government has been the most significant developer and disseminator of new technology, either directly or through state-owned enterprises or para-governmental institutions. It has also been the major funder of the development of new technologies of special interest to Canada, as for example in the energy area - not as in the US through its defence establishments, but rather through other departments and agencies such as the National Research Council, and through grants, loans and tax incentives in support of industrial development.

While in the 1983 survey referred to above, a majority of Canadians (60 per cent) agreed with spending more tax dollars on technological innovation to increase the standard of living (compared with 33 per cent who disagreed and 6 per cent who didn't know), it is interesting and illuminating to note where Canadians thought research dollars should go. Improving health care was the area which most Canadians (31 per cent) felt should be the primary object of increased scientific and technological research. This was followed by developing new energy sources (17 per cent); reducing pollution (14 per cent); developing or improving methods for producing food (13 per cent); improving productivity in industry (6 per cent); developing or improving weapons for national defence (4 per cent); developing more efficient mass transit (2 per cent); improving science education (2 per cent); exploring outer space (1 per cent); and improving communications systems, such as telephones (less than 1 per cent); didn't know or couldn't decide.

While the Canadian public apparently approves of the role of government in Canada as supporting research, it prefers that government be a supporter and regulator, and does not entirely trust government to become too much of an actor in this area. For example, a 1981 public opinion poll asked Canadians to indicate which of three broad groups - government, universities or private companies - would be likely to conduct research which was most efficient, careful, innovative and in the public interest. Almost half (48 per cent) said that universities would be most efficient in conducting research, while fewer (28 per cent) said private companies would be most efficient, and fewer still (21 per cent) said that government would be most efficient (3 per cent said they didn't know). Universities were also cited by 44 per cent as being more careful in conducting research than were government (26 per cent) or private companies (24 per cent); 6 per cent said they didn't know. Forty-three per cent of Canadians thought universities would be the most innovative, followed by private companies (31 per cent) and government (17 per cent); 9 per cent said they didn't know. Universities (41 per cent) were also thought to conduct research which was most in the public interest, followed by government (35 per cent) and private companies (16 per cent); 8 per cent said they didn't know.

One exception to the above was that French-speaking Canadians regarded government research as being more in the public interest (47 per cent) as compared to universities (36 per cent) and to private companies (12 per cent); 5 per cent said they didn't know.(9)

Interestingly enough, the state has also been the most active agency involved in assessing the impact of new technologies, whether through Royal Commissions, such as the Freedman Commission on Railway Run-throughs, or the Berger Commission on the MacKenzie Pipeline; through special committees, such as the Canadian Videotex Consultative Committee; or through particular agencies responsible for these tasks, such as the Science Council.

Various government departments have also taken upon themselves the difficult role of mediating the

various concerns arising from the use of
technologies in Canada. For example, the 1970s
witnessed a growing but often ill-defined concern
about the role of science and technology in the
making of policies and decisions in the public and
private sectors, and also about the effect of such
decisions on scientific and technological
development. Varying in scope, and embracing such
diverse issues as the effect on human beings of
long-term exposure to low-level radiation; the use
of saccharin in diets; the effectiveness of seat
belts to reduce fatal automobile mishaps; and the
adequacy of information about Canada's oil and gas
reserves, this concern is said in various ways to
involve 'scientific and technological controversy'.
To deal with that controversy the federal
departments of Energy, Mines and Resources, Health
and Welfare, and Consumer and Corporate Affairs
each invested considerable time and effort to
mediate, resolve and understand the problem and
potential harm to citizens of Canada.(10)

The state in Canada has also been the primary
sponsor of the many advocacy groups which have
developed in response to the various technological
and scientific incursions experienced by the
general public or by more particular publics such
as the native Indian population or the poor.
Various forms of support through direct grants,
contracts or subsidies for researchers and others
have been the major source of financial support for
most of the advocacy groups which have developed to
promote the 'public' interest with respect to
technology, as for example in making presentations
to committees reviewing the possible impacts of new
pipelines, hydro or nuclear generating stations, or
other projects.

It should be noted as an aside that many of
the most significant issue areas having to do with
technology in the last decade have not been
concerned with 'new' technology as such but rather
with relatively 'old' technologies such as
hydroelectric dams and oil or natural gas
pipelines. What is notable and of a special
significance from the perspective of 'public
acceptance', is that these technologies are those
which will have an impact on 'the land'. There
exists in Canada an almost mystical feeling
concerning the land - it is perhaps the only
mythology which is common across the country. Any
technology which will impact on the land will be
certain to find itself embroiled in controversy as

environmental groups and local citizens respond in opposition. In Canada, while the numerical significance of land-based populations (farmers, fishermen, trappers) is small, they still retain significant popular support through their access to the popular imagination and to the media. To take one example, from time to time discussions arise in Canada concerning the sale of water to the United States. It is interesting and remarkable to note that even though much Canadian industry is owned and controlled by foreign corporations, there is an almost unanimous feeling among Canadians that water should not become a transaction in the trade accounts of the United States and Canada. In addition, many of the newer land-based hydro-developed projects have been proposed for land either already assigned to the native Indian population, or to which they retain as yet unsettled legal claims. In the case of many natives living in remote areas the land may still be important as a source of game and furs (for commercial sale) and as the basis for their cultural identity. The participation by native groups in hearings concerned with technology development has had a significant impact on these developments and on the public perception of the impacts of technologies, as well as on the native groups themselves. This participation has been almost completely funded from government sources. There have been numerous instances in recent times where various government agencies have been providing funds for the proponents of a project, the opponents of the project, and for the arbitrators seeking to reconcile the interests of each - as well, of course, as for the courts to which decisions may ultimately be assigned.

It is therefore not possible or realistic to examine the public acceptance of new technologies except within a context at least in part mediated by the state. The interests of the state and its bureaucracy must be seen as central in the analysis of the public acceptance of new technologies, as for example, the concern of the state with resolving pressures for radical decentralisation (and/or independence) from several regions, as well as the on-going concern with issues shared with other developed countries, such as unemployment, increasing international competition and so on. These overall efforts by the Canadian state have probably been a response to the lack of a private sector Canadian capability in the development of new technologies. The net effect, however, is that the state has come to shape the Canadian response to the acceptance of new technologies in a very profound way.

The Media and Public Attitudes Towards the Acceptance of New Technologies in Canada

About 73 per cent of Canadians have access to one or more US television channels. Despite government regulation of cable-TV and Canadian content regulations, the accessibility of US television channels has meant that Canadians watch much non-Canadian programming. This is compounded by the fact that about 70 per cent of all the programming viewed by Canadian audiences on English-language stations is of foreign (primarily US) origin. For entertainment and sports programming the figure rises to 86.5 per cent. About 40 per cent of the viewing on French language television is non-Canadian (mainly dubbed US programmes or from France). Further, US programming is made possible by the extensive use of cable television. In 1979 an estimated 78.3 per cent of Canadian homes had access to cable and 55.6 per cent of Canadian households received the service. By January 1980 an estimated 58 per cent of Canadian households were connected to a cable-TV system, thereby making Canada one of the most heavily cabled countries in the world. Where cable service is unavailable, there has been the very widespread penetration of satellite earth stations for television reception.

What this meant is that, to a very considerable degree, Canada is a component of the US market for information concerning new technology, as well as information expressing concern for and opposition to new technology. Canada is a constituent component of the US media market for all kinds of information, including popular, intellectual, professional and private (such as client-restricted) information. This has included the very large numbers of US citizens or persons of US origin who have moved into and remain as important members of the Canadian academic and intellectual communities.

Thus, Canadian public attitudes to the acceptance of new technologies are combinations as well as a broad assimilation of the media outpourings of a US perception of needs, history and mythologies. This has meant a peculiar kind of schizophrenia with respect to new technologies. On the one hand, in the US new technology is broadly welcomed as supportive of individualistic enterprise and life style: on the other hand, in Canada new technology is mediated by the state and reflects the power of the state in Canadian life. So while the effect may be the same, broad acceptance of the new technology, the supporting structures, social, cultural, economic, and institutional, are radically different. The effect has been that in Canada technology and the acceptance of new technology have been much more pragmatic or instrumental than in the United States, where new technology and the public acceptance of new technology has been placed in a more ideological context. There is in Canada a very broad willingness across party lines both to accept new technology and to regulate its implementation.

These patterns are well reflected in public opinion surveys. In the winter of 1980 41 per cent of Canadians felt that automation would benefit the economy; by the spring of 1982 this number had grown to 49 per cent. By the summer of 1983 more than half (or 54 per cent) of all Canadians were of the opinion that automation would lead to a stronger economy. Similar changes in public opinion were also recorded for the impact of automation on the price of consumer goods and improvements in the quality of producers. In December 1980 41 per cent of Canadians felt that automation would lower the price of consumer goods; by June 1983 more than half of all Canadians (or 52 per cent) felt that automation would lower the prices of consumer goods. In December 1980 47 per cent were of the opinion that automation led to improvements in the quality of products; by June 1983 another 13 per cent, or 60 per cent in all, held a positive opinion on this relationship.

Canadians are very concerned, however, about the relationship of automation to employment. In June 1983 72 per cent of Canadians were of the opinion that automation in the work place would lead to more unemployment. The same survey found that 72 per cent of Canadians felt that automation would lead to depersonalisation of the average worker. This survey also found that 91 per cent of Canadians believed that technological change produced a greater need for manpower training programmes. Clearly, the public is not only conscious of both positive and negative consequences of technological change in the work place, but sees a great need for policies and programmes that will allow people to cope with such change.

As well as threatening their jobs, many Canadians feel that technological innovations pose a threat to their <u>privacy</u>. Two-thirds (66 per cent) of Canadians surveyed in 1981 agreed that the growing use of computers was a serious threat to individual privacy (29 per cent disagreed; 5 per cent didn't know). And a 1982 survey found that 58 per cent of Canadians agreed that the increased use of machines to store information would threaten customer privacy (36 per cent disagreed; 6 per cent didn't know).

The public seems less certain about the amount of control governments have over high-technology industry. When asked whether they favour or oppose greater government control over this sector, 52 per cent say they do favour greater control (while 42 per cent favoured less), the highest number of any of the industrial sectors (including oil and gas, mining and banking) mentioned. Despite broad public opinion trends opposing government intervention in the economy, these findings demonstrate the public's continuing pragmatism on this topic.(11)

Carriage and Context: Canada's Unique Success and Singular Failure

Many of Canada's successes in the field of new 'physical' technology have been in the area of communications, and especially telecommunications. For example, Canada has emerged as a dominant force in the supply of telecommunications equipment, such as Private Automated Branch Exchanges (PABX), data and voice transmission systems, and custom-integrated circuits for the telephone sector. Canadian industry has a leading position in the development and production of fibre-optic cable. The Canadian space industry has earned respect for the innovation design and manufacture of satellite earth stations, satellite antennae and specialised spacecraft components and control subsystems. Canada designed and manufactured the Remote Manipulator System (CANADARM) for the NASA space shuttle Columbia. In addition, SPAR Aerospace Limited has been at the forefront in the development of satellites such as the ANIK-D series. These technologies, as with the earlier areas of our success in technology - railroad building, aircraft design and so on - have been concerned with linking or bridging populations across broad expanses.

Our successes then have been in the area of communications carriage, i.e. the infrastructure for moving signals. The other side of that equation, of course, is communications content, i.e. what is carried along that infrastructure. Canada has for many years been found to be at the top of international comparisons in telephone calling and length of time spent on the telephone. We have no trouble in the acceptance of technology for interpersonal communication. However, in the broader sphere of developing the industrial capacity to generate and market the commercial/cultural content which would provide a financially significant dividend from our telecommunications system, we have been notably less successful.

Our satellite system, the first in the world, languished economically for lack of commercial users until US industry caught up with the increased technological capability and began to lease the excess channel space. Our vast television cable system carries a significant preponderance of US channels, while our over-the-air television networks carry mostly US programming; videotex technology, continues to search for content and appropriate applications, and both government and industry in Canada now expect any commercial breakthrough to occur first in the US. We have not been successful in developing a commercial product sufficient to the carrying capacity we have created.

The reasons for this are obvious: Canada's small and fragmented population; the two languages further subdividing an already small market; Canada's particular history in never having had a significant period of cultural independence. The effect has been that in many cases others have benefited from Canadian initiatives and innovations.

Increasingly, we are seeing a merging of carriage and content as the market for pure carriage gives way to value added services; to telecommunications companies integrating with computer companies to offer intelligent networks; to cable systems becoming the new supermarkets and so on. In this area the sophistication of Canada in telecommunications and the widespread acceptance

of both old and new technologies is of little
consequence, and we consistently lose out to those
who are able to integrate 'carriage' and 'content'
into single corporate packages.

In our view the communications experience for
Canada has been that the carriage is the content.
It is striking to note that the great philosopher
of the impact of communications 'carriage',
Marshall McLuhan, was a Canadian professor of
English literature - communications content: his
most famous utterance was, of course, 'the medium
is the message'.(12)

The Public Acceptance of New Social Technology

Among its many interpretations, technology can be
defined as the application of knowledge to
practical purposes. In the Canadian context
'social' technology or social innovation has been
at least as important as 'physical' technology.
Because of Canada's position culturally (both
internally with respect to diverse regional needs
and languages and externally vis-a-vis the United
States) and geographically solutions to problems
unique to Canada have had to be developed. While
Canadians are often criticised as not being
entrepreneurial or risk-takers, it is clear that in
the development and application of 'social'
technologies, Canadians have been willing to
explore options and take risks relatively
unhampered by concerns for traditional practices or
ideological positions.

The development of a quasi-commercial but
publicly sponsored radio and television system in
two languages, and with some measure of local and
national network capacity throughout the system, is
a significant innovation observed with interest
throughout the world. The development of
institutional mechanisms for linking two languages
and cultural groups into a common union in the
entire range of spheres of state activity has also
been of interest to emerging countries with similar
problems of nation building in a culturally
heterogeneous environment.

Similarly, the development of a variety of
mechanisms for linking together culturally
heterogeneous groups outside of the two founding
cultures into a national 'mosaic', while ensuring
both the retention of group identity and loyalty to
common national system, is a singular achievement.

Canada

The practice has been to experiment with new forms of publicly supported institutions to deal with perceived social needs, always of course within the overall framework of the market economy. This is expressed in a variety of ways. For example, since Confederation there has been an effort made to redistribute income to the various regions of Canada. As various regions joined to make up Canada, subsidies were arranged to bolster the provinces' municipal tax base, since customs and excise taxes had been allocated to the federal government. In addition, each province joined Confederation with the implied hope that it would prosper since internal tariffs would be eliminated and the benefits of internal free trade would make all provinces and regions of Canada wealthier. As the nation of Canada took shape, and as the effects of a common tariff were experienced, it soon became clear that some regions were prosperous more than others. Some regions experienced a decline in economic activity and demands for compensation were made. The Maritime Rights movement represented just such a demand. It resulted in a doubling of the subsidy to the Maritime provinces and led to a new era in federal-provincial relations. In the words of the Royal Commission on Federal-Provincial relations, the Rowell-Sirois Report:

When, as a result of national policies undertaken in the general interest, one region or class or individual is fortuitously enriched and others impoverished, it would appear that there is some obligation, if not to redress the balance, at least to provide for the victim.

The net effect was to put aside purely economic criteria as the basis and the means for the creation and maintenance of the political entity. This led to a wave of regional development projects and agencies such as the Agricultural Rehabilitation and Development Act, the Atlantic Development Agency, the Fund for Rural Economic Development and the Department of Regional Economic Expansion. The attempt to provide for regional equality has become institutionalised, and what was a social innovation to provide the framework for Confederation is now a part of Canadian culture and politics.

Also, in the process of nation-building a unique institution was created: the Crown Corporation. Almost a contradiction in terms, it

279

allows for government activity to take place in a
country which lives by the ethic of free
enterprise. Crown Corporations are legally
autonomous bodies through which the Government of
Canada or the government of a province carries on
certain activities. For certain purposes, such as
the regulation of highly technical activities or
the provision of goods and services in the interest
of national development or national security, it
has been found convenient to set up agencies
separate from the normal structure of government
authority. These agencies, more or less
independent of the executive in their day-to-day
operations, are subject to overriding control by a
minister, by the cabinet, or by Parliament.

As the name Crown Corporation implies, such
agencies are legal persons in their own right,
separate from the rest of the executive but
enjoying certain rights and privileges as
'emanations of the Crown'. A Crown Corporation may
be engaged in the production or sale of goods and
services, such as Air Canada and Eldorado Nuclear
Ltd; it may be an agency providing goods or
services for government, such as the Royal Canadian
Mint or Defence Construction Ltd; or it may be a
multipurpose organisation, such as the St Lawrence
Seaway Authority.

The individual corporate structure of the
Crown Corporation ensures managerial flexibility
and freedom from the rigid framework of government
financial and personnel controls. But it is an
instrument of public policy and must therefore
respond to the final control of Parliament, a
Minister of the Crown and the public. In general,
questions and debates on the corporations are
confined to structure and policy and do not extend
to interference in any of the details of
management.

There are Crown Corporations in every part of
Canadian life. Too numerous to list
comprehensively, they include: Air Canada, the
Industrial Development Bank, Atomic Energy of
Canada, Canadian Patents and Development, Canadian
National Railways, the Canadian Broadcasting
Company, the Canadian Wheat Board, the National
Research Council, the National Arts Center. The
list could go on and on. Leaving financial
enterprises aside, but including manufacturing,
petroleum and natural gas, mining and smelting,
railways and other utilities, and construction and
merchandising, federal and provincial Crown

Corporations make up over one-third of all Canadian-controlled assets.(14)

Other social innovations in Canada cover the whole range of societal activities, including a vigorous co-operative movement, universal health care, and government participation in public utilities of all types.

Conclusions

From a superficial point of view, there is little that is remarkable about the public acceptance of new technology in Canada - innovator in certain limited areas, Canada has been generally accepting of new technology, although with a continuing concern for certain of the effects which it may have on people and the environment. However, on closer inspection there are several items of interest. First, the significance of governmental bodies in all aspects of technology in Canada is certainly remarkable, especially in the broad context of a market economy and the stridently market-driven approach to new technology in the US. This is almost certainly a result of the degree to which Canadian history has forced a public response to a variety of economic, geographical and cultural conditions which could not be met solely by the market. For Canada to survive as a united country has meant that governments (federal and provincial) have been forced to intervene in all aspects of new technology in a manner which is remarkable in the developed world.

Secondly, it should be noted that the unique Canadian contribution in the broad international area of new technologies (or at least in the narrower area of new technology in North America) is its capacity for and acceptance of social innovation and new social technologies as a means for the resolution of publicly perceived problems. Where in other countries social and economic problems have been responded to within the framework of existing institutions and institutional approaches, the uniqueness of Canadian problems and the lack of institutional traditions in a variety of areas has forced the country to seek out unique solutions to such problems as regional economic disparities, rural development, aboriginal populations, bilingualism, unemployment, and credit formation. That Canada

Canada

has been willing and able to experiment with
developments in these areas without resorting to
radical political changes, to ideological
polarisation, or to disruptive social conflict or
to major internal swings in cultural and social
values, is a measure of what has been described as
the 'Canadian genius for moderation'.

From the perspective of the public acceptance
of new technologies, it is perhaps in the area of
the public acceptance of new social technologies
that Canada has most to teach the world. A major
challenge resulting from the widespread
introduction of the new technologies will be to
develop the social technologies required to
accommodate the problems and opportunities which
will result. Canada is only now beginning to come
to grips with the problems posed by shifts in
international competitiveness and the consequent
problems of long-term unemployment, of shifts in
the nature of the labour force, and of the rapid
rise and fall of new industries and occupations.
If Canada is able to develop appropriate social
technologies to respond effectively in these areas,
it will again have provided direction for its more
technologically active trading partners.

Notes

1.   Corporations and Labour Unions Returns Act,
Report for 1981 Statistics Canada, No. 61-210,
March 1984, p. 21 and Table ix at p. 47.

2. The Bottom Line: Technology, Trade and Income
Growth. Ottawa: Economic Council of Canada, 1983,
p. 42.

3. Decima Research, December 1982, Toronto.

4. The Science Council of Canada, Innovation in a
Cold Climate, 1970.

5. These arguments parallel those developed by a
series of Canadian historians such as H.A. Innis,
Donald Creighton and W.L. Morton. See for example
H.A. Innis The Fur Trade in Canada, Toronto, 1964
and Empire and Communications, London, 1950, Donald
Creighton, The Empire of the St Laurence, Toronto,
1956, and W.L. Morton, The Canadian Empire,
Toronto, 1961.

6. Awareness and Attitudes of Canadians Regarding

Technological Change. (A review of available public opinion data on the perceived impact of technology in Canada), Ottawa: Optima Research, August 1983.

7. See James Guillet 'Nationalism and Canadian Science' in R. Russel (ed.), Nationalism in Canada, Toronto, 1966.

8. See for example, G. Grant, Technology and Empire, Toronto, 1969, and Lament For a Nation, Toronto, 1965.

9. Optima Research, Awareness and Attitudes, p. 5.

10. See for example: The Peripheral Nature of Scientific and Technological Controversy in Federal Policy Formation by G. Bruce Doern, Background Study No. 46, Science Council of Canada, August 1981.

11. 'Public Opinion and Technology: Canadians Adjusting to Change' by Raymond Anderson of Public Affairs International Ltd, paper presented to the Canada Tomorrow conference, November 1983, sponsored by Government of Canada, Ottawa.

12. Marshall McLuhan, Understanding Media, McLuhan of course was an intellectual disciple of a lesser known but no less profound historiographer of communications; Harold Innis. See his Communications and Empire and the Bias of Communications.

13. See Herschel Hardin, A Nation Unaware: The Canadian Economic Culture, Vancouver: J.J. Douglas, 1974, pp. 300-15.

14. Ibid., p. 88.

Chapter Ten

PUBLIC ACCEPTANCE OF NEW TECHNOLOGIES IN SWEDEN

Jan Forslin

Introduction

Scientific and technological advances have long
been a fully integrated part of Swedish experience.
The prerequisites for much of the country's basic
industry were grounded in technical innovations.
Since the eighteenth century Swedes have achieved
international fame as inventors, scientists and
explorers, and this has become a part of the
individual Swede's self-image. The engineer,
inventor or scientist has typically been the hero-
figure in Sweden during the last 100 years, rather
than kings, politicians, warriors, artists or
philosophers. The Nobel Prize has become an annual
reminder of the high prestige of scientific
endeavour and the participation of a small nation
in the learned world community. The Swedish public
school system is one of the oldest, and the
ability to read became widespread at an early
stage. Education and learning have traditionally
benefited from the high social regard accorded
them.
    The period after the Second World War,
characterised by what seemed to be an era of ever
growing affluence due to technical progress,
eventually came to a rather abrupt stop while at
its crest. Serious signs of a threatened
environment now led to a questioning of the basic
philosophy of expanding material welfare. Images
of dead lakes, inedible fish and game catches,
falling levels of subsoil water, pollution of city
air and of the higher layers of the atmosphere,
waste of limited natural resources, ecological
catastrophes all around the world, large-scale
accidents, and prospects of global starvation - all
these issues penetrated through to public

consciousness. The concept of the environment also came to receive a broader definition, including its social and psychological aspects. As a result, crime and impersonal urban life, drugs and teenage prostitution came to be seen as environmental problems linked with the modern, and artificial, lifestyle.

Political Behaviour

Although nuclear power and pollution of the natural environment have been in the forefront of the public's attack, the shift in attitudes from a belief in a prosperous and affluent future to anxiety about an almost immediate 'doomsday', has come to affect the whole array of technical innovations, whether biotechnology, gene-manipulation supersonic aircraft, heart transplantation, the pill, new vaccines and tranquillisers, videos or computerisation. The period of anti-technological opposition culminated with the 1980 referendum on nuclear power, and since then the shock waves of opposition have weakened somewhat.

If technology had once been seen as a salvation for practical and material problems and as a source of national pride, it came during the 1970s to be viewed as a mixed blessing by large groups in society. Unfortunately, there were no polls on public attitudes towards technology until about the 1960s. One could fairly assume, however, that earlier opinions were characterised by a marvelling at and admiration for the wonders of technology. The pressure for an improved standard of living was also much more immediate at the time.

It should also be pointed out that the era of protest and questioning accompanied an increased availability of higher education. The educational gap between the ordinary citizen and consumer of technology on the one hand, and the producer of technical innovations on the other, has diminished. Even though technology, in its transition from a mechanical to an electronic base, has become much more sophisticated and even less transparent in its detail, it has also become cheap and available to anybody in the form of consumer products.

Since the Second World War there have been numerous demonstrations in Sweden against new technological ventures. Initially, protests were connected with the anti-nuclear peace movement of the 1950s. Twenty years later it was protection of the environment that was in question. For the first time ecology had become a political issue. The former farmers' movement - The Centre Party - took up environmental protection as its main cause and managed to attract large groups of people who were not involved in agriculture, many of them indeed second-generation city dwellers. Within ten years it had become the second largest political party. This happened during an era in Swedish politics characterised by an unusually high degree of left-right ideological polarisation, but in spite of the resulting strong pressures, the Centre Party managed to win the support of a substantial part of the electorate for an essentially non-class issue. It was as if ecology were winning out over ideology.

Along with this political shift went changes in consumption patterns and living habits. Quality of life - including quality of working life - became a catchphrase. The big cities were abandoned by the educated young in favour of life in new green suburbs and experiments in new rural living - the green wave. Organic food growing, more ascetic habits, vegetarian diets, fasting, bicycling, collective transport, hiking, camping and canoeing were all expressions of a change in lifestyle, sharply contrasting with the mostly unhampered materialism of the 1950s and 60s. Spontaneous local protests against the spraying of forests, regulation of lakes for hydro power and weed killing took place in numerous locations.

Various alternative movements were born, albeit some were very short lived and others were soon transformed or merged with more established organisations. 'Soft', non-masculine values, experiences and emotions rather than possessions were emphasised. The objectives of 'zero growth' and 'basic needs' became part of the economic debate. Spiritual renewal, with strong strains of meditation and psychotherapy, caught the interest of young, well-educated groups, and technical education fell in popularity. Now, 10 years later, there is a serious shortage of technicians.

The established political parties - with the exception of the Centre Party - did not manage to channel these strong sentiments and reactions against modern industrial society. In opposition the Centre Party had taken a strong stance against nuclear power, which became a symbolic as well as a very practical political issue. The Centre Party's position contributed significantly to bring it into power in the non-socialist coalition government of 1976, the first in more than 40 years. Once there it failed, however, to carry through its policy sufficiently to prevent a large proportion of its electorate from feeling betrayed. In the subsequent election it lost 25 per cent of its voters, and since then it has not managed to regain them.

The development of the Centre Party's strength is shown in Table 10.1 as reported in the official Swedish election statistics and in a prediction of their support in a poll in December 1984.

Table 10.1: Percentage of Votes Cast for the Centre Party in National Elections 1960-82 and a Forecast of Their Support in December 1984

1960	1964	1968	1970	1973	1976	1979	1982	1984
13.6	13.2	15.7	19.9	25.1	24.1	18.1	15.4	13.5

Source: Dagens Nyheter, 23 December 1984.

After the peak in the mid-1970s, which brought the party into the non-socialist governmental coalition, it is now back at its level of the early 1960s - before ecology became an issue. This development hardly reflects the swings in the left-right division of the electorate. The non-socialist side is today - early 1985 - stronger than ever, though the Conservative Party is now the second largest, after the Social Democrats, and not the Centre Party.

The recently established Environment Party - which claims to be the Swedish 'green' party - has adopted ecology as its main concern. So far it has participated only in the 1982 elections, when it received 1.7 per cent of the votes. According to the December 1984 forecast referred to above, it would now obtain the necessary 4 per cent to gain admission to parliament. In Finland, a similar party has managed to get two representatives into parliament, and in the recent municipal elections it got a very strong response. Apart from the matter of ecology, the green parties also question the traditional political system, for example, the dominance of professional politicians. Several of the established Swedish parties are now vying to be seen as the 'green' party. It remains to be seen whether these 'green' parties represent a new political order, in which concern for the environment and the proper use of technology is the guiding principle rather than conflict between classes - protection of the environment, that is, rather than of the underprivileged.

Referendum on Nuclear Power

The nuclear power issue created political turmoil during the second half of the 1970s. Opposition to nuclear power was to be found within factions of most of the political parties, and for this reason it was decided to hold a referendum, a rare event in Sweden. (It had been used three times previously, on prohibition of alcohol in the 1920s, on national pension funds in the 1950s, and the change from left-hand driving to right-hand driving in the 1960s.)

The referendum took place in 1980 under the non-socialist government. It was preceded by an intense information and persuasion campaign in the media. During this campaign the anti-nuclear power movement increased in strength. Activists were estimated to amount to 70,000-80,000 people and the badge of the movement was sold in millions. Since then the now more firmly organised movement has settled at some, 5,000 members. In spite of the heated debate and strong emotions generated, only 76 per cent eventually chose to cast their vote, an unusually low figure for Sweden. Part of the reason for this low participation rate may perhaps have been the rather strange options offered on the ballot.

After negotiations, three options were finally formulated;

(1) Power stations already in operation, under construction or planned should be used for their entire planned lifetime, but nuclear power should afterwards be phased out.
(2) This option was similar to the first, except that it specified a time limit of 25 years; it also included the objective of nationalising the energy industry.
(3) There should be no new plants, and those already in operation should be closed within 10 years.

Formulated in this way, the three options managed to become congruent with the traditional political party positions. Conservatives supported line (1); Social Democrats and the Liberal Party were identified with line (2); and finally, line (3) reflected the position of the Communist Party and of the Centre Party. Line three was considered to be the 'no' alternative, while there were two 'yes' alternatives, one non-socialist and one socialist. There was no fully fledged 'yes' alternative, and it was even claimed that the second line amounted to a call for the abandonment of nuclear power, though it was quite close to line one and could thus be identified as a 'yes' line.

Anyway, the results were interpreted as meaning that the 'yes' side had won. Line (1) received 18.9 per cent of the votes, line (2) 39.1 per cent, the 'no' side 38.7 per cent, while 3.3 returned blank ballots. If one adds this 3.3 per cent to the 24 per cent who did not participate, then some 28 per cent were not satisfied with the procedure, or did not consider nuclear power an essential issue, or expressed distrust in the ability of the political system to handle the issue.

Sixty-six per cent were dissatisfied with having had these three options to choose from as against a more clear-cut dichotomy.(1) The information source most trusted during the campaign was technical experts (75 per cent). Secondly came environmental movements (29 per cent) and thirdly official authorities (17 per cent). Politicians, journalists and trade unions were not seen as having had much to contribute (4-8 per cent). The tendency identified here increased during the months before the referendum, and even among the 'no' voters, technical experts generated more confidence - a traditional Swedish response, even when technology as such was being challenged.

Computer Experience

If waste of natural resources, environmental damage, and nuclear power have been the major issues in connection with the quality of life, then computers have become the symbol for changes in the quality of working life. A number of major studies in Sweden have recently touched upon public acceptance of this form of new technology. In a national survey by the National Central Bureau of Statistics during the summer of 1984, several questions were directed to this area as a commission for 'Datadelegationen'.(2) The sample consisted of 17,000 people from a population of some 4 million in the age range 16-65 years. Non-responses amounted to approximately 13 per cent.

The first conclusion which emerges is that the utilisation of computers in one way or another is far reaching and progressing rapidly. Every third person has some experience of computer equipment, and 25 per cent have used computers for at least one year. On the basis of the sample, one can infer that more than 700,000 people have tried some kind of programming, and of these 95,000 have knowledge corresponding to at least six months of full-time study. Twenty-three per cent are using computers in their work, 15 per cent daily. There are 150,000 jobs directly connected with computerisation, in programming, repair, service and operation of computers, or in design and production of equipment, or in research and teaching. A much greater number still are in occupations which existed previously, but which now have computer support, clerical and administrative tasks being the most common category here, followed by production control and technical work.

Computer experience is concentrated in the young age groups, although up to 5 per cent of the oldest part of the labour force also use computer equipment daily. In the youngest stratum, computer usage tends not to come from work-related activity, but even so some 50 per cent of this group too have computer experience. Use of computers at work is most common among those in their thirties, but even among those aged around fifty, 20 per cent use computers in work.

Men dominate among those having genuine computer jobs, whereas women occupy only 17 per cent of such positions. On the other hand, women largely dominate when it comes to applications in clerical and commercial operations. In technical applications there is again a male dominance.

There seems to be a strong response to the challenges of computerisation. More than half the sample consider themselves in need of further training, most often in order to be able to cope in their jobs, or else in order to get another position, but also as a means of influencing the process of computerisation in their work. Training of users is twice as much in demand as training for computer specialists such as programmers, systems designers and computer operators.

There is a difference between attitudes to and

personal experience of computerisation. Fifty-seven per cent believe that computers create employment, and 66 per cent believe that computerisation leads to simplified work tasks. On the other hand, very many fewer have had their own experiences of this kind. A negligible proportion of the population regard themselves as having lost their job due to computerisation. Of these, 81 per cent were again employed at the time of inquiry. Slightly more - 1 per cent of the population - have experienced their job changing for the worse due to computerisation. More than five times as many think that their jobs have improved for the same reason, and many think that the change did not affect their jobs either way.

When it comes to the respondents' own prospects, 42 per cent see computerisation as something mostly positive, and 22 per cent as something more negative than positive. (One in three does not know or is not interested.) One per cent believe that they are going to have jobs requiring less qualifications, but 8 per cent think that their jobs are going to require more qualifications, while 2.6 per cent believe that their own jobs are going to disappear due to computerisation.

To summarise: this very recent, representative national poll reveals, in general, favourable attitudes towards computer technology. A substantial proportion of the public is getting personal and positive experience of computerisation. Even if it is thought in general that computerisation will have a detrimental effect on employment opportunities and the level of qualifications likely to be called for in future, very few see such prospects as applying to themselves. Twice as many have optimistic personal expectations of the new technology.

Attitudes Towards Modern Technology

Another recent study related to this topic has been carried out within the project 'Jobs in the 1980s: Bridging the Gap between People and Work'. This is an international research programme initiated by the Aspen Institute for Humanistic Studies and Public Agenda Foundation. The Swedish Institute for Opinion Research and the sociology department at the University of Umea have been responsible for the Swedish contribution. Apart from Sweden, the

USA, Japan, Israel, West Germany and Great Britain participated. The main international report was published in 1983. The review here is based on a Swedish publication(3) which draws on several of the Swedish Institute for Opinion Research's studies with various Swedish strata.

Some of these studies are longitudinal. Thus, the general attitude towards scientific and technological progress seems to be very much the same in 1981 as it was in 1959. On both occasions, 60 per cent thought that advances in these areas had made life easier. There is therefore no indication that a more pessimistic assessment had developed during this period. Those who protest against technology are to be found among the 40 per cent not so optimistic, this percentage having also remained the same as 20 years ago - the difference seems to lie in the degree of radicalisation and belligerence.

Computerisation is outstandingly the phenomenon most frequently associated with modern technology, with a 60 per cent response-rate. Automation, robots, electronics, nuclear power and space technology each attracts only some 4 to 11 per cent. In this study computerisation is also mostly seen as something favourable, even if there is ambivalence. Among the positive answers, improved work performance comes out as the most frequent factor. A possibility of unemployment is seen as the largest single threat, but that man loses control in a too technical and bureaucratic society is another problem which gets a mention.

Over a 10 year period a little over a third of respondents experienced major technical changes and 15 per cent minor ones. (It must be remembered that 19 per cent of the sample had not been in work 10 years earlier.) Between 60 and 74 per cent of those who had experienced a small or large technical change in their work thought that their work had improved in terms of cleanliness, reduced physical strain, more interesting tasks and more responsibility. Negative effects were mentioned far less frequently. Less than a sixth said that work had deteriorated by becoming more monotonous, socially more isolated, more mentally stressing, more difficult, or through their having become more dependent on others. More than 50 per cent reported none of these negative effects.

In another question people were asked to rate the advantages and disadvantages with various kinds of advanced technology: medical research, satellite communication, microelectronics, computers, home electronics, research in outer space, industrial robots, and nuclear power. In all but two cases, the advantages were seen as greater than the disadvantages. The two areas where the reverse held true were industrial robots and nuclear power.

Swedes are less enthusiastic on the whole about modern technology than are the pooled opinions in six other Western European nations. This becomes even clearer if one takes the question of whether one would find it good or bad if technological development were more emphasised. Here, Swedish opinions were split equally on the three options Good, Bad and Do Not Know/It Does Not Matter, while almost 60 per cent in nine other European nations thought that such an increased emphasis would be good. In addition, computers are much less appreciated in Sweden than elsewhere. As Zetterberg et al. point out, it is an interesting question why in the three countries with welfare states, Sweden, Denmark and Holland, only one-third is positive towards modern technology, as compared to two-thirds in several other European countries.

Worries in connection with new technology are very concentrated on environmental issues: pollution, industrial waste and the safety of nuclear power. Unemployment due to automation, computer security and personal privacy and risks in working with terminals cause little concern. Zetterberg et al. tested several hypotheses in connection with the source of worries about technology. Parents of small children tend to be somewhat more worried than others. The sex of respondents turns out to be an important factor, as women are clearly more worried than men, even when several other factors are controlled for statistically. Furthermore, if the woman is occupied in a teaching profession, then the relationship gains in strength.

On the other hand, personal experience of modern technology does not show a significant relationship with a less anxious attitude. What turns out to have the strongest correlation with worries about modern technology is one's personal value-system. The people most worried about new technology are those with what Zetterberg et al. call 'interior or reproduction values'. They are also the ones least interested in modern technology.

Their values are typical products of the welfare state and they emphasize empathy and personal development. Access to education and social support creates a commitment to one's own and others' personal development and the creation of life quality. The importance of work is then related to work itself, its meaningfulness, and the social relations and personal development associated with it.(4)

Those most interested in new technology are those with 'exterior or production values', in particular of an individualistic kind. These values are products of the industrial era. Production and creation of material prosperity stands in the foreground. A high and justly shared (according to effort) economic growth has highest priority. A third intermediate group, in terms of attitudes to technology, consists of people with values connected to subsistence; hard work, security and survival are the predominant aspects in this group, which has its roots in agricultural society.
In line with previous observations, willingness to accept computerisation is highest in the exterior values group (74 per cent), lowest in the interior-values group (50 per cent), with the subsistence value group coming in between (59 per cent). These attitudes are influenced by personal experience. Thus, there are as many as 26 per cent in the interior-values and subsistence-values groups who say in answer to a hypothetical question, that they would oppose computerisation despite having had direct personal experience of it. This compares with 12 per cent in the exterior-values group. Amongst those without previous experience, the proportion opposed to computerisation ranges between 22 and 41 per cent.
Technology in the work context is presently being dealt with in numerous studies of working conditions and it is not possible to give a full account of this research here. What can be said is that the impact of new technology is currently one of the main working life issues and an area upon whose importance employer and employee organisations agree. Since much of the discussion here centres on changes in attitudes, data from an original study (5) of technology and work organisation and a subsequent follow up will be quoted. This study was carried out in the car industry, workers occupied in machining were interviewed on two occasions in 1974/5 and in 1981.

About 50 per cent of the sample was identical, but the 1981 data have not yet been published.

The question relevant in this context was formulated 'How, in your opinion does automation affect safety in work/workers' participation in decision-making/prospects for young workers/ physical workload/mental workload/job satisfaction/ importance of trade union?' Table 10.2 gives the results.

From a generally favourable attitude towards automation in the mid-1970s, one can observe a negative shift over the years since. A positive effect resulting from automation is now seen as less likely. In no instance is a more positive view taken than was the case 6 years earlier. Belief has changed more dramatically in relation to prospects for the young. Originally, automation was associated with increased opportunity for young workers. At the beginning of the 1980s the reverse view tended to be taken. At the time of the second interviews, unemployment among youth was becoming a widely recognised problem evidently associated with labour-saving automation.

In addition, mental strain is seen as increasing with automation and work satisfaction as decreasing. No strong link was seen between workers' participation and automation in 1974-5 and this was even more true in 1981, in spite of legislation on co-determination. This is just one study and a small one at that. Its main merit is its longitudinal approach. There is no reason, however, not to accept the general tendencies revealed in this case as representative of their time.

In the mid-1970s automation had not developed far. Few people had at that time direct personal experience with the more advanced levels of automated systems. In 1981 the first fully automated production islands began appearing and since than development has been rapid. Sweden is the biggest user of industrial robots per capita in the world, and computer-aided design/computer-aided manufacture and flexible manufacturing systems (CAD/CAM and FMS) are now being implemented on a large scale. This means that automation is becoming more concrete and tangible than it was even 5 years ago. Simultaneously, attitudes are changing.

Automation has probably been at least as common in clerical work as in production. Much of the discussion has nevertheless continued to be

Table 10.2: Opinions about the Effects of Automation
in 1974-75 and 1981

	Decreases (%)	Increases (%)

Safety
in work

Workers' participation
in decision making

Prospects for
young workers

Physical
workload

Mental
workload

Job
satisfaction

Importance of
trade union

70   60   50   40   30   20   10 | 10   20   30   40   50   60   70

Decreases (%)   Increases (%)

☐ 1974-75

▥ 1981

focused on the situation of workers and most studies have been devoted to them. But today a rapid computerisation of most types of white collar work is occurring up to quite skilled levels. There have been few studies in Sweden directed to this. A regional white-collar trade union did recently survey perceptions of new technology, but data have not yet been reported, although a few observations have been published.(6)

In this study 600 people were invited to answer and 562 of them did so. Sixty-seven per cent felt that dependence on computers is going to increase dramatically in the near future. Half of the sample expected this to lead to unemployment. The same number thought they would have negligible influence over the design of systems. About half also saw the risk that they would become more closely supervised through the new technology, and 45 per cent expected computerisation to sharpen the conflict between union and employers' organisations. In general, it was thought that the unions had little influence as compared with the big suppliers of hardware and systems - and in particular with foreign companies. Eighty-one per cent thought it necessary to use legislation to control this development. Attitudes were the same among men and women. Previous computer training tended to diminish anxiety.

Trade Union Opinion

As stated above, new technology has become an issue of common interest for both employers and unions. In Sweden, where union membership approaches 100 per cent, trade union policy is one important expression of public opinion. Trade union activity and policy should be seen in connection with the legislation on co-determination and on the working environment. Both the latter give the unions far-reaching opportunities to influence technical development. The Swedish trade union movement was alert at an early stage to technological development, and it has been the topic of several union congresses, the most recent issue having been computerisation. The congress of Landsorganisationen in Sverige (LO) - Swedish Trade Union Confederation, set up a council in 1976 concerned with the use of computers. Since then a number of investigations have been carried out and policies developed. In 1981 the council produced an LO-policy on computer applications.(7)

After the Second World War the LO attitude towards more productive technology was that it wished to secure an equitable distribution of the surplus created. The overall problem at the time was to raise the material standard of living for Swedish workers. The LO's attitude led to an acceptance of new technology - at that time work study and mechanisation - so long as it was possible to influence the distribution of wealth.

Since the early 1970s and with increasing intensity, the aim is now to influence development as such. While technological progress is seen as a necessary prerequisite for a prosperous and humane society, acceptance is today less unconditional. On the one hand, technological development is promoted and investment encouraged. On the other, the aim is to participate actively in the process of change. Quite apart from the distribution of income, the consequences of technology for workers' physical and psychological working conditions, the development of their competence, and the lessons for industrial democracy are all considered as key topics by the labour movement.

The LO's policy takes as its point of departure that technology can be utilised in various ways, which are detrimental or beneficial to different extents when judged against the social objectives LO wants to promote. In particular, there is an enormous potential in computer technology, provided one can have union influence on its application.

To fulfil the twin objectives of creating a competitive industry and providing satisfying working conditions, two important areas are pointed out:

• the more information people have, the more they can participate in discussions and decisions on production, which increases safety and security, reduces interruptions, and improves planning and production itself;
• tasks can be facilitated physically as well as mentally, so that people who today for various reasons are prohibited from work can participate in production.

Employment opportunity is one of the most basic issues for the Swedish labour movement, and again there are both opportunities and risks. For Sweden, with its unique co-operative traditions, the emphasis is on joint utilisation of the technology beneficial for the nation as a whole:

• investment in computer technology has to take place to secure welfare and development:
• profits from computer applications must be used for productive investment;
• the influence of wage earners on the direction and design of investment must be guaranteed; investment in computers must promote and develop the knowledge of those employed, and be used so as to favour democratic work arrangements;
• taxation must be changed so that society achieves enough income from an increasingly capital-intensive production system;
◦ changes due to new technology must not be faster than is socially acceptable;
◦ investment in computer technology must promote regional balance in the labour market.

The union of white-collar workers is taking a very similar stance to that of the LO. As regards computerisation, which so far has affected white-collar workers more than blue-collar ones, similar areas are emphasised. The point of departure is that 'high tech' is part of the progress of Swedish society, and is linked to the welfare state. Employment is to be secured with supplementary work to offset the effects of rationalisation, and regional aspects are to be taken into account. The benefits of computer technology, in connection with modern communications systems, are seen as providing an opportunity for a regional dispersion of jobs and for decentralisation. On the other hand, the use of home computers and the allocation of work to households is seen as a major threat to the values of the white-collar union, and is something to be fought with all available means. Employee participation, employer-sponsored training and union influence on the research stage of computer technology are all seen as means to safeguard beneficial development of computer applications.
With this as a starting point, the plan of action for the mid-1980s as decided by the union's congress in 1982 is in summary as follows:

the defence of meaningful employment;
• the requirement that organisations provide long-term investment and employment plans so as to secure measures to counter unemployment;

• the aim that employees be guaranteed co-determination on rationalisation and in procurement of equipment: development objectives and directives can then be formulated;
• legislation on working conditions should be expanded to encompass basic demands concerning work content, work organisation, the individual's influence on his own work, and working hours at computer terminals;
• training must be provided for employees in how to utilise computer technology to improve their work situation.

Final Remarks

The acceptance of technology by the individual and by the public needs to be seen against the background of the perception of technology. If one considers technology as it is expressed in a singular item like a new machine in the work place, or in a new electronic gadget, the impression will be quite different from the one that one gets if each item is seen as part of a general transformation affecting work, consumption, media, communication, leisure, political institutions, living habits - and indeed, all sectors of society.(8) When one adds up all the piecemeal changes connected with new technology in all spheres of life, the total amounts to quite a revolutionary picture. If one adds to this the hypothesis that the present crisis in Western society and its many institutions is a reflection of the decaying industrial era giving way to the new electronic age, then the public has a very fundamental change to come to terms with. The scenario might be frightening or optimistic, but so far the public has not been presented with any scenario to react to. The crisis is formulated in the language and concepts of the nineteenth century, while one should rather be trying to solve the problems and issues of the twenty-first.

On the other hand, it might be that the public, with its practical experience, is correctly sensing the change, while those in government have their cognitive maps petrified with the political inheritance from the Industrial Revolution. If this is so, one must expect to witness another transformation of society from below, during which the present power structure comes to be seen as progressively more inadequate, so that finally it is entirely rejected.

Notes

1. Zetterberg, H, Berg, I, Busch, K. and Frankel, G., 'Det svenska karnkraftsvalet 1980. Varderingar, politik och opinionsbildning', Sociologisk Forskning, vol 3-4, 1981, 26-69.

2. Johansson S., Preliminara resultat fran datoranvandningsundersokningen att presenteras pa statsradsberedningens utfragning om teknikoch sysselsattning den 20 aug 1984, Stockholm: Statiska Centralbyran, stencil, 1984.

3. Zetterberg et al., Det osynliga kontraktet. En studie i 80-talets arbetsliv, Stockholm: SIFO, 1983.

4. Ibid., p.44, (my translation).

5. Forslin, J., Soderlund, J. and Zackrisson, B., 'Automation and Work Organization - A Swedish Experience in Forslin', Sarapata and Whitehill (eds), Automation and Industrial Workers - a 15-Nation Perspective, Oxford: Pergamon Press, 1981.

6. Aulin, U., 'Dataangest vanlig bland tjansteman', TCO-tidningen 34/84, 1984, 7.

7. Facklig datapolitik, Stockholm: LO. Rapport fran LOs datarad, 1981.

8. Bell, D., The Coming of Post-Industrial Society. New York,: Basic Books Inc., 1873; Toffler, A., The Third Wave, New York: William Monrow & Company Inc., 1980.

Chapter Eleven

BELGIAN ATTITUDES TOWARDS NEW TECHNOLOGIES

Alain Eraly

Introduction

Western societies, it is often said, are rapidly entering a post industrial era. Scientific and technological advances have accelerated tremendously in recent years, extending their influence to all spheres of everyday life and shaping social relations. Scientists and engineers, however, continue to form an elite, and their number, even in a country as developed as Belgium, is severely limited. There is no doubt that this elite produces and reproduces highly selective images of modern technological realities, or that these images, amplified by the mass media, become part of the consciousness of all members of modern society. But it would be misleading to restrict one's consideration to this minority of intellectuals. Rather, the current technological explosion renders more crucial than ever an understanding of the attitudes and opinions of the public at large towards technology, as well as of the various groups and institutions which take a prominent part in the social formation of public opinion. The hyper-technological world which is emerging around us is increasingly becoming the 'natural', impersonal context of our lives. But does the public really want it, or understand it? Does it correspond to their hopes? Or conversely does it remain the creation of leading firms and social elites?

This chapter has two aims: one is descriptive and the other theoretical. First, we seek to describe public attitudes in Belgium: how do Belgians see the new technologies, what images do they have, and how do they react towards these technologies? Is it possible to identify general

302

differences among their opinions, and can we relate those empirical features to corresponding cultural and economic sections of Belgian society? Secondly, the chapter outlines a general interpretation of empirical data and, hence, a theoretical framework with regard to public acceptance of new technologies.

In Belgium interest in social attitudes towards new technologies is just beginning. There exist only a few limited surveys on the topic. The empirical foundations in which our diagnosis is grounded are of three kinds: first, we have analysed the results of the various surveys which have been recently conducted in Belgium; then, we have examined the coverage of technological phenomena by the media (newspapers, television and advertising) and by influential institutional actors (politicians, trade unions, employers' federations). Finally, we have integrated the conclusions of several sociological studies we have carried out over the last 3 years in large financial institutions. Those surveys have involved analysis of the attitudes and experiences of clerks whose working lives are becoming more intertwined with new technologies. As one might expect, many of our conclusions about Belgian attitudes could equally apply to other Western countries. Nevertheless, we have sought to describe not only the dominant trends but also the peculiarities of the Belgian case.

Attitudes Towards Technology

The Cognitive and Affective Distance from Science and Technological Progress

Belgians, like Europeans in general, see science and technology at a distance.(1) They speak of their lack both of scientific education and of any relationship with scientists and technologists. To illustrate this point let us consider Table 11.1, concerning attitudes towards computer technology:

Belgium

Table 11.1: Attitudes Towards Computer Technology

(n = 1,000)	Yes %	No %	No answer %
Have you heard about data processing?	83	16	1
Have you ever seen computer equipment at close quarters?	46	50	4
Have you ever used computer equipment?	18	78	4

Source: INUSOP, 1984.

It seems that even if a majority of Belgians (over the age of 18) are not ignorant of computers, most of them have never utilised data-processing equipment, and a majority has never even had the opportunity of approaching any sort of computer equipment! This same survey suggests that there exists some interesting relations between, on the one hand, acquaintance with information technology, and on the other hand, several variables: age: young people are more familiar with computers than are their elders; geography: people in large cities have a better knowledge of computer technology than do people elsewhere; and socio-economic class: higher classes are more frequently acquainted with the technology.

There exist several sections of the population - often outside cities, in Flanders and Wallonia - within which 90 per cent have never seen computer equipment! This general ignorance may seem astonishing in 1984. Be that as it may, we must keep this in mind when we interpret public attitudes. A lack of direct contact with technical realities may suggest a public susceptibility to accept any information about technology regardless of source, due to the difficulty of rectifying information through personal experience.

More profoundly, Belgians are under the impression that they have no hold on technological evolution, that they are not actors in this process which nevertheless pervades - and often threatens - their own everyday life. In our own surveys within large organisations we repeatedly observed this feeling of powerlessness. Technology is generally perceived as a vague and anonymous phenomenon, appearing to employees as external and unavoidable, as if it came from 'somewhere outside society'. Technological development is pictured as something like an historical law to which everybody has to adapt. Individually, people seldom believe that they can, in their own sphere, influence technological processes. They imagine a hierarchy of experts within universities and laboratories who are mostly unknown and whose competence is taken for granted, each expert analogous to the cell of an organ, to a piece of a gigantic machine. In the individual's everyday experience, technological development as a whole becomes incomprehensible and non-human, a mysterious set of forces which he can neither control nor comprehend. Because technology is more than ever a collective, or more precisely, a societal production, it appears mostly as impersonal and disembodied. People easily associate the technical advances of the past with the names of their famous inventors (Edison, Pasteur and Ford) but, paradoxically enough, in the age of computers and biotechnology, this is no longer the case, and people can seldom quote the name of one recent inventor.

The Desire to Take a Part in Technological Change

When invited to react to the following sentence: 'in order to guide scientific and technical research, one should take more account of public opinion, that is, the opinions of you and me', 60 per cent of Belgians agreed.(2)

They express, on the average, a need to participate in the orientation of technological development. Yet, the interpretation of these results is not as obvious as it many seem at first glance; this need to participate is more frequently expressed by: less educated groups; lower socio-economic classes (especially workers); and those politically inclined to the left.

We could claim that professional classes and managers agree less frequently with this statement (only 49 per cent) simply because they already do participate in technological processes and have more opportunities to enter into relationships with scientists and technicians. Our own surveys suggest that a somewhat different interpretation can usefully complement the former one: the need to participate in the orientation of technological processes is in fact related to fears regarding the

social consequences of those processes. Briefly
stated, people need to participate because they
would like to control technologies, and even
sometimes to impede their rise, particularly work
automation and nuclear energy. People want to
participate not so much because they morally and
affectively support technical progress, but rather
because they fear its consequences. Let us examine
this attitude in more detail.

The Ambivalence Towards Science and Technology

A profound ambivalence pervades most public
attitudes towards technologies. On the one hand,
66 per cent of Belgians continue to consider
science as one of the main factors in the
improvement of human life.(3) As an example, in
the case of automation, people characterise
computers by their usefulness (75 per cent), their
rapidity (74 per cent), and their precision (50 per
cent),(4) and 86 per cent remain convinced that
there are still many good things to be discovered
by science.(5) There still exists then, confidence
and hope as regards scientific and technological
development.

On the other hand, 62 per cent simultaneously
reaffirm the old distinction between science and
its applications; science is good in itself, it is
its utilisation which brings about numerous
problems.(6) Indeed, confidence and anxiety
characterise this double-sided attitude towards
technological progress and its impact on social
life. Belgians, like their fellow Europeans, have
a sharp sensation of the acceleration of change.
Seventy-six per cent of Belgians (65 per cent in
the overall European Community) consider that life
has changed greatly in the last 25 years, but among
them, only 45 per cent have the impression that
this change has been in the right direction (23 per
cent think that it has been in a negative
direction, and 25 per cent think that it depends on
which aspects of life are taken into consideration).(7) These results are supported by a survey
(8) of young Belgians, aged 17 to 23, which shows
that technology is perceived in the first place as
a threat. Even if young people express the
conviction that technology has made life easier,
this opinion is surpassed by a belief that
technology has not made people happier and that, in
the future, it will represent a danger for mankind.

This latter conviction is equally shared by the public in general: 60 per cent of Belgians concur in the belief that scientific and technological development involves even greater risks for society and that we will hardly be able to control these risks.(9) Belgians feel that such development 'goes too fast', and 54 per cent think many scientific discoveries are put into practice before their consequences have been sufficiently studied (20 per cent reject this assertion and 26 per cent did not answer).(10)

Let us examine, next, the type of concern which has been expressed towards new technologies. What exactly do Belgians dread?

The Threat of Unemployment

Recent surveys lead us to conclude that fear of employment currently dominates most psychological reactions towards new technologies. Obviously, this is connected with the economic crisis in Belgium; and although it can also be observed elsewhere in Europe, concern is so intense in Belgium that it dominates other fears (such as pollution or military applications of technology). One in two Belgians is obsessed with the idea of having to search for a job.(11) The most frequently cited work motivation is security of tenure. The proportion expressing this as a motivation nearly doubled between 1977 and 1982,(12) and it increased proportionally with age, reaching a maximum between the ages of 45 and 50.

The economic situation and the corresponding 'social deregulation' - a very fashionable expression in Belgium nowadays - constitute the background against which we must interpret Belgian attitudes towards new technologies. Indeed, the conviction that there exists an immediate causal relation between new technologies (especially the automation of work) and unemployment is deeply anchored. In Belgium, as well as in Luxembourg, France and Ireland, the majority of the population is primarily preoccupied with this problem. Sixty-nine per cent declare that they 'really worry about the increase of unemployment due to automation', and 38 per cent consider this problem as 'more disquieting than all others' (28 per cent mention pollution, 15 per cent the artificiality of everyday life, and 9 per cent the human risks caused by medical and pharmaceutical discoveries).(13) Young people, even more frequently than their elders, hold technology responsible for unemployment.(14) <u>In the public mind, technology is chronically seen as meaning the replacement of workers by machines.</u>

Thus, we should not be surprised that Belgium seems to yearn for 'nostalgia' more than the other European countries except Italy: 50 per cent of Belgians think that 'it would be nice if we could stop machines being produced and go back to Nature' (only 27 per cent refuse this nostalgic desire).(15) Let us note finally that women, in this respect, seem slightly more anxious than men, and that workers, more than other socio-economic sections of the population, show greater concern about increasing unemployment as a consequence of automation.

The Fear of Pollution

As contrasted with West Germany, Denmark, Great Britain and the Netherlands, Belgium seems more afraid of the impact of new technologies on employment than of pollution:(16) while 71 per cent of Belgians fear the destruction of nature, only 28 per cent give this problem first priority. Now there is no doubt that the ecological sensitivity of Belgian population has increased during the last 5 years. And this trend has been confirmed by most electoral results, local as well as national - ecological parties have made their entrance into the Belgian Parliament, the European Parliament and several communal councils. Furthermore, the recent shipwreck of the French <u>Mont-Louis</u> (whose cargo contained uranium hexafluoride) on the Belgian coast was intensively covered by the media, and undoubtedly affected public opinion. (The specific case of public attitudes towards nuclear energy is discussed below.)

Other Fears

Other anxieties regarding technology are more moderate. Apart from unemployment, pollution and nuclear power, Belgians do not appear particularly concerned about technological risks. As for genetic experiments, centralised data-bases (and their impact on human freedom), synthetic food, observation satellites, road accidents, to all these, Belgians seem more indifferent than, for example, do the Germans, the Danes and the Dutch.

Belgians' attitudes towards new technologies are not homogeneous but differ according to the various types of technology. This is why, below, we distinguish briefly between types, particularly attitudes towards information technology and nuclear energy.

Attitudes Towards New Technologies

A majority of Belgians declare themselves in favour of the continuation of medical and pharmaceutical research.(17) For example, 80 per cent consider that research concerning organ transplants is worthwhile and should be continued.(18) This domain of research - medicine and pharmacology - gives rise to a high-level of support because its goals are socially accepted, and also because people can relate this research to their own life: health remains a profound and dominant preoccupation and people have the opportunity to experience concretely the benefits of medical discoveries and technological advances in this domain. There exists a similar common assent with regard to the development of new energy sources, as well as with regard to the reduction and control of pollution. Conversely, when asked in which sectors they would be in favour of a limitation of research expenditure, Belgians cite such sectors as armament and national defence, or space exploitation. Biotechnology and new agricultural technologies remain nearly completely underlined unknown and their coverage in the mass media is insignificant. In fact, computers and nuclear power largely dominate the images people have of technology.

Attitudes Towards Nuclear Energy

In Belgium, as in other European countries, attitudes towards nuclear energy are a very special case. First, opinions constantly fluctuate, varying from year to year, and they have a high volatility. Secondly there is in this case no such thing as average opinion. Rather, there exist two opposing groups (for and against) with a large undecided mass in between. Several surveys were conducted between 1979 and 1983 which showed that, in general, opposition to nuclear power stations had increased in Belgium and that there existed significant regional differences. But they also showed that the proportion of favourable opinions had also increased everywhere, except in Wallonia. Hence, we clearly observe in Table 11.2 a process of polarisation of public opinion.

Table 11.2: Attitudes to Nuclear Energy 1979-83

	1978	1983
The country as a whole		
unfavourable	29	39
unconcerned	43	30
favourable	28	31
Brussels		
unfavourable	31	36
unconcerned	37	27
favourable	32	37
Wallonia		
unfavourable	33	46
unconcerned	37	30
favourable	30	24
Flanders		
unfavourable	26	30
unconcerned	47	29
favourable	27	41

Source: INUSOP, 1978-83.

In any case, nuclear energy continues to alarm most Belgians.(19) And if they are in favour of new energy forms, it is because those forms offer a possibility of limiting the development of nuclear energy. This attitude towards nuclear energy is peculiar because a contradiction exists between acknowledgement of the economic necessity of and opposition to nuclear power. The fact that one accepts the gravity of the social consequences of a possible lack of electricity due to the under development of nuclear power, does not imply that one will be favourable to nuclear energy.(20)

Belgium

Attitudes Towards Information Technology

Computers are in vogue in Belgium as everywhere
else. Not a day passes without newspapers devoting
articles to computers, universities organising
relevant conferences and so on. For most Belgians,
computers are both a hope and a deep concern. As
already mentioned, in spite of the diffusion of
computers throughout the various spheres of
activity, many Belgians continue to feel themselves
at a distance, both cognitive and practical, from
information technology, and they consider
themselves to be unable to bridge this gap. Here
may be a precursory sign of what sociologists
frequently call the 'dual society', a society
divided into a productive elite and a larger class
which, for various social, economic and educational
reasons, has no access to the 'high-tech' universe.

Computers and Employment

Some 65.6 per cent of Belgians believe that
computers cause a decrease in the number of jobs;
23.2 per cent foresee no impact from them; and for
11.2 per cent computers increase employment.(21)
Another survey (Table 11.3) yields similar results.

Table 11.3: Opinions on the Impact of Computers

(n = 1000)	Agree %	Do not agree %	No answer %
Computer technology			
creates unemployment	78	16	6
increasingly replaces man in the work place	82	12	6
increase the pace of work	71	21	8

Source: INUSOP, 1984.

Although in general the belief that computers
create unemployment is held by a majority, there
exist significant regional - and thus, cultural and
economic differences. Among the Walloons, 78 per
cent are convinced of a negative impact, but only
58 per cent share this opinion in Flanders and in

Brussels. This difference is even more significant if we take into consideration only particular subgroups of managers and employers: 18 per cent in Wallonia and Brussels believe that the introduction of computers will result in increased employment; more optimistically, 51 per cent in Flanders hold the view.

These data are significant: they reflect a basic segmentation which is simultaneously linguistic and cultural, social and economic, and represents Belgium's historical inheritance. The duality between the Flemish north and the French-speaking south affects attitudes such as those towards new technologies, as evidenced in the greater pessimism expressed in Wallonia.

In any case, it is to be expected that opinions on technologies will not be independent of the social and economic conditions of the individuals who express them. The regions which are more directly exposed to the economic crisis (in Belgium, that means all Wallonia and the province of Limburg) evidently tend to be more pessimistic and distrustful regarding new technologies, and to attach exaggerated importance to computers.

The same conclusion applies to social classes. Indeed, if we now divide the population according to socio-economic levels, we observe that it is more frequently women and unskilled workers who are convinced of the negative impacts of computers on employment; economically vulnerable categories thus feel themselves potentially threatened by computer technology. Conversely, skilled workers, managers and employers are significantly less concerned about its impact. In short, belief in the negative role of computers as regards employment is not independent of the circumstances of work. It is not simply a question of public opinion and political debate, of ways of thinking and their shaping by the media. Opinions are always connected to the everyday experience of individuals, the uncertainties they currently experience in their working lives and, their concerns about their own future. For a large portion of the public, computers represent both a subjective and an objective threat. For another portion, economically and (consequently) intellectually privileged, computers are viewed as opportunities, as a means of social advance and economic benefit.

The Impact of Computers on Social Life

How do Belgians see the consequences of computers with regard to human relations, outside and inside organisations? In this matter they certainly display less pessimism.(22) For example, with the exception of an active and informed minority (often those on the political left and ecologists), the man in the street is not particularly concerned about the growing artificiality of everyday life, or about the risks of a restricted freedom implied by the utilisation of centralised data-bases and their connections (in Belgium, more particularly the National Registers and the information system of the police). In short, the average Belgian does not think that Orwell's world is just around the corner.

Table 11.4: The Impact of the Introduction of Computers

	Worsens %	Does not change %	Improves %	No answer %
The relations between workers	35.1	47.2	15.6	2.1
The relations between workers and management	36.6	44.1	15.6	3.6
The relations between workers and the public	28.4	48.4	19.1	4.1

Source: Pourquoi pas?, Chambre belge de la mecanographie, ICSOP, 1984

Table 11.4 shows that even if a significant minority believes that computers have negative impact on social relations, most Belgians reject this. However, 50 per cent feel that in large industries human relations mostly deteriorate as a result of computer technology. Finally, it should also be stated that if Belgians are not particularly concerned about the deterioration of social relations due to computers, they are also not optimistic. They seem reluctant to devote lyrical discourses to the emergence of an Information Age, a great political and economic decentralisation sustained by computers, a 'convivial and transparent society', and so on. In spite of the sensationalism of the mass media, Belgians remain sceptical and do not rest their hopes and illusions on the current data processing 'revolution'.

The Place of Computers in Everyday Life

In which sectors of the economy do Belgians think computer-based progress has been most rapid? Above all, they cite the banks (75.4 per cent of people questioned), large industries (69.3 per cent), commercial firms (54.1 per cent) government services (49.1 per cent), and supermarkets (44.7 per cent),(23) that is either sectors in which technological advances have largely been covered by the media (large industries, banks and more recently, government services), or sectors in which the public directly experiences the new systems (banks, supermarkets).

Further, let us examine in Table 11.5 a set of reactions towards computers.

Table 11.5: Public Opinion Towards Computers

(n = 1,000)	Agree %	Do not agree %	No answer %
Computers			
save time	90	3	7
improve everyday life	50	44	6
offer more advantages than disadvantages	64	28	8
improve the quality of public services	54	39	7

Source: INUSOP, 1984.

We note from Table 11.3 that the majority of Belgians are convinced of the usefulness of computers.

Attitudes towards the quality of public services merit further clarification. The majority of Belgians question the efficiency of government administration. Since the coming to power of a right-wing government, there have been several reports, each concluding that public administration is bureaucratic, rigid, inefficient and poorly managed (those studies concerned, among others, the Office of Employment, the postal services, urban transport, national education and the railways). These conclusions have been widely reported - and sometimes amplified - by the media, reinforcing public distrust of public administration.

Now while Belgians seem convinced that the introduction of computers in administration will improve the quality of public services, it is doubtful whether opinion is grounded on actual experience. People probably just juxtapose computers 'which go fast' with administration where 'everything goes slowly'. In addition, newspapers have reported the results of recent studies showing the lack of data-processing equipment, and more significantly, of overall technological strategy within public administration: government agencies are seemingly under equipped with computer-based systems. All this may explain the apparent trust Belgians place in computerisation of public services.

We see that, at a very general level, Belgians look favourably upon the consequences of computer technology in everyday life. Once again, however, their attitudes contain an ambivalence. This appears in Table 11.6.

Table 11.6: Opinions About Computers

(n = 1,000)	Agree %	Do not agree %	No answer %
Computer technology			
deeply transformed habits	76	18	6
is made for young people	54	40	6
requires long study before one understands anything about it	45	49	6

Source: INUSOP, 1984.

Here we observe the expression of those feelings of exclusion and resignation described above. To a certain extent, to say that computers are the province of the young is equivalent to excluding oneself, consciously or not, from technological progress, leaving it to the rising generation. Similarly, to believe - as 45 per cent do - that only technically educated people can utilise computers indicates a tendency to feel excluded from the technological game, in spite of the current development of personal computer technology.

And what about private life? Do Belgians expect to integrate computers within their family lives and leisure activities? This seems unlikely. Only 4 per cent of the Belgian population has considered the possibility of buying a home-computer in the near future.(24) Fifty-three per cent rule out the idea of using personal computers to amuse themselves during their leisure time (39 per cent think it could be useful). 49 per cent do not imagine utilising computers to manage their domestic budgets (44 per cent think it could be useful).(25) These data indicate that if most people seem ready to admit the utility of computers within the working sphere, they are far from accepting - or even imagining - introduction of them in their own private lives.

Finally, there is a similar ambivalence towards the introduction of computers at school. Here also, distance and ignorance play an important and sometimes surprising role. Parents of school-aged children were asked whether computers had been introduced in their children's schools in order to support teaching activities.(26) There were negative answers from 35.2 per cent, 28.6 per cent answered positively - and the remainder, that is 36.2 per cent, simply did not know! This indicates an important lack of information. This lack is even more significant in what follows. Those whose children went to schools where computers have actually been introduced, were asked whether their children had already received the opportunity to make use of those computers: 13.5 per cent answered yes, 5.4 per cent no - and 81.1 per cent were incapable of answering! More than ignorance, this suggests that parents do not take an interest in the introduction of computers at school. Nevertheless, 49 per cent of Belgians consider the introduction of computers in primary school useful (44 per cent consider it useless) and 70 per cent

consider introduction in secondary schools useful
(23 per cent consider it useless).(27)

The Social Production of Attitudes Towards Technology

Fundamentally, we can study the process by which
images and values about technologies are socially
produced and reproduced at three complementary
levels:

• a socio-psychological level: the analysis in
this can seek to comprehend and define the basic
dimensions which constitute attitudes, and to
connect them with broader psychological systems
such as style of life, personality or ideology;
. a socio-practical level: the focus here is on
the analysis of the concrete social systems in
which the actor participates: the tasks he
performs, the position he holds within an
organisation, his relationships, etc.,
• the societal level: here, we study the global
institutional context shaping attitudes, the
propaganda of the various institutional actors
(employers, politicians, trade unions, etc.),
political responses, etc. In what follows, we will
briefly develop these three levels. But our
conclusions must not be taken as absolute
statements but rather as programmes for further
research.

The Multidimensionality of Attitudes

Technologies, and more specifically computers,
bear large semantic diversity. People
spontaneously associate a wide range of mental
images with the word 'computer'. This word may
refer to a concrete tool, but it can also symbolise
a definite type of society. The diversity of
attitudes towards technology can be understood in
terms of a corresponding diversity of images. And
apparent contradictions among the attitudes of an
individual (e.g. acceptance combined with
hostility) can be due to conflicting images. (We
often observed, for example, employees who enjoyed
using a computer in their daily lives, and , at the
same time feared the rise of a Computerised
Society; as a consequence, those employees showed a
particular ambivalence towards computer

technology.)

Rosseel,(28) for example, shows that attitudes towards computers conceal a diversity of aspects. He distinguishes two main dimensions:

- a horizontal dimension: how do people value computers (for example, unskilled female employees more frequently express negative valuations than skilled male employees);
- a vertical dimension which opposes realistic conceptions to mythical (projective) ones.

The author uses 'Small Space Analysis', a technique for processing data which permits the representation of a spatial configuration. Rosseel obtains the results shown in Figure 11.1.

Figure 11.1   The Results of Rosseel's Analysis

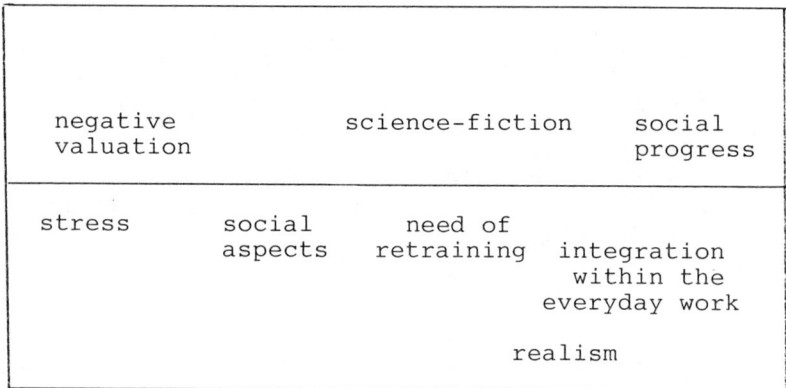

negative valuation		science-fiction	social progress
stress	social aspects	need of retraining	integration within the everyday work
		realism	

The left part of the diagram brings together negative attitudes vis-a-vis computers; one cluster of items expresses the conviction that working with a computer is, in itself, unacceptable; a second cluster expresses the fears of increased stress due to the new systems. In the central part, we find the 'cognitive' representations of computers; at the top, the 'Orwellian' conception of a computerised society; at the bottom, a more realistic conception, which views the computer simply as a powerful machine with, in between, acknowledgement of possible impacts on employment and of the new technical skills to be acquired. The right part of the diagram comprises the positive attitudes: at the top, an enthusiastic support whereby the computer is seen as a means of accelerating one's social promotion; and finally, at the bottom, a positive attitude which seeks harmoniously to integrate computers in everyday activities so as to improve work performance and the quality of working life.

We observe, then, that attitudes towards computers are not simply a question of being in favour or against: there exists a large variety of dimensions. Rosseel also analyses the connections between attitudes towards technologies and broader attitudes towards life in general. As an example, he observes that the negative valuation of computers is integrated within a more global sensation of social powerlessness and pessimism with regard to the future, and that it is associated with the expression of 'traditional' motivations (family life, stable job, etc.). A positive valuation is often connected with the search for new lifestyles or with 'neoliberal' convictions. Rosseel's studies show the critical importance of the wider cultural patterns of specific social groups in determining the genesis of attitudes towards technologies. Yet, one should not overestimate the importance of those broad cultural determinants. Attitudes towards technologies are not simply a question of socialisation, but also of practical experimentation and individual strategies.

The Effects of Practical Experiences on Attitudes

Our own surveys suggest that attitudes towards computer technology depend on whether the individual customarily utilises computers. As compared with uninitiated people, computer users appear significantly more favourable to new technologies. But this is not the sole effect of practical technological experience on attitudes. Among computer users, particularly those who have utilised them for a few years, we frequently observe a process of 'affective neutralisation'. In the course of time people attach less importance to computers within the overall set of their concerns. The new technology is no longer mythological, either in a negative or in a positive sense. The computer becomes one element among others within the working life. In other words, the image of the actual system takes the place of abstract and global representations, and employees utilise their own concrete experience to modify and reconstruct their attitudes.

Organisational Strategies

Our surveys yield another conclusion: the overall experience of computer technology varies very much according to the position (power, status, etc.) that one holds within an organisation. Here, attitudes towards technology can be seen as a sort of 'bargaining relationship over time ' (29); it always involves an aspect of <u>individual strategy</u>. In order to understand an attitude, one has to comprehend the social framework of constraints which the actor cannot easily change and within which he plays a specific game. Differences in attitudes are rooted in different organisational positions and games. The actors become more favourable towards technology when they view new systems as a means to improve their own positions. Conversely, criticism of the technology is often displayed by those who think computers could make them lose their degrees of freedom, or influence or status.

The Societal Context of the Attitudes Towards Technology

Finally, analysis of attitudes towards technology should ideally take into account the social production of the images of this technology. Political parties, trade unions, federations of employers and firms are among the main producers of information about technology. The media echo their discourses and add their own analysis. This is not the place to study the resulting enormous output in detail. One conclusion, however, commands attention: evidently, the overall representations of technology which are constructed by institutional actors are intimately connected with the political and economic strategies and interests of those actors. For example, trade unions lay stress upon the relationship between technology and unemployment, and the reduction of working hours. Employers put the emphasis on the link between technology and economic recovery, as well as on the maintenance of productivity. Private firms themselves contribute actively to the shaping of public opinion. For example, in the case of large contracts with the public sector, they seek to influence public opinion in order to put pressure upon politicians: they insert advertisements in newspapers, organise press conferences, invite journalists to visit their factories, and so on. Within this information battlefield, public opinion appears both selective and volatile. As we have seen, the Belgian population is relatively uninterested in technological problems in general. Nevertherless, it sometimes concentrates on specific issues, selected for their social 'visibility', as for example, the shipwreck of the Mont-Louis, the possible construction of a new nuclear power station, or the introduction of robots in a large automobile factory. The catastrophe of Bhopal led journalists to take an interest, for a few days, in the chemical industry in Antwerp. And currently, the replacement of ancient telephone network centres is the occasion of an important linguistic conflict, the Walloons refusing to abandon to GTE-ATEA and BELL Telephone - two firms situated in Flanders - the whole of this gigantic contract. On each occasion it is not technology in itself which attracts the interest of public opinion, it is rather the sensational character of a specific issue and its connections with deeper cultural and social concerns. In seeking to understand attitudes towards technology, one should always remember the wider cultural and social roots of those attitudes. Different images of technology find their origins in distinctive social experiences.

. To influence the shaping of public attitudes, government has a large range of possibilities, both economic and educational. There exist various forms of public aid for firms investing in new technologies. Public agencies also organise numerous exhibitions of new equipment. The Ministry of Education seeks to shape attitudes by introducing new technologies within various educational programmes, and this policy is also followed by the Ministry of Employment for the retraining of the unemployed. Finally, one should not forget that government is itself a major user of technologies. Provided that it is able to plan for its future needs and that it develops a coherent policy regarding the place of technology within public services in the future, it may enter into long-term contracts with technological firms and universities so as to promote technological research and innovation. In Belgium, unfortunately, such a general public policy has proved to be difficult to develop. Government, as a whole, has never succeeded in adopting any form of general planning regarding its technological needs. Moreover, the current national deficit severely restrains the possibility of the government's engaging in expensive large-scale strategies.

Conclusion

Belgians, as we have seen, are both distant and ambivalent towards new technologies, and many of them are quite pessimistic regarding the future. The fear of unemployment dominates most of their attitudes. Apart from medical care, which is an atypical case, their attitudes are quite negative. Now, this does not mean that Belgians <u>resist</u> technology. On the contrary, up to now no major industrial conflict has been caused by the introduction of new systems. Attitudes do not coincide with behaviour, so that pessimism does not mean hostility. Technologies are currently under development in Belgium as anywhere else. Besides, to resist, one needs the power to resist. The degradation of the labour market following the relative decline of the trade unions has led to a progressive re-centralisation of Belgian society. This trend has possibly rendered social consensus about technological decisions less important to obtain, from the employers' point of view. At this point, a distinction could be drawn between technological investments leading to organisational rationalisation and those creating new products and services. No consensus can be obtained concerning the former, because of their impact on employment. Up to now a large number of technological investments have promoted rationalisation, that is, they have increased productivity <u>and</u> unemployment. To believe that public attitudes play an important role in technological advance is to assume that the public has some power to decide. Obviously, this is true to only a limited extent. Technological decisions continue to be taken by an elite, and the attempt to create positive attitudes often appears as a means of diminishing social resistance.

Although the promotion of new technology is one of the main goals of the Belgian government, the government has never succeeded in defining a general and coherent policy with regard to public acceptance of new technologies and public participation in technological choices. Hence, we can expect that for the majority of the Belgian population, technology will remain, in the future more than ever, a threatening inevitability.

Notes

1.   Commission of the European Communities (ECC), DIMARSO, 1979.

2.   Ibid.

3.   Ibid.

4.   INUSOP, 1984.

5.   ECC, DIMARSO, 1977.

6.   ECC, DIMARSO, 1979.

7.   ECC, DIMARSO, 1977.

8.   INUSOP, 1980.

9.   ECC, DIMARSO, 1979.

10.   Ibid.

11.   SOBEMAP, 1980.

12.   SOBEMAP, 1977, 1982.

13.   ECC, DIMARSO, 1979.

14.   INUSOP, 1980.

15.   ECC, DIMARSO, 1979.

16.   Ibid.

17.   ECC, DIMARSO, 1977.

18.   ECC, DIMARSO, 1979.

19.   62 per cent of young people between the ages of 17 and 23 are opposed to nuclear energy (INUSOP, 1980).

20.   ECC, DIMARSO, 1979.

21.   _Pourquoi pas?_, Chambre belge de la mecanographie, ICSOP, 1984.

22. INUSOP, 1984; ECC, DIMARSO, 1979.

23. Pourquoi pas?, Chambre belge de la mecanographie, ICSOP, 1984.

24. Ibid.

25. INUSOP, 1984.

26. Pourquoi pas?, Chambre belge de la mecanographie, ICSOP, 1984.

27. INUSOP, 1984.

28. E. Rosseel, Quelques etudes sur les attitudes vis-a-vis des technologies nouvelles, Brussels: Vrije Universiteit, 1984.

29. M. Crozier, 'Comparing Structures and Comparing Games' in G. Hofstede and M.S. Kassem (eds.), European Contributions to Organization Theory, Assen: Van Gorcum, 1976, pp. 193-207; M. Crozier and E. Friedberg, L'acteur et le systeme, Paris: Ed. du Seuil, 1977.

Chapter Twelve.

PUBLIC ACCEPTANCE OF NEW TECHNOLOGIES IN THE NETHERLANDS
Jan Berting

Social Change and Public Acceptance of New Technologies

Changes in the public acceptance of new technologies are interdependent with changes of perspective on societal development. For that reason I will discuss briefly the sequence of perspectives on societal development that can be observed in the Netherlands since the end of the Second World War. Four perspectives on societal changes will be distinguished, each connected with a specific type of public acceptance of new technologies.

1945-63 Faith in the Benevolent Long-term Social Effects of Industrialisation

The first period in our description of changing perspectives (1945-63) covers the reconstruction of the empire, when an intensive industrialisation programme was launched in order to create new economic opportunities.

Until well into the 1960s the dominant perspective on societal development in the Netherlands was an optimistic one. Economic growth, based on industrialisation and propelled by science and technology, promised a gradual unilinear evolution towards a more open society with the following characteristics: increasing individual occupational and social mobility together with a growing equality of educational opportunities, a fading away of differences based on class and lifestyles, a concomitant strong growth of the middle classes as a consequence of the increasing demand for highly skilled and professional workers and, consequently, a decrease of collective types of antagonism, especially of class struggle.

In the world of work the majority of the working population accepted the introduction of new technologies as a normal element in the process of modernisation. Technological determinism was the prevailing idea as regards the role of technology in society. As the long-term consequences of technological development on society were generally looked upon as favourable for social life, adaptation to the technological and concomitant organisational changes was seen as a logical requirement. But political and economic elites seemed to be more strongly convinced of the need for workers to adjust to changing conditions than were many categories of blue and white-collar workers. This was exemplified by the plethora of books, seminars and training courses during those years on the general theme of 'worker resistance to change, and how to overcome it'. Nevertheless, resistance to technological change at this time was mild by comparison with the rejection of some technological changes which were to appear later. Technological development went hand in hand with full employment, increasing purchasing power for the majority and the development of the social security system.

1963-72: a Growing Awareness of the Costs of Economic Growth

In the following decade (1963-72) there was a growing awareness of the limits of economic growth. Particularly, city dwellers in the heavily populated, industrialised western part of the country were confronted with a decrease in their quality of life as a consequence of air pollution (SO_2), traffic congestion in the inner cities, and the lagging behind of recreational facilities in relation to the increasing needs. Moreover, some major social problems had not been adequately solved in the preceding period: there was still a severe housing shortage, and social inequality had not been reduced to the extent that many had hoped for. The Netherlands was apparently not on its way to becoming a social paradise. Lastly, many felt that they had too little influence on the decision-making process at both the national and local levels. In this period there was a rather sudden shift from a relatively submissive attitude towards the political authorities to a more aggressive one. The generation that grew to maturity in the 1960s strongly rejected existing patterns of social relationships, including the hierarchy of life styles.

In this period the acceptance of new technologies was not as important public issue, although small, dispersed groups were trying to convince public opinion of the degrading effects on the natural habitat of uncontrolled industrialisation. Faith in the contribution of the natural sciences and technology to the solution of major social problems was clearly waning. In this respect the Netherlands was conforming to the pattern described by Galtung in the comparative ten-nations study Images of the World in the Year 2000, based on data collected in 1967: 'The countries with a high level of technical and scientific development seem somehow to say "so far, but no further"'. It was as if science scepticism had set in at a certain level. 'This science scepticism is particularly focused on the inner sphere of human existence and particularly pronounced in the nations most developed in terms of science and technology.' (1) Moreover, Galtung concluded that: 'Techno-economic development is not reinforced by growing optimism, but rather seems to lead to growing scepticism and pessimism'.(2) In this context it is noteworthy that the publication of The Limits to Growth - A Report for the Club of Rome Project on the Predicament of Mankind in 1972, was followed in the Netherlands by intensive discussion both within the scientific communities and by the public at large. Furthermore, the feeling of being extremely vulnerable as a small country to changes in international relationships was markedly deepened by the energy crisis of 1973.

1972-9: Belief in the Malleability of Social Life

Growing scepticism and pessimism did not apply - at least in the Netherlands - to the role of the social sciences. On the contrary, the period 1972-79 can be characterised by a strong belief in the malleability of social life by political decision-making with the assistance of a contribution of the social sciences.

Although there was a rising awareness, shared by the general public, of important, unintended and undesirable consequences arising from growth, a belief was also growing in the possibility of social solutions to the main problems facing society. Technological determinism and the concomitant need to adapt social life to the requirements of technological development were no longer accepted. The dominant perspective in this period stressed the primacy of social life and the opportunity of adapting technology to its requirements. The conviction prevailed that societal development was controllable through planning. In this period the Scientific Board to the Government (Wetenschappelijke Raad voor het Regeringsbeleid) was founded (1972). This Board was to inform the Dutch government about those developments likely to have long-term effects on society and to formulate policy-oriented proposals related to those developments. Two years later the Social and Cultural Planning Agency (Sociaal en Cultureel Planbureau) was founded. This agency was given as its main task a responsibility to undertake scientific research in order to arrive at a systematic description of the social and cultural welfare situation in the Netherlands, and of expected developments within this field. The tasks of these agencies were clearly connected with the dominant perspective on societal change of this period.

An important goal in this period was an improvement in the quality of democratic life by extending the opportunities for political participation (participatory democracy), especially at the local level, in order to restore the legitimacy of the political system. This development was accompanied by strong growth of the agencies of the welfare state. This latter development is illustrated by the enormous rise of social security expenditures between 1970 and 1980: from 16,756 million florins to 70,920 (i.e. from 15 per cent to 23 per cent of the national income).

1979-84: Declining Faith in the Malleability of Social Life. Restoration of Technological Determinism?

In the latest period, 1979-84, faith in the malleability of social developments has weakened. Since 1982 social security expenditure has been stabilised at 24 per cent of the national income. The number of unemployed has risen from 210,000 in 1979 to 825,000 in 1984 (15.5 per cent of the dependent working population).

The public acceptance of new technologies came to occupy a more prominent place as a political issue than in the preceding period. In 1980 a debate on the societal consequences of microelectronics started after the publication of the report on Microelectronics and society by the Advisory group to the Minister of Science Policy.(4) A considerable number of working documents, reports and books were published to inform policy-makers, professional workers and the general public. In 1983, a nationwide debate was organised on the future energy policy of the Netherlands.

At the present time one can observe the following general trends:

a growing, but reluctant support amongst the working population for the introduction of new types of information technology at work;

a rather enthusiastic acceptance, especially amongst the younger generation of the opportunities provided by personal computers, and a stronger orientation towards educational systems related to developments in information technology;

increased efforts by policy-makers to reinforce the public acceptance of new and controversial technologies; moreover, more attention is now being paid in the mass media to the positive effects of technological innovations;

there has been no restoration of the hard-core technological determinism of 1945-63; efforts are being made to learn from failures of policies which were based on the dominant perspectives of preceding periods.

Introduction of New Technologies not a Major Societal Problem

The introduction of new technologies does not as such show up as a major societal problem in the public opinion polls. Unemployment and housing problems were seen by the public as the major social issues of the 1960s. Environmental problems were barely mentioned. At the beginning of the 1970s, problems connected with the pollution of the land, air and water came to the fore. In the 1970s those types of problems were considered to be the most important ones by 45 per cent of the adult population, compared with housing problems: 42 per cent; wages, prices, inflation: 13 per cent; employment and taxes: 13 per cent; traffic: 13 per cent; political problems: 10 per cent.

Table 12.1: Important Problems and Their Relative Weight 1972-82 (in per cent)

Politically:	1972/3			1977			1981			1982		
	not active	All active	very active	not active	All active	very active	not active	All active	very active	not active	All active	very active
Types of problems												
a) employment	12.5	10.9 *	6.5	41.2	37.8 *	28.5	40.3	34.3 *	24.6	49.6	43.7 *	32.3
b) financial and economic	2.8	3.2	4.0	3.6	4.9 *	8.4	8.5	10.6	11.5	12.8	14.8 *	20.4
c) wages and prices	14.3	13.6	13.4	12.8	11.4 *	10.3	4.5	5.0	4.8	3.1	2.6 *	1.3
d) taxes and social security	3.9	3.6	3.7	1.7	1.6	1.4	0.5	0.8	0.5	0.7	0.5	0.6
e) welfare state	6.6	7.4	5.8	8.3	8.5	10.9	6.8	7.0	6.4	5.7	5.4	6.3
f) education, culture, recreation	1.7	2.1	2.9	1.2	1.4	1.7	0.5	0.7	2.1	0.7	0.5	1.0
g) housing	13.7	12.3	13.1	5.9	6.7	7.6	9.6	8.8 *	9.8	1.8	2.6	2.6
h) environment	22.4	23.9	24.8	5.4	7.3 *	10.7	4.2	5.0 *	7.0	1.3	2.1	2.6
i) law and order	3.0	2.2	2.0	5.2	4.5	2.4	6.0	4.6 *	2.8	5.4	5.4	2.8
j) foreign relations and defence	1.2	2.6 *	5.5	1.8	1.8	0.7	3.0	3.9 *	6.0	7.0	8.9 *	10.8
k) energy	0.0	0.0	0.0	2.2	2.4	2.3	5.8	7.2 *	9.4	2.8	3.2	3.7
l) politics and democracy	9.7	10.3	10.3	2.3	3.0 *	5.0	2.1	2.9 *	5.2	3.5	5.1 *	7.8
m) values and norms	3.5	3.6	4.2	4.4	5.2	7.3	4.4	5.4	6.9	1.9	1.8	3.6
n) (ethnic) minorities	4.7	4.3	4.0	3.9	3.5	2.9	4.0	3.7	3.1	3.6	3.4	4.3
Total	100	100	100	100	100	100	100	100	100	100	100	100
Number of respondents	311	970	96	539	1,370	137	594	1,578	158	506	1,504	149

Note: Significant difference at or above 0.05 indicated by an asterisk. Source: Sociaal en Cultureel Rapport 1984 (Social and Cultural Report 1984), Den Haag: Staatsuitgeverij, 1984, p. 328.
Source of data: Nationaal Kiezersonderzoeken (National Research on Voting Behaviour) 1972/73, 1977, 1982 en 1982.

Table 12.1 gives data concerning the changes in public opinion with regard to the major social or societal problems between 1972 and 1982.

Between 1972 and 1982 the importance of problems connected with wages, prices, housing and the environment decreased. Employment became a very important problem - a reflection of the strong increase in the number of unemployed workers - and aslo more importance became attached to financial and economic problems, problems of law and order, and problems related to foreign relations and defence. The difference between those politically active and those not are seen to be relatively small, and both categories show the same trends. Athough the importance of environmental problems became somewhat overshadowed by the problems generated by economic stagnation, the support for government-financed environmental programmes remains strong, as appears from Table 12.2.

Table 12.2: Support for Government Financing of the Fight Against Environmental Pollution 1970-82 (in per cent)

	1970	1975	1978	1979	1980	1981	1982
Support for drastic intervention against environmental pollution:							
strongly agree	71.3	45.2	43.9	41.2	41.8	41.9	40.3
agree	25.0	43.8	41.7	48.0	49.1	46.4	42.9
neither agree, nor disagree	2.4	7.7	9.0	6.8	6.5	6.7	2.7
disagree and strongly disagree combined	1.3	3.3	5.4	4.0	2.6	5.0	4.3
Number of respondents	1,819	1,755	909	1,797	1,805	1,702	1,749

Source: onderzoek 'Progressiviteit en Conservatisme' (Research on Progressive orientations and conservatism), 1970; 'Culturele veranderingen in Nederland' (Cultural changes in

the Netherlands)' 1958-75,   1978, 1980, 1981, 1983,
reported   in   Sociaal en Cultureel Rapport   1984,
Table 11.14, page 290.

In   1983 89.4 per cent of the respondents   aged
16-24   supported or strongly supported   government-
financed measures against environment pollution, as
did 87.6 per cent of respondents aged 25 or more.

During   the whole period   under   consideration,
environmental   problems were discussed in the   mass
media.   At the beginning of the 1970s attention was
primarily   directed   at   air and   noise   pollution.
After   a   number   of   restrictive   measures,   more
cautious   behaviour by the   industries   responsible
for   some   of   the easily traceable types   of   air
pollution,   for instance,   oil refineries,   and the
setting   up of warning systems together reduced the
number of complaints.   The attention of the general
public   then shifted to the pollution of the   land,
caused   by irregular dumping of   toxic   chemicals,
especially   after   it was discovered that   several
newly   built   living-quarters were built   on   sites
that   were so contaminated that the health of   the
people   living there was endangered.   In the   same
period   the   attention   of the general   public   was
several   times   attracted   to   the   dangerous
consequences of the alarmingly increasing pollution
of   the   main   rivers,   especially   the   Rhine,   by
industrial   waste   from French,   German   and   Dutch
industries.   Since   the   1970s the quality of   the
Rhine's water has improved somewhat.   Nevertheless,
the contribution of French potassium mining to   the
pollution   of the Rhine is still considerable,   and
intensive   political   negotiations have   failed   to
produce   an   effective solution   to   this   problem.
Another   serious   problem which still bothers   both
the   authorities   and   the general   public   is   the
poisoning   of   the Rhine's waters by   heavy   metals
such   as   mercury   and   cadmium,   pollution   that
seriously   affects the life of many types of   birds
and   seals in the Dutch Shallows,   one of the   most
important   nature reserves in Europe.   The surface
of   the   water is also exposed to   the   detrimental
effects   of   phosphates   resulting   from   use   of
detergents.

Until   now   the   measures   undertaken   by   the
authorities to protect the environment in the areas
mentioned   have not been very effective.   And   the
feeling   prevails   in the Netherlands that there has
been   a   negative development with   regard   to   the
environment.   Many   people have reported that they

have been troubled personally by water pollution (see Table 12.3).

Table 12.3: Percentage of Respondents in Different Age Groups Claiming to have been Personally Troubled by Water Pollution in 1982

18 to 24 years of age: 22	45 to 54 years of age: 5
25 to 34 : 11	55 to 64 : 8
35 to 44 : 4	65 and over : 4

Source: NIPO, Public opinion poll 1982 (samples: 992 and 1094 interviews respectively).

Nevertheless, the legislation to protect the environment has been extended since the beginning of the 1970s, new taxes have been imposed, the yield from which is used to counter detrimental effects on the environment, and the research effort connected with environmental problems has been augmented (Table 12.4).

Table 12.4: In-house Expenditure on Research on Environmental Problems 1973-81 (in millions of guilders)

	1973	1974	1975	1976	1977	1978	1979	1980	1981
Industry	−	−	−	−	39	42	36	33	50
Research institutions (predominantly governmental or semi-governmental)	38	49	61	63	63	85	85	91	87
Total (excluding universities)	−	−	−	−	102	127	121	124	137
Research on environmental problems as a percentage of total government-financed research and development	−	−	2.5	2.9	3.2	3.3	3.8	2.9	2.3

Source: Centraal Bureau voor de Statistiek, Kwartaalbericht milieu (Central Statistical Office, Quarterly communication on environment), 1984, 1,. vol 1, no. 1, 12.

The sum of total expenses on environment in 1982 was 3,915 million guilders; and the total estimate for 1985 was 5,025 million guilders.(5)

It is apparent from Table 12.5 that the environmental costs of industries are increasing. In 1982 about 40 per cent of the total expenses on environment were made by industry (excluding building industries).

Table 12.5: Environmental Cost of Industries (Excluding Building Industries) in Millions of Guilders

	1979	1980	1981	1982
Water pollution	491	565	582	627
Air pollution	387	456	494	453
Waste	154	174	182	179
Other	203	205	243	254
Total	1,234	1,400	1,500	1,513

Although many initiatives to protect the environment have been institutionalised, the role of groups of activists, on both national and local levels remains very important for the general public's awareness of the environmental effects of pollution and of the potential dangers from the industrial application of new industrial activities, such as offshore drilling, the installation of Liquid Natural Gas Terminals, and recombinant DNA-research.

In this context, one very successful outcome of organised protest against the potential effects of a large scale project should be mentioned. After the flood disaster of 1953, the Dutch parliament came to a number of very drastic decisions to safeguard the population of the south west regions of the Netherlands. One of these decisions was the closing of the branches of the Scheldt estuary (the Western Scheldt excepted) to the tidal influences of the North Sea. However, it seemed that the planned construction of a huge dam in the Eastern Scheldt would not only destroy the important oyster culture, but would also cause drastic changes in the biological equilibrium of the whole of the region. After a year-long struggle it was finally decided to build open surge dam that could be closed in the event of tidal waves endangering the region. This decision compelled the development of new technologies to surmount the huge problems connected with construction. The cost of this mega-project rose steeply after this decision was originally taken in 1974 (estimated cost by 1984 was 7,859 million guilders).(6) The majority of the Dutch population approved of this solution in 1975 (53 per cent thought it a good solution; 19 per cent not a good solution; 28 per cent did not know).(7)

Public Acceptance of New Technologies: Energy

After the energy crisis of 1973 the Dutch public's awareness of the vulnerability of the Netherlands as regards its energy resources was very acute, notwithstanding the considerable supplies of natural gas that were available. In the 1980s this apprehension became considerably reduced, as is indicated in Table 12.6.

Table 12.6: Opinions About the Energy Problem Between 1979 and 1983 (in per cent)

	1979	1981	1983
a) The chance that the Netherlands will be confronted with energy problems is:			
very big	13.0	15.6	4.1
big	42.4	34.9	19.8
neither big nor small	11.4	13.1	11.0
small	23.2	24.2	35.2
very small	10.0	12.3	29.9
Number of respondents	1,796	1,779	1,749
b) Did you yourself take measures to reduce energy cost?			
yes	59.3	73.5	76.1
no, not yet	40.7	26.5	23.9
Number of respondents	2,022	2,012	1,894

Source: Culturele Veranderingen in Nederland, 1979, 1981, 1983.

Understandably, the greatest problems with regard to the public acceptance of new sources of energy have been related to nuclear power. Since the middle of the 1970s a series of confrontations has taken place between groups of activists on the one hand and local authorities and police forces on the other. Several demonstrations and actions had as their objectives the prevention of transport of nuclear waste and the drawing of the general public's attention to the dangers of the production of nuclear energy. In 1981 53 per cent of the Dutch public said that demonstrations and discussions had helped elucidate for them the issues involved in nuclear energy and nuclear weapons: 12 per cent said the issues had been blurred by those actions, 31 per cent said they made no difference and 4 per cent did not respond.(8)

In 1983 a nation-wide debate was organised after the public had been given opportunities to get systematic information on energy sources, between September 1981 and October 1982.

Several activities as proposed by groups holding different and opposing points of view were subsidised by the steering committee for the Nation-wide Debate on Energy. Approximately 42,000 persons participated in the meetings, which were organised by the steering committee throughout the country. One hundred and eighty social organisations sent in their comments on this issue. These participatory procedures mobilised a greater number of people, representing a very wide range of opinions, than had earlier participation procedures which had been used for other issues.(9) However, this nation-wide debate on energy sources did not have a significant effect on the distribution of opinions on nuclear energy. The Netherlands Institute for Public Opinion (NIPO) reported in 1983 that 50 per cent of the population was in favour of shutting down the nuclear power plants (men 43 per cent; women 53 per cent). The majority of the population does not expect that the existing nuclear power plants will actually be shut down (84 per cent). Eighty one per cent of those who are in favour of shutting down the nuclear power plants nevertheless expect that this will not happen, and 93 per cent of those who endorse the opposite point of view do not think it will happen either (10) - whereas in 1981 69 per cent of all respondents expected the nuclear power plants to be shut down.

Table 12.7 shows opinions about alternative sources of energy.

Table 12.7: Some Opinions with Regard to Alternative Energy Sources in 1981 and 1983 (in per cent)

	1981	1983
a) Which energy source will become important?		
windmills	70.9	71.9
nuclear energy	66.5	69.5
solar energy	77.7	75.3
Number of respondents	1,700	1,700
b) Shutting down of nuclear power plants:		
we need more nuclear power plants	7.0	7.5
we need to maintain the existing nuclear plants	53.5	57.4
we have to shut down existing nuclear plants	39.4	35.1
Number of respondents	1,945	1,882

Source: Culturele Veranderingen in Nederland, 1981, 1983.

At the present time the public is waiting for a reaction from the political system. Recently, some indications have been given that the minister responsible for energy policy is considering an extension of the number of nuclear power plants. However, analyses by specialists of the present energy resources of the Netherlands and of the prospects for the 1990s, considering several policy-options, do not lead to the conclusion that a decision on the future of nuclear energy is urgent,(11) although its postponement may have important consequences for research and development in this field.

Military Technology

The production and export of military technology in the Netherlands have been the target of small groups of activists and some weekly journals since the beginning of the 1970s, but their activities have not really aroused the interest of the public at large. In a few cases public opinion has definitely opposed the export of military technology to countries with political regimes considered to be immoral. Workers in these industries - the construction of frigates, submarines, and destroyers, equipped with advanced electronic systems, the military aircraft industry, military optical instruments, electronic systems for fire-detection, advanced systems for tracing submarines - tacitly accept these industrial activities. In the case of South Africa the export of military technology - submarines - was blocked by the Dutch parliament.

The most massive opposition against military technology is related to the question of whether the Netherlands, as a member of NATO, should accept the installation of new Cruise missiles on Dutch territory. The peace movement succeeded in organising some impressive mass demonstrations against the installation of these Cruise missiles in 1982 and 1983. The effect of these demonstrations on the public acceptance of nuclear weapons is not at all clear. It may be pointed out that even in 1975 the majority of the Dutch public was in favour of a reduction in the nuclear responsibilities of the Dutch armed forces. The peace movement started its campaign against nuclear weapons and the nuclear responsibilities of the armed forces in 1977. Since 1975 there has been a majority in favour of the reduction of nuclear responsibilities (Table 12.8).

In 1983 a majority expected the new Cruise missiles to be installed in the Netherlands (October 1980, 80 per cent; September 1982, 70 per cent; 1983, 76 per cent). From the available data 'peace-researchers' conclude that the stability of the basic attitudes with regard to this issue is considerable: 35-40 per cent of the population can be considered convinced adherents of the anti-nuclear movement and 25 per cent, as outspoken followers of Nato's nuclear deterrence policy.(12) The latter category is strongly overrepresented in the defence and foreign relations elite. Hence, its societal power position is relatively strong.

Table 12.8: Opinions About the Nuclear Responsibilities of the Dutch Armed Forces 1975-83 (in per cent)

	1975	1978	1979	1983
Yes, in favour of fulfilment of our nuclear responsibilities	23	20	26	29
Not in favour	60	53	56	50
Don't know	17	27	17	21
Total	100	100	100	100

Source: See Note 12.

After the national peace demonstration in September 1983, the number of unconditional opponents of the installation of Cruise missiles increased (depending on the nature of the question asked) some 10 to 20 per cent. But the effects of the demonstration were rather different when we look at Table 12.9.

Table 12.9 and responses to similar questions indicate a decreasing resistance by the public at large towards the installation of Cruise missiles. Politically, the installation of Cruise missiles and the reduction of nuclear responsibilities is, although pending for the moment, still a very hot issue which may easily lead to strong polarisation as soon as the political debate is resumed.

Table 12.9: Endorsement of the Government's Decision to Install Cruise Missiles on Dutch Territory (in per cent)

	Before the peace demonstration (6.9.82)	After the peace demonstration (15.11.83)
Endorse	26	33
Not endorse	56	56
Don't know	18	11
	100	100

Source: See Note 12: J. Brinks et al., p. 163.

The Netherlands

Information Technology

The debate on the development of information technology and its societal consequences started rather late in the Netherlands. This statement holds both for the political system and for the general public. Specialists who were informed about what was going on in research and development warned policy-makers in the political system repeatedly, but their warnings and advice seem to have been neglected at the beginning. Before the 1980s almost no systematic attention was paid in the press to the new developments in information technology, although earlier types of computerisation and the development of data-banks had been an issue, especially in connection with the privacy of personal data. Even though the privacy of data is still not regulated by effective legislation, and the issue is repeatedly brought to the fore and discussed by representatives of the scientific community and participants in the political system, it does not have a high salience with public opinion.

The debate on the societal consequences of microelectronics started after the presentation of the report of the Advisory Board on Microelectronics, chaired by Mr Rathenau, former head of the Philips Laboratories in Eindhoven. The Board recommended that the government create urgently a centre of expertise in this field of activities in order to extend and intensify efforts such as the development of specific programmes. Moreover, in order to catch up lost ground in the production and application of microelectronics in professional and other products, the Board advised the government to give financial support to advanced designs of microelectronic components and systems. The Board's forecasts of the effects of the application of microelectronics in industry had particular repercussions both in the mass media and within networks of experts connected with the political system. After the publication of the Advisory Board's report, several decisions were taken by the government in order to develop specific programmes. These decisions and proposals set in train many activities by research institutes, management circles, labour unions, different types of advisory boards, and the educational system.

From the beginning of the 1980s, a considerable number of studies and reports have been

published on the role of information technology in
Dutch society and on the opportunities and problems
created by its development and application. It was
explained above in which way the interdependence
between technological development, on the one hand,
and social conditions and consequences of this
development, on the other, is handled. Recently, a
Note was presented to Parliament by the Minister of
Education and Science concerning the integration of
science and technology in society.(14) In this
Note a programme and policy for technology
assessment was proposed, based on experience
derived from several fields of research and
application (information technology, environmental
technology, computers in education, energy
technology, technology and work, technology and
health). Moreover, research concerning the public
acceptance of technological innovations is seen as
an important element of the proposed science
policy. The Cultural and Social Planning Agency
has already started on this research programme.

In contradistinction to energy and military
technology, the problems connected with the
introduction of new information technologies have
not aroused much public attention. This lack of
interest on the part of the general public is
perhaps related to the fact that the introduction
of new information technologies at work has taken
place at a much slower rate than was expected by
many specialists in the field in 1980. Moreover,
the workers who are confronted with technological
innovations mostly see these developments as
unavoidable. The labour unions have published
several reports, but have not been able to
influence developments in a noticeable way.
Traditionally, their influence has always been
strong at the national level (e.g. in wage
negotiations), but they are not equipped to
intervene effectively on the shop floor. In this
context, it is important to note that
representatives of the labour unions were not
allowed to join the consultations concerning the
introduction of a national banking system, although
this system will have important consequences for
workers in both the postal services and banking.

In 1981 an analysis was made of`the contents
of news items on automation and microelectronics in
papers, weeklies and professional journals.

Table 12.10: Automation and Microelectronics Issues Treated in Newspapers, Journals and Weeklies (in per cent)

Issues:	Office automation	Automation	Micro-electronics
1) Technological (e.g. development of software)	42	29	43
2) Economic (e.g. employment, situation of branches of industry)	6	29	34
3) Social-cultural (e.g. quality of labour, training and education)	17	25	14
4) Political/ management (e.g. policy problems)	1	6	7
5) Organisational (e.g. changing structure of organisations, role of management)	34	11	2
Total	100	100	100

Source: T. Wentink, (Kantoor) automatisering en microelectronika in de pers. Inhoudsanalyse van artikelen in dag-week-en vakbladen in 1981 (Office Automation and Microelectronics in the Press. Content Analysis of articles in newspapers, weeklies and professional journals) VIFKA-rapport 1981: Media over Technology. Adapted from Table IV, p. 49.

This analysis showed that the issues discussed were most frequently of a technological nature. Table 12.11 indicates the most important generators of such issues. Government agencies played an important role as generators in the case of microelectronics.

Table 12.11: Who are the Generators of Issues on
Microelectronics and Automation? (in per cent)

Generators:	Office automation	Automation	Micro-electronics
Government (agencies)	1	18	40
Enterprises	25	27	20
Labour unions	7	1	1
Professional organisations	3	5	1
Branch of industry organisations	33	8	1
Research organisations	1	4	10
Organisations of employers	0	0	1
Universities	12	7	15
Educational institutes	0	8	1
Other special interest organisations	3	4	2
Individual scientists	1	5	7
Consultants	8	8	1
Other senders	10	0	0
Total	100	100	100

In most cases, the contributions on office
automation, automation and microelectronics show a
positive orientation towards technological
developments (Table 12.12).

Table 12.12: Nature of Orientation of Contribution
(in per cent)

	Office automation	Automation	Micro-electronics
Very positive	19 ⎤	24 ⎤	32 ⎤
	⎥ 66	⎥ 44	⎥ 57
Positive	47 ⎦	20 ⎦	25 ⎦
Neutral	24	37	38
Negative	7 ⎤	17 ⎤	5 ⎤
	⎥ 10	⎥ 19	⎥ 5
Very negative	3 ⎦	2 ⎦	0 ⎦
Total	100	100	100

In 1982 578 persons at managerial and supervisory level in industry, banking, insurance, trade, education, public health and civil service (at both state and city levels) were asked to report on the consequences of automation for their organisation and for their own position in it. As Table 12.13 shows, most of the managers and supervisors reported that organisational changes played a more important role with regard to their own position than automation as such. The higher the managerial level and the higher the educational level, the more the respondents held the opinion that automation would change their position within the organisation. The majority, however, was convinced that automation would not change the existing hierarchy.

Table 12.14 shows the managers' replies to the question as to whether automation would affect their positions in the future. More than 50 per cent of the managers thought automation would not affect their positions, some 30 per cent expected a positive influence on their own position, and 20 per cent expected negative effects.

Table 12.13: Relationship Between Managerial Level
and Perceived Types of Changes in the
Organisation's Structure as a Consequence of
Automation (in per cent)

Types of organisational change	Supervisory level		
	Higher	Middle	Lower
Quantitative effects on functions: less jobs Qualitative effects on functions: the contents	27	29	15
Effects on organisation: a more rigid structure	20	21	21
Effects on supervision: decrease of middle level, more strict types of supervision	4	7	3
Effects on technical nature: increase in number of computers etc.	17	16	15
Centralisation within the organisation	2	1	0
Decentralisation within the organisation	4	1	0
Other types	17	15	28
Total	100	100	100

Source: B. van Dam, T. Wentink en H. Zanders, Management en automatisering: kansen en bedreigingen. Een opinieonderzoek bij managers over de gevolgen van automatisering (Management and automation: opportunities and threats. Opinion research among managers on the consequences of automation), VIFKA-publicaties 1982, p.62.

Table 12.14: Expected Effects of Automation on Managers' Positions (in per cent)

Effects on:	Positive influence	No diff- erence	Negative influence	Don't know
Contents of own function	35	51	9	5
Execution of own function	35	50	10	5
Flow of information	74	15	5	6
Working conditions	31	46	16	7
Productivity	74	20	1	5
Communication	42	37	14	7
Social relationships within organisation	8	65	21	6
Commercial performance	70	21	1	8
Internal consultation	23	59	13	5

Source: As Table 12.13.

The level of acceptance of new information technologies by managers and supervisors in different industries and services has been quite high, although in most cases there has been no explicit technology policy within their organisations. On a national level it is evident that the most important interest groups have still not developed a clearly defined policy with regard to the development and applications of information technology. Nevertheless, much is going on related to this issue, but these activities have scarcely been co-ordinated. As has been remarked above, many of these activities have an affinity with the problems that are generated by a changing perspective on social development. A number of seminars, workshops and conferences have been organised, and analyses have been carried out, in order to enhance the understanding of policies which may be adopted so as to avoid the failures or undesired consequences of former policies which were (implicitly) based on techno-economic deterministic criteria, on social voluntaristic convictions, or on social-deterministic approaches.(15)

Some Concluding Remarks

It is obvious that the total range of issues connected with the public acceptance of new technologies is far too broad to be dealt with in a short review of the present state of affairs in the Netherlands. This contribution contains a rather restricted selection of issues and data in this field.(16)

The following tentative conclusions may be drawn from the preceding sections and from the sources on which this contribution is based:

1) Changes in the public acceptance of new technologies are connected with changes of perspective on societal development within the population. Since the end of the Second World War several shifts in the 'dominant' perspective on societal change can be discerned. These shifts seem to be related to changes in the public acceptance of new technologies.

2) Moreover, in the period under review the situation with respect to new technologies changed in another important respect. Since the beginning of the 1970s an increasing number of professionals have been recruited by the main social institutions in the Netherlands. The development of interlocking networks of professionals in ministries, semi-governmental organisations, labour unions and other organisations of employees, organisations of the scientific communities, political parties, consumer organisations, and many voluntary environmental organisations etc., has changed the present situation drastically in comparison to the situation before the 1970s. Experts and professionals have come to play a more important role in the preparation of policies bearing on new technologies since the beginning of the 1970s. This development seems to have contributed to a growing distrust of experts by the general public, as it has became evident that experts operating in the same field may confront the general public with sharply conflicting views on the acceptability of new technologies.

3) The acceptance of new technologies may also be dependent on the positive effects of strategies followed in connection with the adoption of new technologies in an earlier period, and on the institutional arrangements which resulted from these strategies. An example is the Dutch

agriculture and cattle-breeding industry. The
preservation of a stong position in the world
market of the products of the Dutch agricultural,
horticultural, meat and dairy industries is
strongly dependent on an ability to use the
potentialities of new technological developments
(in biotechnology, enzyme technology, genetic
manipulation, immunology, soil biology and ecology
etc.).

The present position of Dutch agriculture is
the result of a strategy followed by the Dutch
government and producers in this economic sector
after Dutch agriculture was severely hit by
economic crises between 1880 and 1900. Although
the authorities were convinced adherents of a
laissez faire policy, a new strategy was developed
by the government and the producers in order to
find those markets in which Dutch agrarian products
could compete successfully. The new strategy
envisaged the adaptation of farmers and cattle-
breeders to the new economic conditions. New
technologies were adopted, such as new types of
transport, cold-storage technology and hot-house
technology in an export-orientated horticulture.
Important reorganisations took place in the
agrarian sector, such as the development of an
efficient co-operative auction system. Moreover, a
specific educational system emerged (elementary and
middle-level agricultural schools) and together
with an agricultural college (the Agricultural
University at Wageningen which is a centre of
research on technological innovations in this
economic sector) and a network of agriculttual
consultants.(17)

In this model, state and agrarian circles
pulled together in order to make strategic choices,
to build a new infrastructure, to develop a new
educational system and to programme research in
relation to the changing needs of this sector.
Since the beginning of the 1970s multi- and
interdisciplinary research and development have
been planned and co-ordinated in national five-year
programmes.(18) This example demonstrates how in
such cases as the 'Dutch agrarian model'
technological innovations are channelled by the
existing institutional arrangements. The public at
large is generally unaware of the introduction of
new technologies in this type of industry.
However, in several cases public opposition rises
when the effects of such technical changes become
evident (for instance, in bio-industries).

4)   The potentiality of new technology to
348

contribute to the solution of problems that traditionally rank high in the public mind is also an important determinant of its acceptance. An example is provided by the Dutch system of safe-guards against the major disasters that can be caused by the North Sea and the main waterways. New technological applications connected with this protection system are generally accepted, although they may be extremely expensive, as in the case of the surge dam described above. It is interesting to note that in this case there has been strong opposition by interest groups against the negative consequences of this project as regards both the biological equilibrium and the important oyster culture in the area. However, a technical solution was accepted by the parties concerned. This solution stimulated the development of new technologies in off-shore construction.

The protection of the country against the dangers of the sea and main waterways is organised in long-established institutions, of which the state department for the maintenance of dykes, roads, bridges and the navigability of canals (Rijkswaterstaat) is of paramount importance. In several respects Rijkswaterstaat may be considered as a 'state within the state'. Thus, the remarks made above in relation to the Dutch agricultural system also pertain to this protection system. Nevertheless, it is evident that the importance given within the Dutch culture pattern to this type of problem affects the public acceptance of new technologies in the area. Looking at this issue from a wider perspective, it may be suggested that differences between the culture patterns in different parts of Europe may partly explain differences in the acceptance of certain new technologies such as nuclear and military technologies, and other technologies that have consequences for the prevailing convictions about the dignity of man.

5) The analysis of the data on the Netherlands indicates that several conditions specific to this country are of importance for its public acceptance of new technologies. In particular, there is the fact that the Netherlands is a densely populated country, traversed by several of the main and many polluted European waterways, a fact that contributes to the public's awareness of the dangers of the increasing growth of certain types of industry. Another rather specific element in the Dutch situation is the presence of a relatively

large number of important multinational enterprises
with their head offices in this country. This
specific element heightens the public feeling that
several important technological developments and
their social and economic consequences are very
difficult to control on a national level. Here we
have a very important difference from the
conditions prevailing in agriculture at the end of
the nineteenth century. The strategy adopted at
that time to improve the position of Dutch
agriculture in the world market could be developed
because then there was no class of big landowners
trying to control future developments in accordance
with their own interests. In the case of new
technological developments, such as the development
of information technology, the power balances
between the state, small and medium-sized
enterprises on the one hand, and multinational
enterprises on the other, and their interrelation
with the labour unions and the educational systems
are quite different. A consequence of this
situation is that it is not very likely that a
similar strategy can be developed in this area
comparable to the above-mentioned agricultural
model.

6) Looking at the development of information
technology, we note that the public acceptance of
information technology shows characteristics that
are rather different from those characterising the
acceptance of new sources of energy and of military
technology. In the latter cases the authorities
are confronted with the firm rejection of some
options by considerable sections of the public, and
thus far they have not been successful in changing
the acceptability of these options. Major
decisions may, therefore, be a threat to the
legitimacy of their authority. Such a legitimacy
problem does not exist with regard to the
application of information technology. In fact, in
this case, the authorities, especially at the
governmental level, are worried by the lack of an
organised response to proposed policies.

7) The introduction of information technology at
work tends to be positively accepted by those who
occupy managerial positions. Those who are not in
command tacitly accept the resultant technological
and organisational changes. This acceptance is in
most cases an adjustment to developments that are
considered to be unalterable by them. This tacit
acceptance at work stands in contrast to the
acceptance of personal computers by private
consumers.

Notes

1. J. Galtung, 'The Future: a Forgotten Dimension' in H. Ornauer, H. Wiberg, A. Sicinsky and J. Galtung (eds), Images of the World in the Year 2000, The Hague-Paris: Mouton/Atlantic Highlands N.J. Humanities Press, 1976, pp. 62-3.

2. Ibid. p. 73.

3. D.L. Meadows, The Limits to Growth - A Report for the Club of Rome Project on the Predicament of Mankind, New York: Universe Books, 1972.

4. Rapport van de adviesgroep Rathenau, Maatschappelijke gevolgen van de Micro-electronica (Report of the Rathenau Committee, Societal consequences of microelectronics), 's Gravenhage: Staatsuitgeverij, 1980.

5. Wetenschappelijke Raad voor het Regeringsbeleid, Beleidsgerichte Toekomstverkenning (Scientific Council to the Government, Policy Orientated Exploration of the Future), Deel 2, 1983, p. 204.

6. G. Goemans en H.N.J. Smits, 'Kostenbeheersing van een megaproject: de Oosterscheldewerken', (Cost control of a megaproject: The Eastern Scheldt Surge Dam), Economisch Statistische Berichten, vol. 69, 1984, p.991.

7. NIPO, 23 and 24 December 1984.

8. NIPO Public opinion poll 23-24 November 1981 (1,101 interviews).

9. Rapport van de Brede Maatschappelijke Discussie Energievraagtuk, 1984, p. 343 ff.

10. NIPO Public Opinion Poll 7-8 February 1983 (1,015 interviews).

11. Wetenschappelijke Raad voor het Regeringsbeleid, Beleidsgerichte Toekomstverkenning, Deel 2, 1983, p.238 ff.; Wetenschappelijke Raad voor het Regeringsbeleid, laats en Toekomst van de Nederlandse Industrie (Position and future of Dutch Industry), 1980.

12. J. Brinks et al., Wapens in de peiling. Opinieonderzoek over internationale veiligheid (Arms and safety), Den Haag: Staatsuitgeverij, 1984, p. 29.

13. Ibid., p.32.

14. Nota Integratie van Wetenschap en Tevhnologie in de samenleving (Note on the intergration of science and technology in society), Tweede Kamer, vergaderjaar 1983-1984, 18421/2.

15. T. Huppes en J. Berting, (eds), Op weg naar de informatiemaatschappij... Maatschappe lijke gevolgen en determinanten van technologische ontwikkelingen (Bound for the information society ... Societal consequences and determinants of technological developments), Leiden: Stenfert Kroese, 1982; M.L.A. ter Borg et al., De maatschappelijke beoordeling van technische kennistoepassing (The Societal Assessment of Technical Knowledge), Werkdocument 2 van de WRR., January 1984.

16. In this contribution several topics were omitted or mentioned only briefly, e.g. biotechnology (e.g. in agriculture), new medical technology, introduction of CAD/CAM systems, space technology, the effects of pesticides on the environment, acid rain, changes in the educational system related to information technology, the changing position of (working) women as a result of changes in technology, the role of transfer-points between universities and industrial organisations, the effects of multi- or transnational enterprises like Philips, Unilever or Shell on a small country like the Netherlands, and the role of the mass media.

17. W.O.C. thoe Schwartzenberg, Ruim 100 jaar Nederlandse agrarische export in vogelvlucht (A Bird's Eye View of Netherlands Agricultural Export Over More than One Hundred Years, Agricultural Economic Institute), Mededeling 248 van het Landbouw-economisch Instituut 1981.

18. Nationale Raad voor Landbouwkundig Onderzoek TNO, Meerjarenvisie landbouwkundig onderzoek 1982-1986 (National Council for Agriculture Research, Medium-term Perspective on Agriculture Research 1982-1986), Studierapport 10, Den Haag.

Chapter Thirteen.

PUBLIC ACCEPTANCE OF NEW TECHNOLOGIES IN SPAIN

Rafael Lopez-Pintor and Luis Ramallo

It is the aim of this chapter to describe the evolution and present situation of Spanish public opinion on new technological developments, as well as to elucidate the sense and direction of public policy in the fields of science and technology.

The process of opinion-formation in Spain and, also, of policy-making and implementation have been deeply affected in this area, as well as in most fields of public policy, by the general recession in the West since the early 1970s, and particularly by the specifically Spanish political transformations of the last decade. For those who are not familiar with the Spanish scene, it seems advisable, therefore, to summarise some major structural features of the Spanish context both economic and political.

At present, Spain is the world's eleventh industrial power, but it is one whose development has come late as compared with the major industrial powers of the West. Although Spanish industry started developing during the last quarter of the nineteenth century, its fastest and most definite growth took place from 1955 to 1975 during the authoritarian period of the Franco regime, the last twenty years of a much longer dictatorship. Spain's political economy at the time was overprotectionist, and a welfare state developed out of economic affluence and rather paternalistic social policies. Presently, Spain having ended its negotiations with the EEC, entered the Common Market in January 1986.

It follows from this that the democratisation of the policy in the second half of the 1970s took place at a time of economic recession, and yet at the same time as expanding welfare and redistributive policies. The public sector of the economy amounted to about 29 per cent of GNP by 1976, and reached 39 per cent in 1984. This increase in public expenditure at a time when democracy was being re-established was not coupled with economic policies aimed at responding to the most recent industrial reconversion, the development of new technologies and so on.

Spain is an energy-dependent country to the extent of 65 per cent of its energy budget; with regard specifically to oil, its foreign dependence is over 90 per cent. An almost absolute dependence also exists in respect of the most recently developed technologies, most notably, biomedical, computer and microelectronic technologies. Among industrial countries, Spain is the leading buyer of modern technologies.

A corollary as well as a cause of this situation is the fact that the level of national investment in science and technology is one of the lowest among industrial countries. Only 0.5 per cent of GNP is devoted to science and technology, in contrast to a European average, after excluding Portugal and Greece, of 2 per cent. Nevertheless, there have been some recent public initiatives to improve national standing in this area, and they will be referred to later.

Four key topics are covered in this chapter, and they are considered in terms both of public opinion and of public policy-making. They are: the energy crisis and new sources of energy, where a special concern exists with regard to nuclear power and nuclear power plants; industry and the environment; the development of science and technology; and consumer concern with food processing and its control.

The Energy Crisis and Nuclear Power

There are sufficient opinion-research data available to allow us to look at this and other topics, but with different scope and intensity in each case. Sometimes, data are available only with regard to the general public, while at other times data also exist in respect of industrial owners, managers, scientists and politicians.

As result of the energy crisis of the early 1970s, concern about a future repetition of this crisis has been growing amongst the Spanish public. Although there has never been massive awareness of the energy problem, an increasing concern with energy topics has been periodically illustrated by

the surveys which have been undertaken by the
Centre for Sociological Research since the late
1970s.(1) As might have been expected, this
concern started sooner and grew faster among
entrepreneurs and politicians, but they were not
enthusiastic about transmitting their concern to
the public at large since restrictive economic
policies were not feasible during the uneasy
equilibrium of the late 1970s, while democracy was
being restored.

A National Energy Plan (PEN) was drawn up in
1978 and revised in 1983. Given Spain's heavy
dependence on foreign oil for energy supply, the
plan provides for an increasing dependence on
nuclear power, while dependence on oil is to be
curtailed. (Spain is both a producer and exporter
of uranium.) The nuclear policy, nevertheless,
also implies a moratorium on the opening of new
nuclear plants. The nuclear moratorium was decided
upon by the Socialist Government in 1983. It has
affected the opening of seven new plants, some of
which were already built, while some others were
still in the planning stage. (One of these plants,
that at Lemoniz in the Basque country, has
repeatedly been the object of terrorist attacks by
the Basque separatists of the ETA organisation.) A
motion to maintain the nuclear moratorium was
approved at the last national conference of the
Socialist Party in December 1984. At present there
are seven nuclear plants working in Spain, and
together they provide about 10 per cent of energy
supplies.(2)

Public opinion is becoming knowledgeable about
nuclear energy and nuclear technology: around 30
per cent of the population know something about it.
Nevertheless, a majority of the public is worried
about the problems and dangers which may result
from nuclear plants. In fact, 47 per cent of those
in a national sample described nuclear plants as
'very' or 'quite' dangerous. A similar percentage,
45 per cent, would favour state ownership of
nuclear plants, and only 5 per cent would favour
the status quo of private ownership by the electric
companies. When a question was posed regarding
substitutes for nuclear energy, 41 per cent
mentioned solar energy.(3) At the elite level,
politicians are less suspicious of nuclear energy.
In a sample of deputies from the Lower House of the
Cortes (parliament), a very large majority favoured
increasing public investment in civilian uses of
nuclear power (69 per cent compared with 24 per

cent among the general public) and said they felt
confident about scientific advice in the
hypothetical situation where a group of scientists
were prepared to recommend the opening of a nuclear
plant to produce electric power as being without
danger (84 per cent compared with 37 per cent among
the general public). Public mistrust of nuclear
technology is most widely spread among the middle
classes, and it is from these that the basis of the
ecology movement largely comes. As might be
expected, there is also an ideological correlation
in rejecting nuclear technology. Yet there is not
a strong correlation with age: in almost any age
group under 50, trust and mistrust of nuclear
energy are found equally.(4) In the early 1970s
there were clear-cut generational and class
components in the movement against nuclear
technology - though that movement was broader than
this suggests, and also worked towards a social
basis for the redemocratisation of the country.
That basis was made up of the urban upper-middle
class aged below 30. But today, the generational
and class components have dissolved throughout the
middle classes, whether urban or not, and with
respect to all age groups under 50.(5)

The Development of Science and Technology

The Spanish public is ambivalent with regard to new
scientific and technological developments. On the
one hand, the majority tends to have positive
feelings about the social functions of scientific
and technological research, as well as about the
intentions and goals of individual scientists.
Most Spaniards also think that the government
should spend more on scientific research, even if
this means raising taxes. On the other hand, there
is little knowledge about new technologies, and
there is a feeling that scientists may become
dangerous because of their esoteric and specialist
knowledge. This latter opinion is shared by 67 per
cent of the general public, but it is a minority
view among politicians from the Lower House of the
Cortes, only 23 per cent. Most people also think
that technological progress has made work safer,
easier and less boring. However, at the same time
they think it may lead to unemployment - this
opinion is shared by 68 per cent of the public but
by only 24 per cent of Cortes deputies. These
empirical data relate to 1982, but there are also

data on some of these points which allow comparison of attitudes over the period since 1974. In that time the public has become less supportive of science and technology, and perhaps more suspicious of them. Yet this does not necessarily mean that it has become more hostile. In 1974 62 per cent of Spaniards supported the idea that more money should be spent on scientific research. In 1982 only 48 per cent of the general public still held this opinion, but 100 per cent of deputies did so. Those who supported more research expenditure, even if a tax increase were necessary to cover it, amounted to 67 per cent in 1974. By 1982 the percentage backing that opinion, though still large, was only 50 per cent (among deputies the view was held by 87 per cent).(6)

The need for improved scientific and technological development is felt, then, by the general public, but even more by elite sectors of the society which are more closely related to the matter. Stronger support for scientific development is found not only among politicians, but also among industrial managers, professionals and scientists themselves. Spanish entrepreneurs are very aware of the relationship between national capacity for technological autonomy and ability to compete internationally, as can be deduced from existing surveys.(7) Regarding scientists themselves, more than 50 per cent of a national sample of researchers in 1974 felt that scientific research in the country should be oriented towards technical development (20 per cent), or to meeting the needs of economic development (11 per cent), to getting more technological autonomy (19 per cent) and to supporting the requirements of national industry (10 per cent).(8)

The ambivalence towards technology mentioned above has repeatedly been found amongst youth. In a sense, young people dream about all the goods resulting from scientific advance while at the same time feeling very suspicious of contemporary technology. This conclusion can be drawn from a re-analysis and evaluation of research on youth in Spain from 1960 to the present.(9) Although as a whole, contemporary Spaniards should not be considered very anti-technology, a decade of economic recession and energy shortages, as well as the need for re-industrialisation, seem to be feeding an expansion of anti-technological feelings. An anti-technology factor, in the statistical sense of the word, was already there in

the early 1970s as was pointed out above. It was
particularly strong among certain sections of the
urban middle class, especially males under 35 and
females under 25 in 1972. At that time the
'cluster' who were anti-technology amounted to
about 15 per cent of the national population. By
1978 it had increased to about 22 per cent.(10) In
the Spanish case at least, and maybe in some other
European countries as well, the anti-technology
factor cannot be disassociated from fear of
(nuclear?) war. In Spain this seems to be particularly acute, according to comparative opinion
statistics gathered by the Atlantic Institute.(11)

The restructuring of the steel and
shipbuilding industries has been and still remains
a main focus for political and labour conflict in
Spain. There is, on the one hand, much active
protest by the unions and on the other much
hesitation on policy by the government. But the
majority of the public is revealed in surveys to be
in favour of re-industrialisation even if this
brings more unemployment - this according to the
same source, the Atlantic Institute. (Spain's
unemployment rate was over 20 per cent in 1985.)

There are several recent institutional
responses to the needs for new technological
development which deserve to be pointed out.
First, and in addition to the National Energy Plan
which has already been mentioned, there is a
National Electronic and Computer Science Plan
(PEIN) whose main targets for 1987 have already
been fulfilled. This is now being revised to make
it viable up to 1988. Performance was better than
had been predicted following an unexpected growth
of both investment and exports during 1984.(12)

Secondly, legislation has recently been put
forward to foster and promote the electronics and
computer industries through tax benefits, credits
and research subsidies. This may affect over 140
firms. Production of software products is also
included under the umbrella of this new technology
policy.(13)

Thirdly, the Ministries of Education and
Industry are engaged in preparing a bill for a Law
of Science and Technology as well as a Scientific
Research National Plan covering the period from
1986 to 1990. Both the law and the plan are aimed
at increasing public and private investment in
science.

Finally, a Centre for Industrial and
Technological Development (CEDETI) has been in

existence since 1984 for the promotion and financing of research projects, with a priority in the areas of microelectronics, biotechnology, health and the food industry. These are areas in which Spain is more heavily dependent on foreign supplies.

Industry and the Environment

Public concern about the environment has kept growing in Spain since the early 1970s. At the beginning of that decade, a third of the people living in cities larger than 50,000 agreed with the statement that 'there is much exaggeration about the problem of air pollution', very few people (18 per cent) went along with the idea of 'paying more taxes on cars and petrol in order to help with the pollution problem'.(14) Yet around half of the Spanish public was to some extent concerned with air pollution. In a national sample in 1971 58 per cent of the public described the problem of pollution as 'very important'. In 1974 54 per cent of the population held to the opinion that the problem was 'serious', but fewer people considered that it was 'urgent'.(15)

At the time, the advocates of a better environment came mainly from the urban upper-middle class. They were those who could expect to suffer least from pollution, but they were also those who ideologically sensitive to the 'issue of the time' throughout the West. Cars and industry were mostly thought to be to blame for the evil. Interestingly, local authorities were perceived as less concerned and yet more responsible for the pollution problems than were national authorities.(16) This was a strategy of the power game during the Franco regime: channelling protest towards local authorities in order to protect the national office-holders closest to the dictator.

By 1982, after a decade of cultural and political changes, environmental concerns had spread gradually throughout the entire society to the point where ideological (left to right) or urban-rural differences had become increasingly blurred, at least at the most superficial level of opinion expression (75 per cent of the general public were now prepared to agree with the idea that 'many industrial factories frequently contribute to increasing pollution of the environment'(17)). Almost as many people among the

general public, and a still larger percentage among
deputies, were in favour by this time of increasing
public investment in environmental research. Only
investment in health research came as a higher
priority - 86 per cent among the general public and
99 per cent among deputies. Lower priorities were
given to public investment in consumer goods (53
per cent and 69 per cent respectively); civilian
applications of nuclear power (24 per cent and 69
per cent); aeronautics (30 per cent and 56 per
cent); space research (23 per cent and 49 per
cent); and military research (14 per cent and 30
per cent).(18)

Consumer Concerns About Food Processing and its Regulation

This is an area of values, behaviour and policy
where Spanish society comes out looking impotent
and pessimistic. This conclusion may seem obvious
enough to those familiar with the Spanish scene,
though not much opinion or policy research has been
conducted on the topic, or at least not much has
been available. Spanish legislation on food
treatment and product quality control is sometimes
overcasuistic, and sometimes obsolete. Policy
enforcement tends to be weak, and consumer
associations and lobbying activities in this area
are neither very strong nor very effective.
Against this background, the climate of public
opinion is one of alienation. There is a sense of
impotence coupled with an expressed demand for
total control.
    Some scarce opinion-data from the period
between 1972 and 1982 illustrate the point. More
than a decade ago there was almost unanimous
feeling that the government should be less tolerant
of manufacturers responsible for adulterating food
products (86 per cent felt this among the general
public), and that it should have total control over
prices as well as product quality (85 per cent).
Only a tiny minority (4 per cent) of the people
questioned were against these ideas.(19) Ten years
later, only one-third of the public believed that
'most products on the market are not excessively
dangerous to one's health'. And coupled with this,
as pointed out above, is a belief that public
investment in scientific research dealing with
consumer products should be increased.(20)

Conclusions

Overall, as regards the acceptance of new technology in Spain, it seems quite clear that the mood of Spanish public opinion is more supportive than negative, and that public policy is also very positive - new technologies are being given a special priority, though policy is somewhat late in the day compared with the major industrial powers. But a partial exception must be made regarding nuclear power plants, where both public opinion and governmental action are more restrictive and less optimistic.

As for the role of the mass media, this basically reflects the mood of society, and the style of political decision-making. Media coverage of science and technology has been slowly improving since the early 1970s. More attention is usually given to advances in health and communications technologies than to any other fields of technology and science. Media influence on the public acceptance of technology clearly fits into a reinforcement model: enhancing trust in those technologies which people fear least (health techniques and computers); and encouraging mistrust in that technology which people fear most (nuclear power and its military or civil applications). This is the more visible bias of the media - the written word, radio and television - although nuances and variations occur according to the ideological stand of each medium, except for TV. This is public, and therefore closer to the governmental positions of the time, fitting better the French than the British style of public television.

Spain cannot escape the hopes and fears aroused by the most recent technologies of this industrial era, in the midst of recession and with an urge for re-industrialisation. As before in history, social aspirations for progress mix with some of the persistent shadows of human experience: fear of joblessness and of injuries to life from unknown technical evils. This cultural scenario is common to most Western societies of the day. But there are, in addition, specifically Spanish traits, such as the country's late and more dependent industrialisation, and the political changes over almost a decade of transition from a decadent system of personal power to representative government. These are circumstances which largely account for a rather vacillating policy in the realm of industry and technology, as well as for a certain lack of social experience and capacity for organisation, lobbying, responsible mass action, and effective control of policy implementation. Yet on the whole, the balance today is more positive than negative, as Spain faces the immediate future within the broader social, economic and political system of the EEC.

Notes

1. See the opinion appendices that are periodically issued at the Revista Espanola de Investigaciones Sociologicas, REIS (different years) of CIS, a social research centre which is attached to the Spanish Premier's office.

2. Manuel Garcia Ferrando, 'El debate publico sorbre el uso de la energia nuclear', Revista Espanola de Investigaciones Sociologicas, 16, 1981, 57-90.

3. Ibid., 78-9.

4. Manuel Garcia Ferrando, 'Imagen de la ciencia y de la technologia en Espana', forthcoming in Revista Espanola de Investigaciones Sociologicas.

5. For a picture of public opinion in the early 1970s see Rafael Lopez-Pintor and Ricardo Buceta, Los Espanoles de los anos 70. Una version sociologica. Madrid: Tecnos, 1975.

6. Manuel Garcia Ferrando, 'Imagen de la ciencia'.

7. Manuel Garcia Ferrando, 'Actitudes de los empresarios Esponoles ante el cambio social y la crisis economica', Revista International de Sociologia, January 1984, 93.

8. Pedro Gonzalez Blasco, El investigador cientifico en Espana, Madrid: CIS, 1980, p.11.

9. Miguel Beltran et al, Informe sociologico sobre la juventud espanola, 1960-1982. Madrid: S.M., 1984.

10. Francisco Andres Orizo, Cambio socio-cultural y comportamiento economico, Madrid: CIS, 1979, pp.202-5.

11. Comparative surveys by the Atlantic Institute for International Studies in Paris, as published by the Madrid daily newspaper El Pais, 7 June 1984.

12. There was a public report by the Ministry of Industry in February 1985. This information was published by the Madrid paper Diario 16, 2 February 1985, 19.

13. There is a Royal Decree of 24 January 1985.

14. Rafael Lopez-Pintor and Ricardo Buceta, Los Espanoles, pp.125, 170.

15. Elena Bardon Fernandez, 'Contaminacion y medio ambiente' (a research note), Revista Espanola de la Opinion Publica, 38, 1974, 314-15.

16. Ibid., 320.

17. Manuel Garcia Ferrando, 'Imagen de la ciencia.

18. Ibid.

19. Rafael Lopez-Pintor et al., Los Espanoles, pp. 125, 170.

20. Manuel Garcia Ferrando, 'Imagen de la ciencia'.

Chapter Fourteen

NUCLEAR POWER : AN INTERNATIONAL COMPARISON OF
PUBLIC PROTEST IN THE USA, BRITAIN, FRANCE, AND
WEST GERMANY (1)
Wolfgang Rüdig

Introduction

Nuclear energy is the most controversial civil
technology ever devised. No other has been the
subject of such forceful and widespread protest
movements. Anti-nuclear movements stretch across
national boundaries; they often do not just demand
changes in design or political control, but aim at
no less than the phasing-out of nuclear energy
altogether.

There are, however, striking differences in
the numerical strength, political relevance and
actual achievements of these movements.
Substantial opposition has emerged, for example, in
West Germany, France, and the USA, yet the extent
of anti-nuclear protest in Britain has remained
more limited. Nuclear construction has virtually
stopped in the USA, while France continues to order
further stations at a reduced but steady rate.
West Germany and Britain have retained their
commitment to nuclear energy, but in both major
programmes have been delayed or halted.

Public interest in the peaceful uses of
nuclear energy has in many countries in recent
years given way to concern over nuclear weapons and
the escalation of the arms race.(2) The resulting
lull in activity directed against nuclear power
provides a good opportunity for an interim
evaluation of nuclear-opposition movements. Why
have these movements developed in different ways in
different countries? Have they had any significant
effect on nuclear development? To what degree has
their political effectiveness been conditioned by
particular national institutions, cultures, and
other circumstances? Is the current absence of
forceful protest temporary or permanent, and what
can be learnt from past experience about the

emergence, form, and likely results of any future conflicts about nuclear technology?

These questions are dealt with below in three parts. The chapter details first the emergence of anti-nuclear movements in different countries. How exactly did nuclear power become an issue of widespread public protest in the 1970s? Here, there is both the historical development of expert dissent on nuclear-safety issues and local opposition to be considered. Secondly, the chapter analyses the patterns nuclear conflicts have followed in particular countries. Which forms of action were employed by anti-nuclear movements, and how did the nuclear industry and the various political administrative institutions respond? The chapter reviews specifically the different licensing systems for nuclear installations, the role of public participation, and the responses of trade unions and political parties. It examines finally the impact of anti-nuclear movements on the development of nuclear energy: to what extent has the recent world wide contraction of the nuclear industry been due to anti-nuclear opposition?

The chapter draws mainly on the experience of the USA, the United Kingdom, France and West Germany. Since these four countries have the most important nuclear industries, it is reasonable to expect that nuclear energy's fate will be significantly influenced, if not wholly determined, by the future aceptance of nuclear technology in them.

The Politicisation of Nuclear Energy

The peaceful use of nuclear energy always has been a subject of political debate. In the early stages, issues like the size of research and development programmes, the role of private industry, the scale of public subsidies, and the choice of reactor line, dominated the debate. In some cases nuclear energy even became a party political issue. But the main actors were then government institutions and regulatory agencies, electricity supply companies , and the nuclear construction industry. By the middle 1970s, however, all this had changed in the USA and also in most Western European countries. A previously marginal issue, the safety of nuclear installations, now became central. In addition, new actors emerged who, starting from criticism of individual nuclear projects, increasingly moved on to make demands for a moratorium or even the abandoning of nuclear energy altogether. The force of this new onslaught took most industries and governments by surprise. Decision-makers in Europe might perhaps have been less taken aback had closer attention been paid earlier to the growth of nuclear opposition in the USA, because not only did public opposition first emerge there, but also nuclear debates in Western Europe, when they eventually began, drew heavily on the data and arguments coming from across the Atlantic.

The Role of Expert Dissent

A crucial precondition of anti-nuclear opposition has been the emergence of expert dissent over key aspects of nuclear safety. Apart from reactor safety, dissent over virtually the entire fuel cycle, from uranium mining to waste disposal, became increasingly prominent from the late 1960s, starting in the United States.

USA. Expert dissent on nuclear safety in the US can be tracked back at least to 1956, when an application was made for a licence to construct a 100 MW Fast Breeder Reactor (FBR) 25 miles from Detroit. The official Advisory Committee on Reactor Safeguards (ACRS) expressed its concern to the licensing authority, the Atomic Energy Commission (AEC), and this contributed significantly to the first case of local opposition to a nuclear project, in this instance predominantly carried through by Detroit trade unions.(3)

Professional concern about nuclear waste found its first public expression in 1957, when a National Academy of Sciences report drew attention to the unsolved problem of waste disposal, claiming that its solution was crucial for the future of the nuclear industry.(4) The first comprehensive critique of the nuclear industry focusing on the waste problem was published in 1963 by David E. Lilienthal,(5) a former chairman of the AEC, but his book made no real impact.

Not until the late 1960s did dissent over nuclear safety become significant. The issue was then the health effect of low-level radiation. There was already a long-standing scientific debate on this but not in connection with the nuclear industry, rather with fallout from nuclear weapons tests.(6) In 1963, with the signing of the Partial

Test Ban Treaty, public concern over fallout died away. The scientific debate, however, continued, and in 1969 was revived by the findings of Dr Ernest Sternglass concerning the effect of fallout on child mortality. Two senior AEC scientists, Dr Arthur Tamplin and John Gofman, now went on public record to the effect that Sternglass was wrong only in his estimate of damage, and not in his basic thesis that low-level radiation could have adverse health effects. By implication, the AEC was being criticised here as regards routine emissions of radiation from civil nuclear power stations, and Gofman and Tamplin went on to demand that AEC radiation standards be reduced by a factor of ten.(7) Their critique had an important impact on the development of nuclear opposition. Here were two distinguished AEC scientists publicly criticising AEC policy on low-level radiation and the Commission's general approach to nuclear safety: it followed that concern could henceforth no longer be easily dismissed as irrational and lacking any scientific base.(8)

Gofman's and Tamplin's intervention also came at a critical time in the development of the US nuclear industry. The enormous expansion of nuclear energy in the US in the late 1960s confronted many localities for the first time with nuclear power, and in several, groups formed with the aim of contesting the project.(9) In addition, concern over general environmental pollution peaked in 1970 with Earth Day, which followed Rachel Carson's book Silent Spring, and a number of environmental catastrophes such as the Santa Barbara oil spill of 1969.(10)

It was mainly the issue of thermal pollution which brought local preservation groups to join other local interests in contesting particular projects. To cut costs utilities were at this time planning 'once through' cooling systems (which would lead lakes and rivers to warm up, with undesirable effects) rather than the more costly cooling towers or ponds.(11)

By 1970, although dissenting experts, local groups and the environmental movement had made several challenges to the nuclear industry, there was still little really fundamental opposition. Scientists, local groups and environmentalists almost all started with demands for specific reforms, and the industry in many cases moved to meet them. Thus, new safety standards were adopted by the AEC in 1971 which went a long way towards complying with the changes requested by Gofman and Tamplin, and utilities moved away from 'once through' cooling systems.(12) New issues, however, continued to emerge.

The most important single issue in the early 1970s was that of the integrity of the Emergency Core Cooling System (ECCS). ECCSs were designed to avoid the reactor core heating up dangerously in the event of a Loss of Coolant Accident (LOCA). This problem was particularly serious in the type of reactor predominantly used in the USA, the Light Water Reactor (LWR). Since in this ordinary water is used as both coolant and moderator in one system, a simple pipe-rupture could lead to the water flowing out, leaving no heat sink; the core as a result heating up rapidly and eventually melting down, releasing a large amount of radiation into the environment. The ECCS problem became particularly pertinent as the sales drive of the late 1960s came to depend on ever larger reactors needing far more reliable ECCSs - ECCSs which had never been tested. Some debate about these systems had taken place within the AEC in the 1960s, but any suspension of licensing had been rejected to avoid jeopardising the commercial breakthrough nuclear power was then enjoying. Since the nuclear industry's case for the reliability of its ECCS designs rested solely at this time on computer calculations, the AEC initiated research projects on LOCA in which, inter alia, experiments into the reliability of ECCS models were carried out. But ECCS models failed in LOCA experiments, and the results became publicly known in May 1971. By then, a group of MIT scientists, the Union of Concerned Scientists (UCS), had become interested in the ECCS issue. UCS contacts with AEC scientists disclosed that the ECCS problem was serious. With local opposition groups about to take this up to challenge individual projects, the AEC decided to clear the issue with a rule-making public hearing. But, starting in December 1971 and lasting for more than a year, this proved damaging to the AEC when some of its own scientists gave evidence challenging the Commission's position that existing ECCS designs were safe.(13)

The ECCS issue was also joined at the time by two other major concerns. In 1971 the AEC's attempt to solve the problem of nuclear waste by identifying a central site for permanent storage backfired. State scientists demonstrated that the chosen site, a Kansas salt mine, was not leakproof, and also that several pieces of crucial evidence had been overlooked. Following strong opposition in Kansas to the plan, the AEC was forced to abandon the site in 1973, leaving its nuclear-waste policy in rather a shambles.(14)

Second was the issue of nuclear safeguards and weapon proliferation. Dr Theodore Taylor, a former nuclear-weapons designer, had pointed out the ease with which nuclear bombs could be manufactured if fissionable material were available; and popularised by a journalist, his revelations aroused fear of the implications of a widespread use of fissionable material, particularly plutonium in an energy economy dominated by FBRs.(15) Taylor himself demanded stronger safety precautions to reduce the chances of fissionable material being diverted by terrorist groups, but there was public apprehension that a plutonium economy might require security measures whose strictness would seriously infringe civil liberties. And if diversion of nuclear materials at home was a problem, then clearly the use of civil nuclear projects in other countries for nuclear weapons programmes could not be excluded either. This was underlined by the explosion of an Indian nuclear device in 1974, using fissionable material produced in a research reactor supplied by Canada, a development which brought into question the whole subject of nuclear exports.(16)

From what has been said here it is evident that the AEC was unable to control nuclear installations to the extent necessary to retain the confidence of the public, or even of some of its own scientists. And once serious doubts about nuclear safety had been raised within the AEC, independent scientists found little difficulty in gaining access to further information, thanks to the relative 'openess' of the American political system. The latter also provided an abundance of possibilities for disseminating this information and publicly challenging the nuclear industry.

The US experience thus suggests that expert dissent was instrumental in making nuclear energy a political issue, though other factors such as local opposition and the 'open' US political system also played a crucial role. US expert dissent was a crucial influence without which anti-nuclear movements elsewhere would not have emerged in the strength in which they did, since the nuclear sectors of most European countries are more homogenous and less 'open' than American, making it more difficult for internal dissent in them to surface.

Some of the criticisms of nuclear energy voiced in the US were applicable to nuclear energy in general, but other key topics, such as the ECCS controversy, applied only to LWRs. By 1970 most European countries had already adopted the LWR. While the choice of reactor line was thereafter not an issue for them, the new evidence coming from the US was perceived as casting doubt on the entire nuclear enterprise. France and West Germany were such countries, but each lacked the openness of the US political system, and they also had no established procedures comparable to US hearings or the UK public inquiry to take up the new issue. It was groups of 'alternative experts' in these countries who processed the US debate and were instrumental in the emergence of powerful anti-nuclear movements in the 1970s.

While French and German critics of nuclear energy could rely almost entirely on US material, the US safety debate was clearly going to have less impact in countries which had not committed themselves to the LWR, in particular Canada and Britain.

UK. Britain was in 1970 the only major country in Western Europe which had not adopted the LWR. There had been a long debate in the UK between protagonists of the British Advanced Gas-cooled Reactor (AGR) and of the LWR.(17) Criticisms of LWR safety thus tended to be perceived as arguments against the LWR and for the AGR.(18) At the same time, there was no dissent from within the British nuclear establishment about AGR safety which could have induced a wider debate on nuclear power.(19) The greater homogeneity of the British nuclear sector, aided by the generally greater secretiveness of British government, and the perception of the US safety debate as an indictment of LWRs rather than of nuclear energy in general, contributed to the delayed arrival of anti-nuclear protest activity in the UK.

In Britain it was not primarily nuclear reactor safety, but rather other issues associated with the nuclear-fuel cycle which eventually made nuclear energy an issue. This happened when plans for a Thermal Oxide Reprocessing Plant (THORP) became widely known, following an article in the Daily Mirror on 21 October 1975. This article asked if Britain was to become the 'nuclear dustbin' of the world, in view of plans to reprocess and store spent nuclear fuel from Japan. In Britain, it required the authoritative opinion of a Royal Commission to establish criticism of nuclear development as respectable. The <u>Royal Commission on Environmental Pollution</u> published its report on nuclear power in September 1976, and in it expressed doubts about the handling of nuclear waste and the dangers of a plutonium economy. With the US president deciding to abandon reprocessing and the publicising of a major silo-leak at Windscale, both in December 1976, the British public rapidly became very sensitive to nuclear issues. With a decision to be made about THORP, the conditions for the formation of a national anti-nuclear movement and wider public debate had at last arrived in Britain, too.(20)

The Role of Local Opposition

Expert dissent seems to have been a necessary precondition for the emergence of serious protest in the nuclear case, but it is not in itself sufficient to explain the extent and form of protest. The second key element appears to have been local opposition, which played a major role in both protest-group formation and issue-making in France, West Germany and the US. The conditions for the emergence of local protest therefore represent important variables in the explanation of the development and impact of anti-nuclear movements in different countries. The evidence(21) suggests that there are a number of factors which inhibit or promote local protest:

1) <u>Perceived threats to economic interests</u>: Local resistance to nuclear power is most common in agricultural areas where farmers, fishermen, wine growers, and similar groups fear economic deprivation, either because they believe that nuclear plants would endanger the value of their goods, or else because they think such plants would be the first step to a broader industrialisation process. Sites in areas where there is already industrialisation are less likely to be opposed, especially if regional unemployment is high.
2) <u>Number and social integration of affected people</u>: Nuclear plants sited remotely in areas with very low population densities are unlikely to be strongly opposed, particularly if the number of people directly affected is very small. The more homogenous affected communities are and the better

the communication structure within them, the more likely is the emergence of local protest.

3) Political socialisation: Experience of previous regional and ethnic deprivation can lead to the incorporation of the anti-nuclear case into existing conflicts, and will thus increase the likelihood of a strong collective response. In France particularly, the regionalist element in anti-nuclear opposition has been very strong.

In addition to these three factors, there is a fourth important variable: the effect of 'clustered nuclear siting' and 'existing-site' policies. There is evidence from several countries that strong local opposition is unlikely if nuclear plants are built on established nuclear sites. Surveys in the US show the host communities of operating nuclear power stations have a favourable attitude towards their plants. Opposition is most frequent among potential host communities. There have been very few cases where host communities have actively protested against further plants on an established site.(22)

UK. The importance of siting policy in the emergence of local protest is amply demonstrated by the British case, where clustered siting combined supportively with other factors, in particular the remoteness of sites and the few nuclear projects which came up in the early 1970s after the safety of nuclear power had first been seriously questioned. A policy of remote and clustered nuclear siting has been followed in the UK over many years, most reactors in the 1950s and 1960s having placed in remote coastal areas of low population density.(23) These had attracted some local opposition by farmers and fishermen, but there had been no wider impact. (24) During the 1970s there were few stimuli for local resistance, nuclear power having by then entered a period of relative stagnation.

Of the three nuclear installations which fell to be sited in that decade, two were placed on already developed nuclear sites. It was provided that one reactor would be built at Heysham in Lancashire, where another AGR was already under construction. There was no genuine local opposition to this. The new reprocessing plant for AGR and LWR fuel, THORP, was scheduled for Windscale, a site which since the 1940s had had various nuclear installations for both military and peaceful purposes. The local population having always welcomed the employment brought by nuclear installations, opposition to THORP had practically no local base either. The only greenfield site to be developed in Britain in the 1970s was Torness, in the Lothian region east of Edinburgh. The area around the site has a very low population density. The main opposition group, the Scottish Campaign to Resist the Atomic Menace, was formed by environmentalists mainly based in Edinburgh. Torness became the most important case of local opposition in the 1970s, but even so the conflict there was still muted by continental standards.(25)

In the early 1980s there was some potential for stronger local opposition movements, but once again government policy and the electricity utilities' preference for existing sites largely defused this. In this context, the government's decision to postpone test drilling for nuclear waste-disposal sites was crucial, as there had been strong indications that this might lead to vociferous public opposition involving direct action.(26) As regards reactor sites, the Conservative government's 1979 nuclear programme, which envisaged the construction of several nuclear stations, starting with a PWR, gave an impetus to anti-nuclear opposition generally. But strong local protest was again avoided by choosing an existing site (Sizewell) for the first PWR. After strong local feeling in Cornwall had led to acts of civil disobedience by inhabitants of the village of Luxulyan, one of the new sites then being considered, the Central Electricity Generating Board (CEGB) appeared to move even further towards an existing-site policy. Planned nuclear stations are now all to be constructed at the sites of Magnox and AGR reactors, except, that is, for Druridge Bay in Northumberland.(27) This remains under consideration although organised local opposition has been materialising.

Local opposition contributed far more decisively to politicising nuclear energy in the US, France, and West Germany. Many new sites were developed in these countries thoughout the 1970s, and strong local opposition was encountered at several, though not all of them.

USA The ubiquity of local opposition was very important in the United States. Many scientists previously unconnected with nuclear issues became involved only because a nuclear project was being

planned near their own university. The UCS's
attention, for example, was aroused only by plans
to site a nuclear power station south of Boston.
Furthermore, the AEC decided to stage a public
rule-making hearing on ECCSs only because it feared
the issue would be successfully taken up by local
opposition groups in licensing hearings.(28)

The widespread nature of local opposition in
the USA was also closely connected to siting
policy. With siting entirely a matter for
individual utilities, and with utilities relatively
small, many had a demand for just one large nuclear
power station. The wave of nuclear construction in
the late 1960s and early 1970s thus led to a
proliferation of new sites.(29) Subsequent
experience in the US supports the thesis that
dispersed siting has a higher probability of
attracting local opposition than does a policy
confining nuclear capacity to fewer sites.

France. Local opposition was even more important
for the emergence of nuclear power as a political
issue in France and West Germany. France was the
first European country where significant opposition
to nuclear power emerged. Starting with strong
local and regional protest against the Fessenheim
LWR project in Alsace in 1970, conflicts afterwards
centred around particular sites, involving both
local people and environmentalists.(30) They were
encouraged by the great size of the 1974 French
nuclear programme, which sought to replace the
previous maximum use of cheap imported oil with a
similarly vigorous and, in the critics' view, one-
sided commitment to nuclear energy. The action of
the riot police against anti-nuclear protesters
often galvanised opposition against particular
projects. At many sites the conservative rural
population initially had little sympathy for the
long-haired ecologists from Paris and their anti-
nuclear festivals. At Creys Malville, the site of
the French prototype FBR, for example, the local
population was at first not especially worried
about the reactor, responding in a reserved fashion
to the young demonstrators coming into the area.
But the local attitude changed when riot police
brutally cleared the site of demonstrators.
Fleeing protesters were sheltered by stunned
farmers, and the police action mobilised the local
population more than had any anti-nuclear arguments
and pamphlets previously.(31) With few
opportunities for discussing nuclear policy at the

national level, local opposition became of central importance in catapulting nuclear energy into political prominence. The fortunes of the French opposition movement appeared to decline somewhat after the incidents at Malville in 1977, but the escalation of other local conflicts, particularly at Plogoff in 1980, gave the movement a new lease of life and helped to keep nuclear energy on the agenda. On the other hand, systematic attempts by the utility EDF to 'buy off' local opposition by generous financial help to areas hosting new nuclear power stations may have limited the potential for further local conflicts.(32)

West Germany: While in France the issue-making process extended over years, in West Germany there was one incident above all which made nuclear energy a national issue: the conflict between police and local people at Wyhl early in 1975. There had been some scientific dissent towards nuclear energy in the early 1970s, but except for some small environmentalist groups, nobody took much notice.

All this changed with Wyhl in 1975.(33) Wyhl is near the Kaiserstuhl, a prosperous wine-growing area, and while the inhabitants of the village of Wyhl itself were willing to accept the proposed power station, the wine-growers of the surrounding area were afraid of wider industrialisation that might destroy their livelihood. They were also fearful about changes in the micro climate which might adversely affect their product. Opposition was aroused as well by the insensitivities of the utility and of public officials. Citizen action groups sprang up following a radio announcement that Wyhl had been chosen as the site of a nuclear power plant. Contacts were established in France and Switzerland where, just miles from Wyhl, similar conflicts had developed. The opposition to Wyhl became particularly inspired by the successful action taken by the local population of Marckolsheim (Alsace) against plans to construct a lead factory: the withdrawal of this proposal had been achieved with the help of a site occupation. On 18 February 1975, local people spontaneously occupied the site of the planned nuclear power station at Wyhl. Police removed them forcibly two days later. TV pictures of policemen dragging away farmers and their wives at once turned nuclear energy into a major national issue, and the Kaiserstuhl population now received strong support from all over Germany, particularly from the nearby university town of Freiburg. On 23 February about 28,000 people re-occupied the site, plans to remove them being abandoned by the state government in view of the number involved and the adverse publicity the previous action had received. On 21 March, an administrative court withdrew the construction licence.

This event was followed by extensive debate in Germany. Initially, it was the state government's handling of the affair and police behaviour which was at issue, but interest in nuclear safety and associated questions was also stimulated. In addition, Wyhl encouraged the population at other planned nuclear sites to organise citizen action groups. Many other anti-nuclear groups in support of these local struggles formed elsewhere, and existing citizens' action groups widened their aims to include the nuclear issue. Thus it was that a national anti-nuclear movement evolved.

Local opposition continued to play a major role in the national anti-nuclear movement after 1975. In 1976 and 1977 mass demonstrations took place at Kalkar, the site of Germany's first FBR, and at Brokdorf, north of Hamburg. The circumstances at Brokdorf were not unlike those at Wyhl, and again the behaviour of the police was crucial. The authorities had rushed through the licensing process, and police occupied the site hours before the first construction licence was granted in order to prevent a repetition of Wyhl. Demonstrators trying to enter the site a few days later got harsh treatment, and all this helped consolidate the population in opposition.(34) Local opposition was also central in a campaign against the combined reprocessing waste-disposal installation planned at Gorleben in Lower Saxony. Continued local resistance at Brokdorf, Wyhl, Gorleben and, most recently against Germany's reprocessing plant at Wackersdorf in Bavaria then kept the nuclear energy issue alive during the late 1970s and early 1980s.(35)

Conclusions
To sum up, expert dissent in the US having made it possible for nuclear safety to become a political issue worldwide, the four political systems considered here then provided different opportunities for political actors to raise the issue. While the US was the key 'open system', it was, by contrast, far more difficult to raise the

376

nuclear safety issue there through established political channels. Despite the 'import' of criticisms of LWR safety from the US, the anti-nuclear case in each instance was not brought to wider attention until major local opposition groups emerged who, beause they were determined physically to block any nuclear construction in their localities, therefore clashed with police forces. Local opposition was thus pivotal, but the heavy-handed response of state authorities also helped make nuclear energy an issue. Of these four countries only in the UK did nuclear energy become an issue without major local opposition.

It is also noteworthy in this context that nuclear accidents and actual safety problems (as against projected ones) played little or no role in the initial politicisation of nuclear energy. Only when concern over the safety of nuclear energy had been raised, did nuclear accidents begin to receive much more attention, further deepening the case against nuclear energy.

The way in which nuclear power was politicised had a profound effect on the character and development of the various anti-nuclear movements. In Germany and France, anti-nuclear protest was from the outset a radical cause, the case against nuclear power being linked to issues such as police repression and centralisation, thus leading on to broader social critiques and demands for fundamental political change. In the US and UK, anti-nuclear activity began with a more moderate, reformist outlook, confining itself to working 'within the system', though some radicalisation took place in these countries too in the later 1970s. Finally, local opposition and political responsiveness were not only important factors at the issue-making stage; they also continued to play a major role in the saliency of the issue and the subsequent development of the anti-nuclear movements.

Patterns of Nuclear Conflict

Once anti-nuclear opposition movements had formed, they were faced with different possibilities of influencing nuclear decision-making. Their choice, and with it the character of the ensuing nuclear conflict, was largely determined by two kinds of factor: the socio-political characteristics of the groups involved, and the objective opportunities provided by the respective political systems.

The Main Anti-nuclear Actors

In all four countries one can find essentially equivalent elements in the anti-nuclear movement at national level, although their relative importance varied significantly.

The Environmental Movement. The various environmental organisations took on nuclear energy as one among other issues from about the late 1960s. In each of the four countries two types of environmentalist groups can be distinguished. First, there were the traditional nature conservation societies - groups such as the Sierra Club in the US, the Council for the Protection of Rural England (CPRE) and the Town and Country Planning Association (TCPA) in the UK.(36) In general, these groups were not opposed to nuclear energy per se, but campaigned rather against particular sites or for design changes. They chose to work through established channels and avoided involvement in more radical opposition entailing direct action or party political activity. In Germany and France, 'old' conservation societies of this sort, not having aquired the same standing as their counterparts in the US and UK, played a far smaller role in the anti-nuclear movement.

The wave of environmentalist protest in the late 1960s also produced in all countries of the Western world 'new' groups, their formation sometimes influenced by the unwillingness of the 'old' groups to take on new issues like nuclear power. In both the US and the UK Friends of the Earth (FOE) began in the early 1970s with radical demands and actions. However, after establishing themselves as groups to be reckoned with, in both countries they became integrated into the political process, and pursued their aims by working 'within the system' via participation in hearings, public inquiries, lobbying, court actions and so on. In the US, FOE was also joined by a broad array of other national environmentalist groups specialising in particular issues or particular forms of action, such as litigation.(37)

The French Amis de la Terre (Friends of the Earth) were rather different. The Amis were a less well-organised pressure group, more a loosely structured network of local groups, and with a more radical outlook in terms of both ideology and political action.(38) In Germany the leading 'new' environmentalist group of the 1970s was the Federation of Environmental Citizen Action Groups (BBU - Bund Bürgerinitiativen Umweltschutz), formed in 1972. From the time of Wyhl in 1975, the BBU was engaged in non-violent direct action and civil disobedience.(39)

The New Left   Another important force in the anti-nuclear movement was the 'New Left', by which is meant the non-Moscow-orientated left which emerged as a political force with the student movements of the late 1960s. The involvement of the New Left in the anti-nuclear movement occurred at different stages in the four countries. In France it became strongly involved in the energy movement as early as 1971, adding to the anyway more radical outlook of the French environmental movement.(40)   In Germany the New Left was not involved in any environmentalist or anti-nuclear activity before February 1975. Until then it had dismissed environmental issues as an establishment plot to divert attention from more pressing questions, but afterwards the whole of the New Left, from Maoist avant garde parties to anarchist drop-outs and left social-democrats, took on board the nuclear issue. To some extent, this embracing of the nuclear issue was a result of the failure of the New Left's previous campaigns and of the dearth of alternative issues.(41) In the US, the New Left became involved only from about 1976 onwards.(42)

The US, France and West Germany had all had strong student movements, and their remnants eventually became important participants in the conflict over nuclear energy, but the student movement in Britain was always less important.(43) In Britain it took until 1978-9 before nuclear energy was seriously considered as a potential 'left' issue, and even then it never acquired the same importance for radical politics as it had in the other three countries.

The previous experiences of members of environmentalist groups and the New Left served as starting points governing much of their perception of the nuclear issue and their initial reaction to it. But it was the state's response to anti-nuclear demands which had the strongest single influence on the development of the anti-nuclear movements. The responsiveness of a political system to any protest is largely governed by

379

features particular to that system. In the nuclear case there is, to speak generally, first the licensing process, often governed by the general principles of administrative law and procedure and providing different degrees of delay or obstruction. Secondly there is the general framework for lobbying legislatures and executive organs, influencing the policies of political parties, or in the extreme, setting up new anti-nuclear parties.

The Licensing Process
USA   In the US, licensing procedures offered a major opportunity for delaying nuclear projects. Electricity utilities have to apply for construction and operating licenses to the licensing authority. (Before 1974 this was the AEC, but its regulatory functions were afterwards taken over by the Nuclear Regulatory Commission (NRC).) After detailed discussion between the regulator and the applicant, a local hearing takes place in which objections to the construction application are heard before an Atomic Safety and Licencing Board (ASLB), whose members are appointed by the licensing authority. The decision of the ASLB is then open to appeal to the regulator, and can also be challenged in the courts. The whole procedure is repeated for operating licences except that local hearings are not then compulsory.(44)

Local opposition groups were often formed specially to contest a hearing. At first their motivation was mainly to secure better safety standards, design changes and so on, but from the early 1970s some of them began to oppose nuclear projects altogether. Local groups usually had few illusions about their chances of having an application turned down completely, the applicant and the AEC/NRC having generally reached a consensus beforehand. Where there were serious doubts, the application were normally withdrawn before any hearing took place.(45) Objectors thus tended to be faced at hearings with a fait accompli: with the AEC/NRC having basically accepted the application, there could be little hope that the ASLB as part of the AEC/NRC would throw it out, and indeed, there has not been a single case where ASLBs have done so. Local hearings nevertheless gave objectors some leverage. Their main strategy was delay, drawing out hearings to cause financial problems for the utility. In some cases, utilities agreed to carry out design

changes or to strengthen safety standards beyond
AEC/NRC norms after private negotiations with
objectors, who then withdrew from the hearing so
that construction could start.

This strategy of delay required substantial
financial resources. Specialised lawyers with
detailed technical knowledge had to be hired, often
at substantial cost. Also essential were
'alternative experts' to challenge applicants.
Objectors were generally disadvantaged here as they
could hardly match the expertise mobilised by the
nuclear industry.(46) Local conflicts were thus
largely carried on by specialised lawyers and
experts, local opposition groups often being just
support organisations collecting funds. There was
the occasional demonstration or march, but these
were mostly marginal.(47)

The hearing over, a further challenge could be
mounted in the courts. At federal level,
environmentalists had increasingly been using the
courts, and specialised groups of enviromental
lawyers had emerged, especially the Environmental
Defense Fund and the Natural Resources Defense
Council. In the nuclear field this was encouraged
by the Calvert Cliffs court ruling in 1971 which
forced the AEC to include the environmental impact
of nuclear power (e.g. thermal pollution) in local
hearings, in accordance with the National
Environmental Protection Act (NEPA) of 1969. This
ruling did not actually stop any nuclear project,
but it did lead to a licensing hiatus and thus
long delays in the implementation of projects.(48)
In the early 1970s courts also became more liberal
about legal standing to sue, allowing litigation in
the public interest whereas previously personal
deprivation had been required.(49)

Overall, the licensing process provided anti-
nuclear groups with some opportunities to use
established channels to make their views known, but
it did little to improve the legitimacy of nuclear
policy, since objectors usually emerge from
hearings with stronger opposition than before.
Local hearings were also unable to contain the more
radical anti-nuclear demands of the later 1970s.
With growing frustration over the limited
participatory possibilities offered by the
licensing process, anti-nuclear campaigners started
to turn to more overtly political forms of
campaigning.

France. In contrast to the US, the French

licensing system proved virtually impenetrable to anti-nuclear groups. Its main features in the 1970s (minor changes were made in 1981) were as follows. An application from the state electricity utility Electricite de France (EDF) was initially processed within government, the application then being sent to the prefect of the relevant region, who passed it on to the responsible local authority. Here, the documents were displayed in the town hall and a register laid out to give the local population the opportunity to express views. No presentation of expert evidence or adversarial forms of oral evidence were admitted. In charge of this 'enquête d'utilite publique' (inquiry into the public utility of the project) would be an inspector or inquiry commission named by the prefect, and at the end of the required period this inspector would collect the register, note possible objections, and write a report to the prefect, including a policy recommendation. The prefect then sent everything back to Paris, where after several other intra-governmental steps, the 'declaration d'utilite publique' (DUP) was granted, giving the project final approval - the DUP having the legal status of a government decree and being practically immune to court action. No application was rejected throughout the 1970s, irrespective of the amount of opposition expressed at the 'inquiry'. Environmentalists have repeatedly sought to challenge DUPs, but no license has ever been cancelled, since only the correctness of the formal procedure is checked. (50) This system incensed many local populations, and this, together with police action, was instrumental in precipitating much of the violent reaction seen at some sites.

West Germany In West Germany the licensing system proved somewhat less impenetrable, and the modest accomplishments of the German anti-nuclear movements were largely due to this.
The German licensing authorities are the individual state governments. The electricity utility applies to its state government, which then starts a complex examination process involving, among other bodies, the Reactor Safety Commission attached to the Federal Ministry of the Interior. The only form of public participation is a local hearing. Its conduct is, however, more similar to the French than to the US local hearing, or even to the British public inquiry. As in France, there is

no adversarial procedure or cross-examination of witnesses, and only local inhabitants can raise objections. Hearings often last only a day and, since a licence has never been refused on the basis of one, they have tended to be regarded as farcical.(51)

More important has been the right of local citizens to challenge nuclear projects in the courts. Court actions had been pursued before 1975, but with the exception of a ruling in the 1950s prohibiting uranium mining in the Black Forest, they had never been successful. The decision of the administrative court of Freiburg in March 1975 to throw out the construction licence for the Wyhl plant was thus the first legal success of the anti-nuclear movement.(52) But despite this, there was never much hope in the anti-nuclear movement that the nuclear programme could be stopped by litigation alone. In the event, this view has proved correct. Refusals of licences by lower courts have usually been overturned by higher courts, and the Federal Constitutional Court has also ruled that the German Atomic Law, the main legal basis of the nuclear programme, involves no infringement of constitutional rights.(53) So while courts on several occasions have delayed nuclear construction, they have not proved an unsurmountable barrier to further nuclear development.

UK In Britain licensing of nuclear installations has several separate aspects, and one of them can involve a process which has become the main focus of opposition to nuclear projects, the public inquiry. Following essentially the same procedure as for other planning applications, an application is lodged with the local authority. Should the authority refuse a licence, the applicant can appeal to the Secretary of State. Alternatively, the latter can take his own initiative and 'call in' an application even if the local authority has approved it. In either case, the minister appoints an 'inspector' to conduct a public inquiry into the application and write a report. The 'inquiry' is formally intended to provide additional information for the Secretary of State. The public inquiry was originally conceived as a forum for the discussion of purely local matters, and certainly not of government policy, but latterly public inquiries with wide terms of reference have been conducted for projects with national significance and the

function of the local inquiry has been
substantially widened.(54)
    The public inquiry has also given anti-nuclear
campaigners a clear context in which to campaign.
Their main demand in the middle 1970s was not that
THORP be abandoned outright, but rather that a
public inquiry be conducted into the proposal. The
Labour Government having 'called in' the
application, objectors had high hopes of making a
significant impact: they accepted the public
inquiry as a suitable venue for the 'nuclear
debate' and put their efforts into constructing a
good technical case against the application. There
was at this stage no discussion by them of site
occupations or other protest action.(55) Their
reconsideration of the justness and legitimacy of
the decision-making process occurred only after the
inspector's report had summarily dismissed all
their objections. Calls for civil disobedience
were then heard for the first time.(56) However,
in the ensuing years much attention has been given
to ideas for reforming the public inquiry process
to make it more suitable for the consideration of
major projects of national importance.(57)
    After THORP, the more traditional
environmental organisations continued to see the
public inquiry as their main chance of influencing
nuclear energy policy, but a more radical faction
of the anti-nuclear movement proposed boycotting
inquiries, or else participating in them only to
expose their unfair character as a prelude to
subsequent protest action. Government reaction
went some way towards pre-empting this more radical
approach. First, following local public inquiries
into proposals for test drilling for nuclear waste-
disposal sites,(58) the entire test-drilling
programme was abandoned, an outcome which the more
moderate environmentalists were able to claim as a
confirmation of the effectiveness of working within
the system. And secondly, the inquiry into the
plan to install a PWR in Britain was given the
broadest possible terms of reference by the
Secretary of State, thereby allowing a detailed
discussion of every aspect of UK energy policy.
Radical anti-nuclear groups were already in decline
by the beginning of this latter inquiry, and its
detailed and lengthy character left little scope
for more radical campaigning outside the inquiry
process. Again in this case, therefore, the
institution of the public inquiry appears to have
been able to integrate the main force of anti-

nuclear criticism into the licensing process itself. On the other hand, this effect was possible only because of an accompanying 'low key' policy which avoided the emergence of any severe local conflicts. When nuclear conflicts failed to escalate, this contributed to a general decline of public interest in nuclear energy questions and, in turn, a re-orientation of the more radical elements of the anti-nuclear movements towards nuclear weapons issues, both of which developments helped to contain the nuclear debate largely within the confines of the public inquiry process.

The licensing process for nuclear installations has thus played a crucial role in the development of anti-nuclear opposition in all four countries. However, the licensing procedures in France and West Germany were widely perceived as repressive as soon as strong local opposition movements emerged, and these procedures thus contributed to the rise of radical protest movements concentrating on direct action and civil disobedience. In the US and UK, by contrast, the licensing systems appeared to provide more scope for influence, although in the end environmentalists in these countries tended also to be disappointed and disillusioned, so that they then looked for other opportunities to influence nuclear decision-making.

Political Campaigning

Many forms of anti-nuclear action have been employed outside the licensing process. There has been political campaigning within established political institutions, including pressure-group lobbying. Where possible, referenda have been sought. Attempts have been made to influence political parties, and the electoral arena has also been entered directly. Finally, there has been direct action and civil disobedience, both as symbolic protest and with the explicit aim of stopping nuclear power through physical obstruction. Violence has sometimes been used, both at demonstrations, particularly in conflicts with police forces, and in terrorist attacks upon nuclear installations.

The pattern of these various forms of action has varied considerably between the four countries. In France and West Germany direct action was the first major form of protest, and more conventional campaigning came to the fore only when that had failed. In the UK and the US traditional pressure-

group tactics were first employed, and it was only
after disillusionment with their effectiveness that
a more radical anti-nuclear movement relying on
direct action emerged.

USA In the US, impatience had grown over the
existing procedures by the middle 1970s,
particularly as regards local licensing hearings,
which were seen as a charade whose only positive
function was clearing a path for subsequent legal
action.(59) But the period of judicial 'activism'
was also passing and courts became less responsive
towards anti-nuclear litigation.(60) Protest was
now increasingly geared at influencing decision-
makers directly, a switch of emphasis assisted by
the Administration's attempt to formulate an energy
policy for the US following the oil-price crisis of
1973-4.
    Lobbying efforts were able to benefit from
enhanced co-ordination among anti-nuclear groups.
In 1974 a first national conference of anti-nuclear
groups was organised by Ralph Nader, well known for
his consumer protection activities. Nader's
initiative was mainly geared at improving co-
ordination among local groups and scientific
support organisations so as to mount a stronger
challenge to nuclear projects at all levels. The
Critical Mass Energy Project, as part of the wider
Nader network, thus became an important clearing
house. Nader's network was also active in
lobbying, and other 'public interest groups' which
had been set up in the early 1970s, such as Common
Cause, also became interested in energy issues.(61)
    Lobbying activities by 'public interest
groups' are a rather elitist way of political
action, with grass-root involvement basically
confined to financial support of research and
lobbying staff in Washington. These activities
nevertheless had some impact in the nuclear field,
their peak achievements occurring during the Carter
Administration, when they helped to secure passage
of the Nuclear Proliferation Act of 1978, which
imposed strict conditions on the export of nuclear
facilities.(62) Carter also stopped commercial
reprocessing, tried to halt the FBR project at
Clinch River, and initiated major research
programmes into solar energy and other renewable
energy sources. On the other hand, Carter sought
as well to 'streamline' the nuclear licensing
process and to reaffirm the commitment to LWRs, a
policy which was to fail in the wake of the Three
Mile Island (TMI) accident.(63)

An important consequence of the change in emphasis by groups towards more political methods were the various 'citizen initiatives' of 1976. In 17 states, mainly in the West, laws can be passed by a referendum vote, and using this procedure, an initiative arose in California proposing a law with complex environmental and health restrictions. This initiative was supported by both FOE and the Sierra Club, but was rejected by the electorate (by a margin of 2 to 1), as were similar referenda in Arizona, Oregon, Washington, Montana, Colorado and Ohio. The referenda campaigns nevertheless provided an opportunity to bring anti-nuclear arguments to a wider section of the population, and the California state legislature was sufficiently impressed by the defeated initiative that it passed additional legislation prohibiting further nuclear construction until the problem of nuclear waste had been solved, thus in effect imposing a moratorium. (64) Further citizen initiatives in Hawaii and Montana were successful in 1978, and many other state legislatures also proceeded to impose additional restrictions on reactor construction at this time,(65) adding to regulatory uncertainty.

Anti-nuclear campaigners also found they could challenge nuclear policies in public utility commissions - regulatory bodies set up at state level to control pricing policies. As electricity demand began to stagnate following the oil crisis, utilities found it much harder to finance new construction projects. They therefore tried to increase electricity prices in order to finance new power stations, but this policy proved vulnerable. In Missouri a referendum in 1976 stopped the process of financing nuclear power stations through a surcharge on electricity bills. Anti-nuclear groups in many other states, allied to consumer protection groups and campaigns against energy poverty, mounted successful challenges in public utility hearings, arguing that energy-conservation programmes could keep demand down and avoid price increases.(66)

All in all, there were thus plenty of opportunities for anti-nuclear groups to challenge nuclear projects through established decision-making procedures, and working within the system brought the anti-nuclear activists some rewards. But to some sections of the movements this was not enough. Nuclear power had been hindered and delayed, but not stopped. Some parts of anti-nuclear movement including other New Left groups previously not engaged in environmental issues, now started to look for more radical action.

At Seabrook, New Hampsire, all legal interventions and lobbying had failed, and this state did not allow citizen initiatives as an alternative campaigning possibility. In this situation some sections of anti-nuclear groups in New England, pacifist groups such as the American Friends Service Committee (Quakers), anarchists, drop-outs following 'alternative lifestyles', and other leftist groups from Boston and New York, joined together to form the Clamshell Alliance, with the intention of using direct action and civil disobedience to prevent the start of construction on the Seabrook site. Encouraged by Wyhl in 1975, various acts of civil disobedience were carried out from August 1976 onwards. The occupation of the site by more than 3,000 protesters on 30 April 1977, followed by the arrest of 1,414 the following day, hit the national headlines and made nuclear energy for the first time in the US a prime issue of radical politics.(67)

The Clamshell Alliance model quickly spread, and by 1978, according to one estimate, about half of all planned nuclear power stations were opposed by similar alliances. Civil disobedience became the main form of action for this new generation of anti-nuclear groups.(68) Forms of action first practised by the pacifist movement in the 1950s and later taken over by the civil rights movement(69) had thus found their way into the anti-nuclear movement.

The TMI accident in March/April 1979 naturally provided an enormous boost for the US anti-nuclear movement. By this time the new and more radical type of anti-nuclear activity based on civil disobedience had become dominant, and most new recruits after TMI were drawn to forms of political action modelled on the Clamshell Alliance.

The various factions of the anti-nuclear movement came together to organise a major national demonstration on 6 December 1979 which attracted more than 100,000 people. But their unity was limited, and the heterogeneity of anti-nuclear movement increased even further after TMI. With the emergence of the Alliances, a split had already occurred with groups such as FOE, the Sierra Club, the Nader organisation and other public interest groups, which preferred to continue fighting nuclear energy through established channels. There was also friction between national groups and local opponents following a 'Not in my backyard!' approach.(70) On top of this, the new Alliances proved highly unstable. By 1980 the Clamshell Alliance had itself been torn apart by internal ideological splits between those seeing 'civil disobedience' as a symbolic act of protest, and others wanting to employ direct action as a means of actually stopping nuclear construction, so risking direct and possibly violent conflict with the police.(71)

Other sections of the New Left had also begun to become very critical of the 'direct action' approach, and sought instead to re-direct attention to more traditional forms of political campaigning involving citizen initiatives, trade unions, and political parties.(72) In addition, groups like Environmentalists for Full Employment sprang up in the late 1970s to argue against nuclear power on employment grounds, and to build up links with the trade unions.(73) This strategy had some success, with the miners, autoworkers and other small unions showing some sympathy with anti-nuclear objectives.(74) But it did not really succeed in making trade unions a powerful new ally for the anti-nuclear movement.

Other new initiatives focused on the electoral process. There had already been successes in making nuclear energy an issue in state elections and Democratic Party primaries.(75) Established environmental groups had been active in this for some time, and the rise of Political Action Committees (PACs) as a method of channelling additional funds to candidates provided another opportunity.(76) These attempts to intervene in the electoral process have had some limited effects at state level, but are seen as not far-reaching enough by others, who have argued instead for direct participation in elections by an ecology party. A first such attempt was in fact launched in 1980 when the new Citizens' Party put forward one of the most famous of US environmentalists, Barry Commoner, as presidential candidate on a platform combining anti-nuclear and ecologist demands with other New Left issues.(77) While such activities were rejected by the direct-action faction of the movement as establishment politics, environmentalists mostly found the Citizens Party to be too much to the left, and it has not been able to make an impact either electorally or among the anti-nuclear and environmental movement as a whole.(78) A new initiative to form a decidedly non-marxist 'green' party has been launched still more recently.(79)

All these new initiatives have, however, been of very marginal importance compared with their European counterparts. Even the more radical anti-nuclear movement of the late 1970s could never mobilise more than a few thousand people and, most importantly, it generally lacked the support of the local populations. The more traditional forms of action employed by the established environmental groups and other public interest groups were in the end more significant in terms of impact than were those of the direct action groups. Given the broad range of entry points for pressure-group activity in the US, as well as the general fragmentation of the political process in a federal system, and the successes in delaying nuclear construction, it is not really surprising that the more radical nuclear challenges remained marginal.

In the 1980s the movement against nuclear energy has largely disappeared from public view, with many groups having shifted their attention to issues concerned with nuclear weapons and the arms race. The successful integration of anti-nuclear movements into the political system would no doubt have been far more difficult had nuclear construction continued in the face of TMI, but the collapse of US nuclear construction after this in effect robbed the movement of essential stimuli for continued action.

France   France displays many political features which are diametrically opposite to the US system. Not only did the French licensing system provide no scope for influence by environmentalists, but on the political level, too, opportunities were very limited. The structure of nuclear decision-making in the 1970s is exemplified by the Commission PEON, an advisory group with representatives of government departments and the energy industries. Its advice frequently became government policy. Parliament, political parties, trade unions and other groups had no access and were not consulted. Lobbying activities were unlikely to be successful in view of the broad consensus between government officials, the public utilities, and the business community, based on the practice of technocratic government and common education in elite schools.(80) This strong elite consensus enjoyed the further advantages of a centralised political system able to implement its results.

It was precisely because of these features of the French political system that direct action was so widely adopted. The charade of the 'enquête d'utilité publique' and the powerlessness of local authorities led to much anger on the part of local populations, and they and environmental groups therefore practised 'direct action', mainly in the form of site occupations. This had proved successful in the fight against extension of a military zone in Larzac in 1973-4, and also at Marckolsheim in 1975. Site occupations therefore became the most familiar form of action between 1975 and 1977, being met by determined police counteraction, and this reaction, together with the unresponsiveness of the government, contributed significantly to the radicalisation of the movement. Direct action reached its peak in 1977, when about 60,000 demonstrators marched to the FBR site at Creys Malville. There was a violent conflict between the police and some demonstrators, and a man was killed by a police grenade. Much of French public opinion was critical of the police on this occasion, but the impact of Creys Malville on the movement was shattering. It made it clear that mass demonstrations and site occupations could not stop nuclear power. There were intense discussions, particularly questioning the violent tactics pursued by some groups, and the movement entered a crisis, so that in 1978 there was hardly any anti-nuclear activity.(81)

The failure of direct action forced the movement to consider other options, such as alliances with other groups, especially trade unions. The mainly communist Confédération Général du Travail (CGT) and the more conservative Force Ouvière (FO) were solidly behind the French nuclear programme, but elements in the predominantly socialist Confédération Française Democratique du Travail (CFDT) had voiced criticism in 1974. Some employees in the nuclear sector itself were unhappy about the size of the programme, its implications, and the safety of nuclear workers, particularly at the reprocessing plant at La Hague. Furthermore, the issue of reprocessing and crack-detection in the containment vessels of PWRs had led to strikes in 1979 in which the CFDT had played a dominant role.(82)

Among the parties, only the Socialist Party (PS), its moderate ally, the Movement of Left Radicals (MRG), and the small left-socialist Unified Socialist Party (PSU) made any criticism of

nuclear energy. The PSU expressed total opposition
to nuclear power and sought to portray itself as
the vanguard of the anti-nuclear movement. The MRG
demanded a nuclear moratorium and, at one stage,
sought an electoral alliance with the ecologist
groups. But for the PS, it was mainly the size of
the programme and the way it was implemented
without public debate which gave offence rather
than nuclear energy as such.(83)

The efforts of parts of the ecology movement
to enhance co-ordination with other political
forces led to joint petition in 1979 involving Amis
de la Terre, CFDT, PS and PSU as the main
signatories. The document criticised the
government's 'tout nucléaire' approach, demanded a
full public debate, and favoured alternative energy
programmes based on conservation and other energy
sources, but it avoided opposition in principle to
nuclear power.(84)

The 1979 statement was controversial within
the movement, and there was even stronger debate
about electoral participation. Ecological
candidates had been participating in elections
since 1973, and in 1977 ecological lists achieved a
first 'breakthrough' when in local elections they
topped 10 per cent of the poll in parts of Paris
and in constituencies close to planned nuclear
power stations. Ecological lists were also put
together for elections to the National Assembly and
the European Parliament, and in the presidential
elections of 1981.

Despite the discrimination of the French
electoral system against small parties, the 1981
presidential elections brought the ecologists close
to major influence. Their candidate polled 3.87
per cent of the votes, and with the difference
between Giscard amd Mitterrand within one or two
percentage points, the ecologist vote might have
been decisive in the second round of the ballot.
But there were no formal negotiations and no
recommendation for the second ballot. Mitterand's
promise of a reconsideration and full public debate
on the nuclear programme seems to have been
sufficient to secure him most of the ecologist vote
and guarantee his election.(85)

Some ecologists had high hopes of a
termination of the nuclear programme under
Mitterrand, but they were to be disappointed. Only
one project, Plogoff in Britanny, where local
opposition had been exceptionally strong, was
cancelled. Otherwise the construction schedule was

simply stretched, and in particular neither the extension of the reprocessing facilities at La Hague nor the FBR programme was cancelled. The delays in the nuclear construction programme would probably have been necessary in any case, as large overcapacities in electricity generation were building up and EDF was getting into severe financial difficulties.(86)

The promised public debate on nuclear power was confined to a two-day session in the National Assembly, and there were minor reforms of the licensing system.(87) The main concession was the launching of a major conservation programme, a move apparently aimed at appeasing the CFDT and other critics of French energy policy.(88) In the face of these disappointments, the anti-nuclear movement largely disappeared as a force to be reckoned with. Mitterand had for many been the last hope, and with all alternative forms of action having already failed, there was no momentum left.

A revival of nuclear opposition only on the level of 'green party' politics is also unlikely to materialise at present. A party organisation, the Greens (Les Verts) was finally formed in the run-up to the 1984 European Elections, although the MRG and PSU also make efforts to portray themselves as 'green'. While France moved towards a new system of proportional representation with a 5 per cent hurdle for March 1986 elections to the National Assembly, the anti-nuclear green party could not pull off more than 1.22 per cent and made no impact at all.

West Germany. In West Germany direct action dominated much of the early opposition to nuclear energy. Following Wyhl in 1975, there was a widespread consensus among anti-nuclear groups that site occupations were the form of action most likely to succeed. But by 1977 serious doubts had emerged and alternatives were then considered.

It was above all the outcome of the Brokdorf demonstrations in 1977 which led to these doubts. Most of the anti-nuclear movement had always adhered to the principles of non-violence - this was the line taken by the main environmentalist organisation, the BBU, as well as by the youth organisations of the Social-Democrat and Liberal parties. Other groups, however, wanted another attempt to occupy the site at Brokdorf, with no commitment to non-violence - this was the position taken by various revolutionary avant garde parties of broadly 'Maoist' persuasion (the so-called K-groups) and by other groups with more anarchist views and a belief in 'spontaneous' action (the so-called 'Spontis'). Both K-groups and Spontis recruited from the remnants of the radical student movement of the 1960s. The resulting split in the anti-nuclear movement manifested itself in two separate demonstrations against Brokdorf in February 1977.

These internal conflicts had serious consequences. The BBU was attacked by federal and state governments for allowing itself to be exploited by 'extremists', and as no understanding could be reached with the K-groups and Spontis on a non-violent strategy, the BBU decided to abandon mass demonstrations and site occupations for the time being: as with the French anti-nuclear movement after Malville, initial hopes of stopping nuclear power by non-violent direct action had had to be abandoned.(90)

On the other hand, as compared with the US, there were few alternatives available. The central government had responded to the upsurge of anti-nuclear feeling with a participatory programme, and various public debates, seminars, research projects, and conferences were organised.(91) But this so-called 'dialogue with the citizen' produced no concrete policy-change, so that anti-nuclear activists came to regard it as a propaganda exercise.(92) The 'Gorleben hearing' which the prime minister of Lower Saxony organised in 1979 in response to public protests against the plans for a combined reprocessing waste-disposal facility also had little standing among radical anti-nuclear groups.

One major effort at influencing policy in the late 1970s was geared at the policy-making process within the established political parties. Such a strategy was potentially very effective, not least because the fragmentation of the licensing process left each state government with a decisive say in the development of nuclear energy in its area.

The Christian-Democratic Union (CDU) and its Bavarian sister party, the Christian-Social Union (CSU), were staunch supporters of nuclear power, but there appeared to be some chance of influencing both the Social Democratic Party (SPD) and the Liberal Party (FDP). SPD and FDP youth organisations had already campaigned against nuclear energy, and some prominent members, usually toward the left of the political spectrum, had pressed for more critical positions on nuclear energy, so that the party leaderships had to find compromises to avoid major embarrassment. The feeling in both parties was strong enough for significant concessions to be made. The formula each came up with involved two critical conditions: nuclear power was to be developed only if there were an additional energy demand which could not be met by other means, and only if the problems of nuclear waste-disposal had been solved. These conditions remained open to differing interpretations, and the first was mainly used by the SPD to re-assure the coal industry and coalminers' union.(93)

The circumvention of the waste-disposal problem was more difficult: the federal government essentially saw this condition as being fulfilled provided planning permission were given for the combined reprocessing waste-disposal plant at Gorleben in Lower Saxony. Anti-nuclear critics on the other hand argued that it was possible no satisfactory solution on waste would ever be found. This debate automatically made the Gorleben plant a crucial plank of the entire SPD/FDP approach to nuclear development. However, the licence to allow construction of the plant had had to be granted by the CDU-led government of Lower Saxony. To complicate matters further, political pressure not to grant the licence came from strong local opposition groups, a burgeoning 'green' party, and also the SPD in the state, which committed itself to opposing Gorleben. In addition, the prime minister appeared to be in the running for the post of CDU party leader for the forthcoming general elections. And finally, in the wake of the TMI accident in March/April 1979, anti-nuclear feelings were running higher than ever. The decision required from the prime minister of Lower Saxony was thus fraught with political risks for himself, while the chief beneficiary appeared likely to be the SPD/FDP Federal Government. These considerations may have been relevant in the prime minister's decision in 1979 to refuse a licence for the reprocessing plant, not because of any outstanding technical problems, but because the project, in the prime minister's words, was not then 'politically feasible'. This was a partial success for the anti-nuclear movement, the German programme appearing for a while to be in a shambles. Frantic negotiation took place between federal and state governments until an interim solution was found. Under this, nuclear waste was to be stored at reactor sites for longer than planned, a network of interim storage sites would be established, and one or two small reprocessing plants were to be built.(94)

All in all, the anti-nuclear movement's strategy of working through existing parties had really achieved little. Repeated attempts to commit the federal SPD party organisation to oppose nuclear energy failed, and only in those states where the SPD was in opposition, Schleswig-Holstein, Lower Saxony, and Baden-Wurttemberg, did it come out with a clear anti-nuclear position.

Attempts to influence trade union policy also failed. The main German trade union federation (DGB) had adopted a strong environmental policy in 1972, but it mellowed its stance in the wake of a worsening economic crisis. In 1977 counter demonstrations in favour of nuclear energy took place, and the trade union movement became a major proponent of nuclear development.(95)

These various experiences of the anti-nuclear movement with site occupations, litigation, lobbying, trade unions, and established political parties all contributed to the move towards an ecological party. Fears that the anti-nuclear movement would become discredited by violence led first to an environmentalist party being launched in Lower Saxony in May 1977, and this, after much internal wrangling, eventually led on to the formation of the federal party, Die Grunen (The Greens) in 1980.(96)

Since the formation of the Greens, the salience of nuclear energy has declined. Most former participants of the nuclear conflict, including the K-groups and Spontis, have been integrated into Green Party political activity. The extra-parliamentary citizens action group movement has largely vanished. The BBU has increasingly concentrated on nuclear weapons issues and has played an important role in the German peace movement,(97) but has neglected more mundane environmental issues as well as nuclear energy and can thus not be regarded any more as the primary national environmental and anti-nuclear pressure group.

Although nuclear construction has slowed and the anti-nuclear movement has declined, controversial issues remain. The questions of nuclear waste and reprocessing are still not solved, and although the Greens have spent most of their energy on internal wrangles and issues outside the realm of environmental politics, nuclear energy has remained at the top of their political priorities. In the long-drawn-out negotiations between the Siocal Democrats and the Greens, in Hesse particularly, energy policy has been one of the main issues. With the Greens refusing to condone any further nuclear construction, a potentially important electoral obstacle to nuclear development has thus emerged.

UK. Political campaigning against nuclear power has centred much less on direct action and 'green' political activity in Britain than on the continent. When environmental groups re-examined their possibilities following the Windscale report in 1978, civil disobedience was considered, but it never became the major form of action. Apart from the absence of sustained local opposition, and the continuing attraction for mainstream environmentalists of working within the established system, there was the fact that nuclear energy was the focus of radical campaigning for only a relatively brief spell: from 1978 to about the end of 1981. The local conflict at the Torness power station in Scotland provided a first chance to try 'continental' tactics of site occupations and civil disobedience, but far fewer people became involved than at German or French sites, the genuinly local element was small, and escalation was avoided. The British police adopted a far more patient approach than had their continental counterparts,(98) and they were not faced with anti-nuclear forces determined to confront them at all cost.
    After Torness, the central target for the growing anti-nuclear movement was the plan of the Conservative Government which came into office in May 1979 to make a major commitment to nuclear energy and, in particular, to the PWR. This government's programme was for 15,000 MW of nuclear plant to be installed up to 1992, with at least the first a PWR. Opposition formed in November 1979, the Anti-Nuclear Campaign (ANC) being set up with a congress in London. The ANC was intended to combine all anti-nuclear forces in Britain, and a major figure behind it was the then leader of the Yorkshire miners, Arthur Scargill. The approach which the ANC followed was internationally unique: an umbrella group covering national and local environmental organisations, political parties, and trade unions was formed to campaign against nuclear

energy. The more traditional environmental organisations did not join, but otherwise FOE was the only major national anti-nuclear organisation not involved. FOE did not want to be affiliated with an outrightly political organisation involving political parties.

FOE's attitude widened the split between the moderate and radical wings of the anti-nuclear movement, which had been developing since FOE's refusal to support the Torness occupation. The ANC appeared at first to prosper, and by the end of 1980 about 200 local groups had been formed. But delays in implementing the nuclear programme and the abandonment of plans for nuclear waste-disposal test-drilling robbed the organisation of momentum, and by 1982 the revival of the Campaign for Nuclear Disarmament (CND) was in full swing, drawing activists and resources to its campaign against the stationing of Cruise missiles. The ANC's two main strategies both in the end failed: attempts to win the support of the trade union movement, and to link the civil and military use of nuclear power. (99)

The concept of building an anti-nuclear force through the trade unions had been tried earlier by the Socialist Environment and Resources Association (SERA), formed in 1973. SERA saw its main aim as being to commit the Labour Party to an anti-nuclear policy, and it therefore set out to campaign within the trade union movement. However, while some trade unions had long opposed the introduction of US nuclear technology, most were staunch supporters of British reactors. SERA was nevertheless moderately successful, and some unions, particularly in the service sector, expressed opposition to nuclear power in conference motions. A main success of the trade union campaign was the stopping sea-dumping of low-level nuclear waste after the National Union of Seamen had refused to handle such waste. But key industrial unions, particularly those with members in the nuclear industry, withstood all attempts to commit them to anti-nuclear positions, and repeated efforts to get anti-nuclear motions carried by the TUC also failed. The National Union of Mineworkers (NUM), however, did commit itself to total rejection of nuclear power in 1980. This followed leakage of a cabinet document in October 1979 in which it was stated that 'a nuclear programme would have the advantage of removing a substantial proportion of electricity production from the dangers of industrial action by coalminers, and transport workers'. The Transport and General Workers Union turned anti-nuclear in 1985, allowing a radical anti-nuclear motion to be carried at the Labour conference, but this failed to win the two-thirds majority necessary to make it party policy. (100)

When an extension of the ANC's scope to cover nuclear weapons was suggested in November 1979, it was rejected in order not to jeopardise the affiliation of as broad a range of organisations as possible. CND was then struggling to keep going and, in fact, seeking to raise nuclear weapons through the campaign against nuclear energy, actually affiliating to the ANC. The roles were quickly reversed. The issue of cruise missiles led to a major revival of CND. Suddenly, nuclear energy was totally passe, and with a swelling of CND ranks, many ANC activities moved over to the peace movement, with the ANC affiliating to CND. But this strategy ran into trouble: association with the peace movement alienated the more moderate environmentalists, such as the Conservation Society, which left the ANC on mid 1980. On the other hand, the issue of nuclear energy was controversial within CND. Since nuclear energy had no mobilising appeal, there was no incentive for CND to pursue an issue which could spark off an internal conflict. CND choose to participate in the Sizewell Inquiry to draw attention to what it saw as military implications of the nuclear energy programme, but the peace movement did not pursue nuclear energy as a major campaigning issue and largely ignored the ANC.

As to other forms of action by the British anti-nuclear movement, direct electroral participation is bound to remain ineffective with the present electoral system: the Green Party was formed in 1973 but has made little impact.(101) Apart from participation in public inquiries, most effort will probably be directed at the Labour and Alliance parties, particularly in view of the crucial importance of central government for policy-making in the UK. There remain anti-nuclear forces in the Labour Party, and also a strong anti-nuclear element in the Liberal Party.

Conclusions

The pattern of anti-nuclear opposition in the four countries shows some striking similarities but also some major differences. In all countries,

essentially the same types of group took up the
nuclear issue, and the same range of political
actions was pursued. The relative importance of
the various types of action has however, differed
considerably.

Only in the UK has the licensing process
remained the dominant focus of anti-nuclear
activity. The public inquiry system appears to
have been very successful at integrating anti-
nuclear opposition and avoiding any escalation of
conflict, in conjunction with a general 'low key'
approach. The licensing process has remained
important in the US and UK, where various options
are open to delay if not to block the construction
of nuclear installations. In France, however,
opposing groups have not had any trust whatsoever
in the responsiveness of the licensing process to
anti-nuclear criticism.

As to political campaigning, the American
political system offered environmental groups most
opportunities to influence nuclear development. In
Germany and France lobbying activities have been
relatively unsuccessful, but impact has been
achieved at the party-political level. Although
attempts to commit the social-democratic parties to
outright nuclear opposition failed, they were
forced to seek compromises which led to some
uncertainty and curtailed the likelihood of nuclear
expansion. In Britain political campaigning
against nuclear energy has largely been
ineffectual, however with the important exception
of the trade union campaign which at least on one
occasion brought an important success.

The biggest contrast between France and
Germany on the one hand and the US and Britain on
the other can be found in the relative importance
of direct action and 'party' political activity.
While both these forms of action can also be found
in the US and UK, they have remained very marginal
in both, but they are perhaps the key phenomena of
French and German approaches to nuclear politics.
Civil disobedience and direct action played an
important role in politicising the issue in both
countries, but early hopes that they could form an
effective strategy for deciding the nuclear
conflict in favour of the opposition proved
illusory. In both countries, the green parties
appear to be the main political carriers of anti-
nuclear demands. While adverse electoral
conditions make it difficult for the French green
party to make any immediate impact, the fate of the

German Greens could well be the decisive political
variable for the future of nuclear energy in
Germany.

The Impact of Anti-Nuclear Movements

What in the end has determined the impact of the
anti-nuclear movements? Numerical strength has
evidently been important, yet as the French case
shows, this alone was not enough. Certainly, the
actual opportunities open to protest movements in
different political systems have been significant.
In general, the most important effect of the anti-
nuclear movement has been to cause delays, an
effect closely related to the fragmentation of
licensing procedures.

The impact of the anti-nuclear movements has
also depended on the standing and resources of the
nuclear industry. Throughout the 1950s and 1960s
this industry was so well established that
government policies were often a direct reflection
of its interests. In the US, however, the nuclear
'iron triangle' of the nuclear regulatory agency,
the nuclear construction industry, and the
electricity supply companies was swept aside in the
1970s. The erosion of such 'iron triangles' has
been much less marked in France and Germany. All
in all, it was a mixture of a more relaxed energy
situation, high interest rates, problems with
financing nuclear construction through price rises,
a general regulatory uncertainty with increasing
nuclear lead times, and mounting economic,
regulatory and political uncertainty over the
general future of nuclear following the TMI
accident, which led nuclear development to falter
in the US. Anti-nuclear movements have contributed
to that uncertainty. The resulting 'nuclear
stalemate' can, in the eyes of many observers, be
resolved only by a major restructuring of the
entire sector, requiring a major federal
initiative,(102) a policy which is at odds with the
approach adopted by the Reagan Administration.

The British situation provides contrasts to
both the US and to Germany and France. Here, the
nuclear construction industry has been largely in
the hands of central government, and all orders in
the 1970s and 1980s were made mainly to keep the
industry alive. The nuclear complex has suffered
from continuous fragmentation and internal
squabbles. Of all four countries, the anti-nuclear

movement has made least impact in Britain, and delays have been mainly due to the internal problems of the industry. France and Germany established long ago a system of reactor construction management, but this has remained a matter of intra-organisation conflict in the UK. Yet the anti-nuclear movement found itself unable to exploit these difficulties very effectively: delayed nuclear development provides fewer stimuli for anti-nuclear protest in the first place and, paradoxically, the relative weakness of the British nuclear sector has probably helped prevent a strong political challenge to itself. Anti-nuclear activity was as its highest when THORP and the 1979 nuclear programme seemed to provide the image of a dynamic and swift transition to a nuclear future, but thereafter the cumbersome implementation of the reactor construction programme largely killed off the momentum of anti-nuclear opposition.(103)

As far as energy conservation is concerned, Britain stands out again, as its efforts in this area have remained limited. Bigger inroads appear to have been made in France and Germany.(104) Together with stagnant demand and sometimes large overcapacities, energy-conservation programmes may have weakened the electricity market, so that the rationale for further expansion of generating capacity has been reduced and this has naturally limited the speed of nuclear expansion. In the US the large-scale support given to conservation and solar energy programmes in the 1970s has been severely cut by the Reagan Administration.(105) Energy-conservation efforts at state and local level appear to have been less directly affected by the lack of federal support than have research programmes, though their overall contribution so far appears to have been patchy.(106) The main difference from the European situation has been in the policy of electricity utilities: faced with major economic and political uncertainties regarding expansion of generating capacity, many have sought actively to reduce demand.(107)

With the nuclear industry in crisis, one might have expected that the anti-nuclear movement would try to drive its case home with vigorous energy-conservation campaigns, but on the whole this has not happened. This is an indication of one of the movement's greatest weaknesses: although it has made increasing efforts to provide a positive world-view with support of 'alternative' or 'soft' energy paths, with a very high level of consensus among anti-nuclear activists for these aims,(108) its mobilisation potential has been predominantly reactive.

In conclusion, the impact of anti-nuclear movements on nuclear development and energy policy has been slim but not unimportant. By delaying nuclear projects, anti-nuclear movements have damaged the economics of nuclear power, and so have some energy-conservation programmes which have partly been launched in attempts to appease protest groups. On the other hand, anti-nuclear movements have failed to change the commitment of any major nuclear power to nuclear energy, they have not significantly altered the balance of energy research, and the recent worldwide crisis of the nuclear industry has been mostly caused by recession, the stagnation of electricity demand and associated phenomena. Even where electricity utilities have been forced to make major energy-conservation efforts, as in the US, their overall commitment to an eventual resumption of nuclear construction has not been seriously impaired.(109)

If the world economy picks up and electricity demand rises again, one might therefore expect nuclear construction to recommence. As shown, the level of activity of the anti-nuclear movements has certainly decreased in recent years. The movement has not, however, vanished without trace. Its activities have turned to different issues, and other forms of action and organisations have come to the fore. While some of the anti-nuclear organisations of the 1970s and early 1980s have turned out to be somewhat ephemeral, there are other groups which have established themselves permanently Public interest groups in the US and UK and green parties in Germany and perhaps in France are organisational entities whose existence appears more durable. Environmental concern in general has not waned in the recession. New nuclear construction efforts are thus not likely to pass unnoticed, and the final battle about the energy future is still to be fought.

Notes

1. I am indebted to Professor Roger Williams for his editorial assistance and great number of heplful comments on earlier drafts of this chapter. Any remaining errors are, of course, my own responsibility alone.

2. In this paper, I deal only with protest movements against nuclear energy. The term 'anti-nuclear movement' thus exclusively refers to protest against nuclear energy and not to movements against nuclear weapons.

3. On this episode, see R. Curtis and E. Hogan, Perils of the Peaceful Atom: The Myth of Safe Nuclear Power Plants, Garden City, NY: Doubleday, 1969; S. Novick, The Careless Atom, Boston, MA.: Houghton, Mifflin, 1969.

4. S. Fallows, 'The Nuclear Waste Disposal Controversy' in D. Nelkin (ed.), Controversy: Politics of Technological Decisions, Beverly Hills, CA.: Sage, 1979, pp. 80, 108.

5. D.E. Lilienthal, Change, Hope and the Bomb, Princeton, NJ: Princeton University Press, 1963.

6. C. Kopp, 'The Origins of the American Scientific Debate Over Fallout Hazards', Social Studies in Science, vol. 9, 1978, 403-22; R.A. Divine, Blowing in the Wind: The Nuclear Test Ban Debate 1954-1960, New York, NY: Oxford University Press, 1978.

7. E.J. Sternglass, Low-level Radiation, London: Earth Island, 1973; J. Gofman and A. Tamplin, Poisoned Power: The Case Against Nuclear Power Plants Emmanus, PA.: Rodale Press, 1971; H. Nowotny and H. Hirsch, 'The Consequences of Dissent: Sociological Reflections on the Controversy of the Low Dose Effect, Research Policy, vol. 9, 1980, 278-94; A. Mazur, The Dynamics of Technical Controversy, Washington, DC: Communication Press, 1981.

8. W. Patterson, Nuclear Power, Harmondsworth: Penguin, 1976, p. 152.

9. For early cases of local opposition in the USA, see Curtis and Hogan, Peril; Nowick, The

Nuclear Energy

Carless Atom; R.S. Lewis, The Nuclear Power
Rebellion: Citizens vs. the Atomic Industrial
Establishment, New York, NY: Viking Press, 1972;
D. Nelkin, Nuclear Power and its Critics: The
Cayuga Lake Controversy, Ithaca, NY: Cornell
University Press, 1971; A. Gyorgy and friends, No
Nukes: Everybody's Guide to Nuclear Power, Boston,
MA.: South End Press, 1979.

10. R.C. Mitchell, 'Since Silent Spring: Science,
Technology and the Environmental Movement in the
United States' in H. Skoie (ed.), Scientific
Expertise and the Public, Oslo: Institute for
Studies in Research and Higher Education, The
Norwegian Research Council for Science and the
Humanities, 1979, pp. 171-207.

11. E.S. Rolph, Nuclear Power and the Public
Safety: A Study in Regulation Lexington, MA.:
Lexington Books, 1979, pp. 104-7; D. Nelkin,
Nuclear Power.

12. E.S. Rolph, Nuclear Power, pp. 105, 113.

13. E.S. Rolph, Nuclear Power, pp. 79-80, 86-99,
114, 142-3, 147-48; D. Ford, The Cult of the Atom:
The Secret Papers of the Atomic Energy Commission,
Revised and up-dated edn New York, NY: Simon &
Schuster, 1984.

14. E.W. Lawless, Technology and Social Shock, New
Brunswick, NJ: Rutgers University Press, 1977, pp.
346-8; R.D. Lipschutz, Radiocative Waste:
Politics, Technology and Risk, Cambridge, MA.:
Ballinger, 1980.

15. J. McPhee, The Binding Curve of Energy, New
York, NY: Farrar, Strauss & Giroux, 1974; M.
Willrich and T.B. Taylor, Nuclear Theft: Risks and
Safeguards, Cambridge, MA.: Ballinger, 1974.

16. M.J. Brenner, Nuclear Power and Non-
proiferation: The Remaking of U.S. Policy,
Cambridge, Cambridge University Press, 1981.

17. R. Williams, The NUclear Power Decisions:
British Policies, 1953-1978, London: Croom Helm,
1980.

18. J. Surrey and C. Huggett, 'Opposition to
Nuclear Power: A Review of International

405

Experience', <u>Energy Policy</u>, vol. 4, 1976, 291-2.

19.   R. Williams, <u>Nuclear Power Decisions</u> pp. 263-4.

20.   R.   Williams,   <u>Nuclear Power Decisions</u>;   G. Boyle, <u>The Windscale Inquiry</u>, Milton Keynes: Open University Press,   1978;   B.   Wynne,   'Windscale: A Case Study in the Political Art of Muddling Through' in T. O'Riordan and R. Turner (eds), <u>Progress in Resource Management and Environmental Planning</u>, vol. 3, Chichester: Wiley, 1980, pp. 165-204.

21.   These conclusions are drawn from case studies of local opposition movements quoted in notes 9, 33, and 35, as well as my own studies of local opposition in Britain.

22. W Rüdig, 'Clustered Nuclear Siting and Anti-nuclear Opposition', paper presented at the Sixth Annual Scientific Meeting of the International Society of Political Psychology, Oxford, St Catherine's College, July 1983.

23.   S. Openshaw, 'A Geographical Appraisal of Nuclear Reactor Sites', <u>Area</u>, vol. 12, 1980, 287-90; S. Openshaw, 'The Geography of Reactor Siting Policies in the UK', <u>Transactions of the Institution of British Geographers</u> (N.S.)̇, vol. 7, 1982, 150-62.

24.   R.R. Mathews and E.F. Usher, 'CEGB Experience of Public Communications' in <u>Nuclear Power and its Fuel Cycle</u>, vol. 7, Vienna: International Atomic Energy Authority, 1977, 145-55.

25.   There have been no detailed case studies of local anti-nuclear opposition groups in Britain. For overviews of local conflicts, see J. Falk, <u>Global Fission: The Battle Over Nuclear Power</u>, Melbourne: Oxford University Press, 1982; and R. Williams, 'Public Acceptability of Nuclear Power', paper presented at the UK Political Studies Association Conference, Manchester, April 1985. The assessment of the significance of local opposition groups, as well as other features of the British anti-nuclear movement referred to below, are based on interviews with activists.

26.   W.L. Miller et al., <u>Democratic or Violent</u>

Protest? Attitudes Towards Direct Action in Scotland and Wales, Glasgow: Centre for the Study of Public Policy, University of Strathclyde, 1982.

27. Atom, no. 312, October 1982, 221.

28. J. Primack and F. van Hippel, Advice and Dissent: Scientists in the Political Arena, New York: Basic Books, 1974; D. Ford, Cult of the Atom.

29. C.C. Burwell, 'A Policy Based on the Expansion of Existing Sites' in M.W. Firebaugh and M.J. Ohanian (eds), Gatlinburg II: An Acceptable Future Nuclear Energy System, Oak Ridge, TN.: Institute for Energy Analysis, Oak Ridge Associated Universities, 1980, pp. 57-60.

30. C. Guedeney and G. Mendel, L'Angoisse Atomique et les Centrales Nucléaires: Contribution Psychoanalytique et Sociopsychoanalytique a l'Etude d'un Phenomen Collectif, Paris: Payot, 1973, p. 138; Amis de la Terre, L'Escroquerie Nucléaire, Paris: Stock, 1978, p. 320.

31. N.J.D. Lucas, Energy in France: Planning, Politics and Policy, London: Europa Publications, 1979, pp. 195-207.

32. On Plogoff, See Plogoff - La Revolte, Le Guilvicne: Editions Le Signor, 1980; and 'Plogoff: Round One for Survival', agenor, no. 80, 1980; the system of financial incentives for nuclear host communities and regions is described in J. Fagnani and J. Moatti, 'The Politics of French Nuclear Development', Journal of Policy Analysis and Management, vol. 3, 1984, 264-75.

33. On Wyhl, see S. van Buiren et al., Bürgerinitiativen im Bereich von Kernkraftwerken, Bonn: Bundesministerium für Forschung und Technologie, 1975; H.H. Wüstenhagen, Bürger gegen Kernkraftwerke: Wyhl-der Anfang?, Reinbeck: Rowohlt, 1975; N. Gladitz (ed.), Lieber heute aktiv als morgen radioktiv-Wyhl: Bauern erzählen, Berlin: Klaus Wagenbach, 1976; B. Nössler and M.de Witt (eds), Wyhl: Kein Kernkraftwerk in Wyhl und auch sonst nirgendwo-Betroffene Bürger berichten, Frieburg: inform Verlag, 1976; T. Ebert, W. Sternstein, and R. Vogt, Ökologiebewegung und ziviler Widerstand: Wyhler Erfahrungen,

Stuttgart: Umweltwissenschaftliches Institut, n.d.
For accounts of the events in English, cf. D.
Nelkin and M. Pollak, The Atom Besieged:
Antinuclear Movements in France and Germany,
Cambridge, MA.: MIT Press, 1982, pp. 61-4; J.
Falk, Global Fission,pp. 103-8.

34. G. Trautmann, 'Defizitärer Planungsstaat und
polititsche Legitimitat - Der Fall Brokdorf' in B.
Guggenberger and U. Kempf (eds), Bürgerinitiativen
und reprasentatives System, Opladen Westdeutscher
Verlag, 1978, pp. 309-36.

35. H. Kitschelt, Kernenergiepolitik: Arena eines
gesellschaftlichen Konflikts, Frankfurt: Campus,
1980; D. Rucht, Von Wyhl nach Gorleben: Bürger
gegen Atomprogramm und nukleare Entsorgung, Munich:
C.H. Beck, 1980; C. Büchele, I. Schneider, and B.
Nossler, Wyhe - Der Widerstand geht weiter: Der
Burgerprotest gegen das Kernkraftwerk von 1976 bis
zum Mannheimer Prozess, Freiburg: Dreisam Verlag,
1982; H. Meyer, Zur neueren Entwicklung der
Burgerinitiativbewegung im Bereich Kernenergie,
Bochum: Büro für Atomenergieprobleme, 1981.

36. T. O'Riordan, 'Public Interest Environmental
Groups in the United States and Britain', Journal
of American Studies, vol. 13, 1979, 409-38.

37. For the UK development, see S.K. Brookes and
J.J. Richardson, 'The Environmental Lobby in
Britain', Parliamentary Affairs, vol. 28, 1975,
312-28; J. Bugler, 'Friends of the Earth is 10
Years Old', New Scientist, vol. 90, 1981, 294-7; P.
Lowe and J. Goyder, Environmental Groups in
Politics, London: George Allen & Unwin, 1983. On
the US, see O'Riordan, 'Public Interest'; R.C.
Mitchell and J.C. Davies III, The United States
Environmental Movement and its Political Context:
An Overview (Discussion paper D-32), Washington,
DC: Resources for the Future, 1978.

38. C.M. Vadrot, L'Ecologie, Histoire d'une
Subversion, Paris: Stock, 1978; M. Chaudron and Y.
Le Pape, 'Le mouvement écologique dans la lutte
anti-nucléaire' in F. Fagnani and A. Nicolon
(eds), Nucléopolis: Matériaux pour l'Analyse d'une
Société Nucléaire, Grenoble: Presses
Universitaires de Grenoble, 1979, pp. 25-78; Nelkin
and Pollak, The Atom Besieged

39. Kitschelt, Kernenergiepolitik; Rucht, Von Wyhl nach Gorleben; Meyer, Zur neueren Entwicklung.

40. C. Leggewie and R. de Miller (eds), Der Wahlfisch: Ökologiebewegungen in Frankreich, Berlin: Merve, 1978.

41. For a more detailed account, see W. Rüdig, 'Eco-Socialism: Left Environmentalism in West Germany' New Political Science (forthcoming).

42. E.M. Cohen, Ideology, Interest Group Formation, and Protest: The Case of the Anti-nuclear Movement, the Clamshell Alliance, and the New Left, PhD, Harvard University, 1981.

43. G. Statera, Death of Utopia: The Development and Decline of Students Movements in Europe, New York, NY: Oxford University Press, 1975, pp. 48-9; O. Klineberg, M. Zavalloni, C. Louis-Guerin, and J. Ben Brika, Students, Values and Politics: A Cross-cultural Comparison New York, NY: Free Press, 1979.

44. C.E. Cook, Nuclear Power and Legal Advocacy: The Environmentalists and the Courts, Lexington, MA.: Lexington Books, 1980; D. Okrent, Nuclear Reactor Safety: On the History of the Regulatory Process, Madison, WI.: University of Wisconsin Press, 1981; OECD, Siting Procedures for Major Energy Facilities: Some National Cases, Paris: Organisation for Economic Co-operation and Development, 1980.

45. Several cases are described in Okrent, Nuclear Reactor Safety.

46. S. Ebbin and R. Kasper, Citizen Groups and the Nuclear Power Controversy: Uses of Scientific and Technological Information, Cambridge, MA.: MIT Press, 1974.

47. See, for example, the case study of the Midland nuclear power station conflict in Cook, Nuclear Power.

48. Ibid., pp. 41-5.

49. Ibid., pp. 35-7.

50. J.P. Colson, Le Nucléarire sans le Francais:

Qui decide? Qui profite?, Paris: Maspero, 1977, pp. 101-50; A. Oudiz, Le choix des sites nucléaires in Fagnani and Nicolon, *Nucleopolis*, pp. 161-222; OECD, *Siting Procedures*, pp. 79-98; D. Nelkin and M. Pollak, 'French and German Courts on Nuclear Power', *Bulletin of the Atomic Scientists*, vol. 36, No. 4, 1980, 36-42; R. Macrory and M. Lafontaine, *Public Inquires and Enquête Publique: Forms of Public Participation in England and France*, London: Environmental Data Services, 1982; G. Kiersch and S. von Oppeln, *Kernenergiekonflikt in Frankreich und Deutschland*, Berlin: Wissenschaftlicher Autoren-Verlag, 1983.

51. Kiersch and von Oppeln, *Kernenetgiekonflikt*; Kitschelt, *Kernenetgiepolitik*; OECD, *Siting Procedures*.

52. H.Kitschelt, 'Justizapparate als Konfliktlösungsinstanz? Das Beispiel Kernenergie', *Demokratie und Recht*, vol. 7, 1979, 3-22; Nelkin and Pollak, 'French and German Courts on Nuclear Power'.

53. *Süddeutsche Zeitung*, 7 February 1980.

54. D. Pearce, L. Edwards, and G. Beuret, *Decision Making for Energy Futures: A Case Study of the Windscale Inquiry*, London: Macmillan, 1979; Macrory and Lafontaine, *Public Inquires*.

55. A. Garry, 'The Windscale Inquiry - a Choice to Participate?' *Proceedings of the Joint EASST/STSA Conference Choice in Science and Technology*, London, Imperial College, September 1983, pp. 111-14; B. Wynne, *Rationality or Ritual: The Windscale Inquiry and Nuclear Decisions in Britain*, Chalfont St Giles: British Society for the History of Science, 1982.

56. E. Goldsmith, E. Bunyard, and N. Hildyard, 'Reprocessing the Truth: The Ecologist Analyses the Windscale Report', *New Ecologist*, no. 3, 1978 (supplement).

57. Pearce et al., *Decision Making*; P. Sieghart et al., *The Big Public Inquiry: A Proposed New Procedure for the Impartial Investigation of Projects with Major National Implications*, London: Council for Science and Society/Justice/Outer Circle Policy Unit, 1979; *Deciding About Energy*

Policy: Principles and Procedures for Making
Energy Policy in the UK, London: Council for
Science and Society, 1979.

58.   For   a case study of one of the nuclear waste
conflicts,   see R.A.   Green and G.   J.   Lake,
Repositories of Knowledge: Geologist Test Drilling
and Nuclear Waste Disposal in Northumberland,
Edinburgh: Science   Technology   and   Society
Association, 1984.

59.   Ebbin and Kasper, Citizen Groups, p. 139, pp.
245-46.

60.   N.J. Vig and P. J. Bruer, 'The Courts and Risk
Assessment', Policy Studies Review, vol. 1, 1982,
716-27.

61.   General accounts of the development of the US
anti-nuclear movement are given in Gyorgy et al.,
No Nukes; D. Nelkin, 'The development of the Anti-
nuclear Movement in the United States', Zeitschrift
für Umweltpolitik, vol. 1, 1978, 131-48; D. Nelkin
and S.   Fallows, 'The Evolution of the Nuclear
Debate: The Role of Public Participation', Annual
Review of Energy, vol. 3, 1978, 275-312; S.E.
Barkan, 'Strategic, Tactical and Organisational
Dilemmas of the Protest Movement Against Nuclear
Power', Social Problems, vol. 27, 1979, 29-37; R.
C. Mitchell, The Nuclear Debate, Discussion paper
D-24, Washington DC: Resources for the Future,
1978; J. Shapiro, 'The Anti-nuclear Movement',
Science for the People, vol. 12, no. 4, 1980, 16-
21; R. E. Kasperson et al., 'Public Opposition to
Nuclear Energy: Retrospect and Prospect', Science
Technology & Human Values, vol. 5, no. 31, 1980,
11-23; L. E. Dwyer, 'Structure and Strategy in the
Anti-nuclear Movement' in J. Freeman (ed.), Social
Movements of the Sixties and Seventies, New York:
Longman, 1983, pp. 148-61; R.C. Mitchell, 'From
Elite   Quarrel   to   Mass   Movement',
Transaction/Society, vol. 18, no. 5, 1981, 76-84;
J. Price, The Anti-nuclear Movement, Boston, MA.:
Twayne Publishers, 1982.

62.   Brenner, Nuclear Power.

63.   J.R. Temples, 'The Politics of Nuclear Power:
A Subgoverment in Transition', Political Science
Quarterly, vol. 95, 1980, 239-60; R. T. Sylves,
'Carter Nuclear Licensing Reform Versus Three Mile

Island', <u>Publius</u>, vol. 10, 1980, 69-79.

64. Nelkin and Fallows, 'Evaluation of the Nuclear Debate'; R.E. Kasperson et al., 'Public Opposition'; J.L. Liverman and R.D, Thorne, 'Public Acceptance of Nuclear Power in the United States of America' in <u>Nuclear Power and its Fuel Cycle</u>. vol. 7 Vienna: International Atomic Energy Agency, 1977, pp. 197-217; K.A. Griffith, 'Ninth Circuit Upholds Nuclear Moratorium Provision', <u>Natural Resources Journal</u>, vol. 22, 1982, 689-98.

65. R.T. Sylves, 'Nuclear Power and the States; A Typology of State Regulatory Instruments and How They Have Been Used' paper presented at the Annual Meeting of the American Political Science Association, September 1982; R. Benedict, H. Bone, W. Leavel, and R. Rice, 'The Voters and Attitudes Toward Nuclear Power: A Comparative Study of 'Nuclear Moratorium' Initiatives'. <u>Western Political Quarterly</u>, vol. xx, 1980, 7-23.

66. W.T. Gormley Jr., <u>The Politics of Public Utility Regulation</u>, Pittsburgh, PA.: University of Pittsburgh Press, 1983; E.M. Cohen, <u>Ideology</u>; A. Gyorgy et al., <u>No Nukes</u>, p. 385.

67. For an excellent case study of the Seabrook conflict, see E.H. Cohen, <u>Ideology</u>.

68. Barkan, 'Strategic, Tactical and Organisational Dilemmas', p. 24; Mitchell, <u>The Nuclear Debate</u>.

69. See L.S. Wittner, <u>Rebels Against War; The Peace Movement. 1933-1983</u>, Philadelphia, PA.: Temple University Press, 1984.

70. Mitchell, <u>The Nuclear Debate</u>; Gyorgy et al., <u>No Nukes</u>; Barkan, 'Strategic, Tactual'. This split became particularly visible in the movement against the restart of the Three Mile Island reactors, see E.J. Walsh, 'Resource Mobilization and Citizen Protest in Communities Around Three Mile Island', <u>Social Problems</u>, vol. 29, 1981, 1-21.

71. Cohen, <u>Ideology</u>.

72. See, for example, S. Vogel, 'The Limits of Protest: A Strategy for the Anti-nuclear Movement', <u>Socialist Review</u>, no. 54, 1980, 125-34.

73. R. Grossman and G. Daneker, Energy, Jobs and the Economy, Boston, MA.: Alyson Press, 1979.

74. R. Logan and D. Nelkin, 'Labor and Nuclear Power', Environment, vol. 22, no. 2, 1980, 6-13, 34; L. Smith, 'Labor and the No-nukes Movement', Socialist Review, no. 54, 1980, 135-49.

75. A number of case are documented in J. F. Pilat, Ecological Politics: The Rise of the Green Movement, Beverly Hills, CA.: Sage, 1980; Cohen, Ideology; and M. Barnes and G. Ujufusa, The Almanac of American Politics 1984 Washington, DC: National Journal, 1983.

76. Environmental Action vol. 16, no. 2, 1984, 25; on the phenomenon of political action committees (PACs), see L.J. Sabato, The Rise of Political Consultants: New Ways of Winning Elections, New York, NY: Basic Books, 1981.

77. The party's founding statement is 'Toward a Citizens' Party', Social Policy, vol. 10, no. 3, 1979, 29-31; see also 'The Citizens' Party Considered', Social Policy, vol. 10, no. 3, 1979, 32-40; and F. Smallwood, The Other Candidates: Third Parties in Presidential Elections, Hanover, NH: University Press of New England, 1983, p. 208.

78. Shapiro, 'The Anti-nuclear Movement'.

79. F. Capra and C. Spretnak, Green Politics, New York, NY: E.P. Dutton, 1984; M. Mayer, 'Grüne USA', Die Tageszeitung, 26 January 1985.

80. C. Sweet, 'A Study of Nuclear Power in France', paper presented at the SSRC Energy Panel Seminar, London, April 1981; Lucas, Energy in France.

81. Lucas, Energy in France; Leggewie and de Miller, Der Wahlfisch; Fagnani and Nicolon, Nucleopolis.; Vadrot, L'Ecologie; Nelkin and Pollak, The Atom Besieged.

82. L. Mez, B. Ollrogge and W. Rüdig, Energiediskussion in Europa: Berichte und Dokumente über die Holtung der Regierungen Parteien und Gewerkscheften in der Europäischen Gemeinschaft zur Kernenergie, Villingen: Neckar Verlag, 1981;

J.S. Eisenhammer, 'The French Communist Party, the General Confederation of Labour, and the Nuclear Debate', West European Politics, vol. 4, 1981, 252-66; Lucas, Energy in France.

83. Mez et al., Energiediskussion; D. Nelkin and M. Pollak, 'Political Parties and the Nuclear Energy Debate in France and Germany', Comparative Politics, vol. 12, 1980, 127-41; S. Lauterbach, Die Ökologiebewegung in Frankreich, unpublished manuscript, 1980.

84. T. Chafer, 'The Anti-nuclear Movement and the Rise of Ecology' in P.G. Cerny (ed.), Social Movements and Protest in France, London: Francis Pinter, 1982, pp. 202-20. This move towards increased co-ordination with the CFDT and the PS was also influenced by the work of the sociologist Alain Touraine; see his book A. Touraine, Z. Hegedus, F. Dubet, and M. Wieviorka, Anti-nuclear Protest: The Opposition to Nuclear Energy in France, Cambridge: Cambridge University Press, 1983. For a criticism of Touraine's approach, see W. Rüdig and P.D. Lowe, 'The Unfulfilled Prophecy: Touraine and the Anti-nuclear Movement', Modern and Contemporary France, no. 20, 1984, 19-23.

85. F. Nullmeier, F. Rubart, and H. Schultz, Umweltbewegung und Parteiensystem: Umweltgruppen and Umweltparteien in Frankreich und Schweden, Berlin: Quorum, 1983.

86. Fagnani and Moatti, 'Politics of French Nuclear Development'; T. Chafer, 'Politics and the Perception of Risk: A Study of Anti-nuclear Movements in Britain and France' West European Politics, vol. 8, 1985, 41-54.

87. J.Y. Faberon, 'Le nouveau pouvoir et l'énergie', Revue politique et parlementaire, vol. 83, no. 895, 1981, 51-70; 'Science in France', Nature, vol. 296, 1982, 303; Financial Times, 9 November 1984, Le Monde, 10 October 1984.

88. Rüdig and Lowe, 'Unfulfilled Prophecy'.

89. W. Rüdig, 'The Greens in Europe: Ecological Parties and the European Elections of 1984', Parliamentary Affairs, vol. 38, 1985, 56-72; Daily Telegraph 18 March 1986.

90. Kitschelt, Kernenergiepolitik; Rucht, Von Wyhl nach Gorleben; Meyer, Zur neueren Entwicklung.

91. K. Lang, 'Information on Nuclear Energy in the Federal Republic of Germany: Establishment of a Dialogue Between the Public and Government Authorities' in Nuclear Power and its Fuel Cycle, vol.7 Vienna: International Atomic Energy Agency, 1977, pp. 121-36.

92. W. Rüdig, 'Public Participation and Nuclear Power Politics', Politics, vol. 1, no. 2, 1981, 35-42.

93. Mez et al., Energiediskussion; Kitschelt, Kernenergiepolititz; L. Mez and M. Wilke, Der Atomfilz: Gewerkschaften und Atomkraft, Berlin: Olle & Wolter, 1977; J. Hallerbach (ed.), Die eigentliche Kernspaltung: Gewerkschaften und Bürgerinitiativen im Streit um die Kernkraft, Darmstadt: Luchterhand, 1978.

94. Rucht, Von Wyhl nach Gorleben; Meyer, Zur neueren Entwicklung; M. Sontheimer and U. Scheub (eds), TAZ-Journal no. 1: Ökologie, Berlin: Verlag Die Tageszeitung, 1981, p. 34.

95. Mez et al. Energiediskussion; Hallerbach, Die eigentliche Kernspaltung; Mez and Wilke, Der Atomfilz.

96. The most authoritative account is L. Klotsch and R. Stöss, 'Die Grünen' in R. Stöss (ed.), Parteien-Handbuch: Die Parteien der Bundesrepublik Deutschland, 1945-1980, vol. 2, Opladen: Westdeutscher Verlag, 1984, pp. 1509-98.

97. E. Kolinsky, 'Ecology and Peace in West Germany: an Uneasy Alliance', Journal of Area Studies, no. 9, 1984, 23-6.

98. M. Stott, 'Torness revisited', Vole, vol. 2, no. 2, 1979, 6-7.

99. This account of the ANC is based on interviews and archive work.

100. Mez et al., Energiedeskussion; D. Elliott, Trade Union Policy and Nuclear Power, (Technology Policy Group Occasional Paper, no. 3), Milton Keynes: Technology Policy Group, Open University,

1981; J. Aberdein, 'Labour Go Anti Nuclear', SCRAM Journal, No.51, December 1985/January 1986, p.3.

101. W. Rüdig and P.D. Lowe, 'The Withered "Greening" of British Politics: A Case Study of the Ecology Party', Political Studies Vol. 34, 1986 (forthcoming).

102. M. Lönnroth and W. Walker, The Viability of the Civil Nuclear Industry, London and New York: Royal Institute of International Affairs/ Rockefeller Foundation, 1979; B. R. Weingast, 'Congress, Regulation and the Decline of Nuclear Power', Public Policy, vol. 28, 1980, 231-55, I. C. Bupp, 'The Nuclear Stalemate', in R. Stobaugh and D. Yergin (eds), Energy Future: Report of the Energy Project at the Harvard Business School, New York, NY: Ballantine Books, 1980, pp. 127-66; M. Hertsgaard, Nuclear Inc.: The Men and Money Behind Nuclear Energy, New York, NY: Pantheon, 1983; G. Allison and A. Carnesale, 'The Utility Director's Dilemma: The Governance of Nuclear Power' in D. Zinberg (ed.), Uncertain Power: The Struggle for a National Energy Policy, New York, NY: Pergamon Press, 1983, pp. 134-53; J. Campbell, Can We Plan? The Political Economy of Commercial Nuclear Energy Policy in the United States, PhD, University of Wisconsin-Madison, 1984.

103. Rüdig and Lowe, 'The Withered "Greening" of British Politics'.

104. Comparison of Energy Saving Programmes of EC Member States, COM(84)36 final, Brussels: Commission of the European Communities, 1984.

105. M.E. Kraft and R. S. Axelrod, 'Political Constraints on Development of Alternative Energy Sources: Lessons from the Reagan Administration', Policy Studies Journal, vol. 13, 1984, 319-30.

106. Cf. the literature review of B. A. Cigler, 'Intergovernmental Roles in Local Energy Conservation: A Research Frontier', Policy Studies Review, vol. 1, 1982, 761-76.

107. W. Walker, 'Utilities and Energy Conservation: Implications of Recent Developments in Utility Policy in the United States' in Facing the Energy Future: Does Britain Need New

Energy Institutions?, London: Royal Institute of Public Administration, 1981, pp. 61-9, P.V. Davis, 'Selling Saved Energy: A New Role for the Utilities' in Zinberg, Uncertain Power, pp. 182-98.

108. A.E. Ladd, T.C. Hood, and K. D. Van Liere, 'Ideological Themes in the Anti-nuclear Movement: Consensus and Diversity', Sociological Inquiry, vol. 53, 1983, 252-72.

109. Hertsgaard, Nuclear Inc., p. 149.

Chapter Fifteen

AN OVERVIEW AND POSSIBLE MODEL

Stephen Mills and Roger Williams

It is essential to recognise at the outset that the countries dealt with in this book are members of a privileged group, and that there are at least two other categories of state, in one of which, the socialist societies, Western-style opportunities for the public to resist or question technical change are more circumscribed, and in the other of which, the countries of the Third World, to do so would, with few exceptions, seem an absurd indulgence. Liberal democracy and affluence, in other words, are almost preconditions for the underlying problems with which this book is concerned, and these are therefore, from one point of view, problems we should be glad to have. A second qualification which is also best made at the beginning is that, even in the advanced industrial societies, it may be that only a relatively small number of people are actually doing what they want most of the time - many people, most of their working lives at least, are doing what economically they must. The reason for mentioning this here is that a book of this kind tends inevitably to intellectualise the problem with which it deals, thereby perhaps failing to bring out the real essence and immediacy of the issues involved as they reveal themselves to those most directly affected.

There are also several caveats to be entered about survey data in this area. In the first place, a sharp distinction must be drawn between what people state their positions to be and what their positions are as demonstrated by behavioural responses. Secondly, although there may be a widely shared belief that technology in general is very desirable, resistance may still manifest itself when specific technologies are encountered and real losses addressed. Thirdly, if public opinion surveys show the population of one country to be more optimistic about technology than the population of another, one still cannot assume that the former will in fact get more advantage from technology than the latter, for optimism might correlate with complacency, and it could also be that the more uncertain a people, the more disposed it might be to be self-critical and positive towards technology-induced change. Optimism and pessimism may also be expressed somewhat differently in different cultures, a point which surveys will not necessarily fully pick up. In this context too it would be a mistake to overlook the fact that stated or revealed resistance to technical change can be perfectly rational at particular levels for particular groups in society, even when for the society as a whole such resistance is highly undesirable, and individuals may also not comprehensively appreciate the aggregate effects of their actions.

It was noted in the introduction that while it is obviously a necessary shorthand, it is no less evidently a gross oversimplification, to refer to 'public acceptance of new technologies' as if this were a homogeneous phenomenon. It is equally important not to get the significance of public acceptance of new technologies out of perspective, in that there are always many other factors bearing upon the introduction of these technologies. Thus, in general there are commercial considerations, such as the availability of venture capital, rates of interest, and tax aspects. There are questions to be asked about the calibre, let alone the commitment, of managers and the rest of workforces. Legal requirements and processes may exert an important influence. And to the extent that technical change is a means and not an end, there is the paramountcy of particular governmental and commercial ends to be considered (maximisation of immediate profits, maintenance of existing sales levels, minimisation of industrial unrest, for instance, on the business side; and defence, ideological and electoral motives on the political one).

It may indeed well be, given the scale of what is involved and the astonishing transformations which have been wrought by new technologies in recent decades, that the public-acceptance problems do not in the long run really amount to much. At least this might be true if acceptance is construed in the narrow sense referred to in the introduction: it could hardly be so if the more dynamic and activist construction advocated there is put upon it, for that requires far more from society and from individuals than a shallow and grudging acquiescence.

The public acceptance of new technologies is conveniently thought of as having a number of interleaved layers. Probably the most basic of these is the health and safety dimension. The complexity here lies in the recognition and accurate description of particular threats and their translation into lay and unsensational terms. Next, one might put the categories of direct loss associated with new technologies, arising from a fall in the value of skills, from the undermining of committed capital, or simply from the job-killing consequences some technologies have or are thought to have. After Hirsch, being deprived of positional goods might be seen as a third level of concern - the most dramatic example here is certainly the way in which the motor car and its wide availability have made the countryside accessible to many, so that its rural remoteness, for long the secret of its attraction to the privileged few, is largely lost. (Or, as Gilbert put it more generally, 'When everyone is somebodee/ The no one's anybody'). The implications of new technology for privacy, and the fear of an increasingly oppressive state constitute a fifth aspect of public acceptance, and a sixth would be patterns of social behaviour facilitated by new technologies which come to be widely criticised, such as for instance the displacement of books by TV, or the advent of computer fraud. Beyond this again are still more 'overarching' considerations - nostalgia for the supposedly simpler life and guilt about comfort, perhaps especially in the context of the world's gross inequalities; and some sense of moral outrage precipitated by what comes to be seen by many as an interference by man in the natural order, an assault which, they are convinced, must end in grief. Finally, since technical change can alter the power balance between the interests in a society, it seems to some to introduce a dangerous instability into the very heart of the body politic. Altogether, with technology impinging so heavily on man's environment, his employment and, with the advent of nuclear weapons, his survival itself, it is scarcely to be wondered at that its introduction has precipitated doubts.

The health and safety dimension of public acceptance requires a separate word. What has happened here is that, with quantity of life becoming steadily more assured this century in the advanced countries, public attention has turned increasingly to quality of life. It has been well said that

the overall consequences of all the technological progress of the past century has been to confine death very largely to the aged, whereas throughout all of human history death occurred more or less regularly amongst all ages of life' (J. Urquart and K. Heilmann, Risk Watch, Facts on File Publications, 1984)

But collective risks having fallen to a 'uniquely low value that is without historical precedent', in the last quarter-century, publics have responded rather curiously to existing and new technological and other risks. Thus, the media regularly draw attention to risks which bear only upon small numbers of people, and which often even for them are a low order of probability, while the public more less complacently accepts, presumably as inevitable, other more serious risks, and above all the annual carnage done by the most dangerous of all technological artifacts, the motor vehicle. Steps to reduce the danger presented by the motor vehicle to its occupants and to pedestrians have been painfully slow in coming; they have sometimes, as with seat belts, actually been resisted; and they have often not, as with motorway safety barriers, been pursued in any logical order.

Why the public reacts as it does to particular risks cannot be fully explained, though the literature on the subject is now large. Thus, for example, distinctions have been drawn between the public's attitude towards voluntary as against involuntary risks, and between the attention it gives to frequent events in each of which only a small number of people is affected as against very uncommon accidents involving large numbers. Perhaps three factors deserve special mention. First, the media carry a heavy responsibility for the way they report hazards, focusing upon the possibly small number affected in a way which is often emotional and/ or didactic. Secondly, members of the public sometimes seem to respond to this sensation-seeking reporting in an unhealthily vicarious way. And thirdly, it must be said that

as well as a shortage of unbiased information,
there exists no simple, widely understood, calculus
for comparing the various risks encountered in
life.

In most, though not quite all, liberal
democracies nuclear power was in the 1970s, and by
a wide margin, the technology which generated the
greatest controversy, and indeed even outright
opposition. It was, in short, at the core of
'technophobia'. It is not difficult to adduce
several reasons for this state of affairs. In the
first place, that nuclear power is related to
nuclear weapons has always been widely known but
until the generation, fascinated by Jekyll, the
public seems to have seen no necessary connection
with Hyde. Since then, however, things have tended
to veer to the opposite extreme. In addition, the
characteristics of the possible accidents with
nuclear facilities, and especially the associated
statistical probabilities, are - understandably
enough given their esoteric nature - very ill
understood so far as the general public is
concerned. A second consideration is that the
radiation known to be the inescapable concomitant
of nuclear power has come to be seen as peculiarly
sinister, in part because of its long term and
genetic effects, in part no doubt because it cannot
be recognised by the senses.

These features, and there may have been
others, set nuclear power apart. They were, of
course, all potentially 'there' from the beginning
of the nuclear age, but they stayed virtually
dormant until activated by the environmental
movement of the 1970s. This itself had complex
origins - among them specific environmental
incidents and books and reports about them which
resonated with latent public opinion; a surge of
youthful idealism throughout the Western world,
partly stimulated by opposition to the Vietnam war
and its domestic ramifications in the United
States; and a strong lead from the United States,
always the home of issue politics and in which,
aided by tax and legal developments,
environmentalism now became <u>the</u> issue. In
retrospect, it was perhaps inevitable that nuclear
power should become for a time the central target
of the environmental movement, yet this was a
prospect the nuclear industry in most countries had
until the 1970s contemplated only as a nightmare,
and certainly not as a situation to be planned for
and offset.

How do matters stand in the mid-1980s? On the basis of the chapter in this book, one would have to conclude that opposition to nuclear power internationally has passed a peak. Yet one still cannot be completely confident that this is a technology which in the end will prove publicly acceptable. For the moment any conclusion must be a very tentative one. The largest peak may still lie ahead, and nuclear power could yet prove a technology with an unacceptable sting in its tail. One of the critical factors must be when, if at all, and where, a really serious accident with it occurs.

After nuclear power, as perhaps one would expect, it is the technologies which bear upon the workplace which seem world-wide to have excited most interest. Attitudes here appear to have been shaped by three distinct, though related, perceptions. First, there is, it seems, a resigned acceptance that whether or not new jobs are generated by the new technologies, some at least of existing ones will be lost. Second, there is a reluctant acknowledgement that since other countries and companies will be making the best use that they can of the new technologies, then more or less whatever their effect, it is only reasonable to expect that one's own company and country will also be trying hard to do their best with them. And third, though perception is hardly the right word here, there remains a very great uncertainty and confusion as to what exactly, overall and in detail, the new technologies portend.

Against this background what is really more surprising is that there has been so little documented or apparent workplace resistance to the new technologies, than that there has been so much. After all, it is a quite understandable human trait to be suspicious of change, especially change which may impact directly upon something as vital as one's employment. And whereas one might have understood a relaxed attitude towards job losses in the economically buoyant 1960s, it seems almost perversely passive of those affected to tolerate as readily as they mostly seem to such a situation in the depressed 1980s. Is it that those who work in the shadow of technology-precipitated displacement reconcile themselves to what they construe as economically inevitable, tacitly according to technical change an irresistable impulse and therefore settling for accommodation, perhaps involving compensation and delayed implementation, rather than seeking direct confrontation? Certainly, it is doubtful that unions are anywhere great arbiters of new technologies. Their central organs tend either to be corporatist and firmly supportive of new technologies, or else weak, and weakened further by recession, and so bypassed. Resistance by unions there may be at the periphery, but it would seem to be mostly reactive, ill-coordinated and altogether far short of decisive. Exceptions, and apparent exceptions, of course, make larger headlines.

The mix of new technological innovations in modern society is certainly a rich one. There are infrastructure technologies, process technologies, dominant and minor product technologies, technologies in which consumer sovereignty is critical, technologies which are 'enabling' or 'strategic' in that they open the way to others, and technologies which are effectively cul-de-sacs. Generalising about acceptance of new technologies against such a background can hardly be a very exact or scientific undertaking.

From the perspective of the public it may be helpful to think of two polar models of technological innovation. In the one, new technologies 'trickle down' from their inventors and designers via engineers and sales forces to a public more or less compliant at the base of the pyramid. In the other, new technologies emerge into societies shaped by deep cultural traditions and which values and policies are constantly and vigorously in the process of evolution. There are probably elements of both models in most contemporary societies, with organisational rationales no doubt pulling towards the former, political freedom and educational liberalism towards the latter. And for all its turbulence, and for all the trouble it causes the sponsors of new technologies, who from their point of view could reasonably wish for an easier passage, it is the latter type of model which in the long run one would expect to triumph. This is so not only for moral reasons, though these reasons are valid enough, but ultimately also on grounds of efficiency. Information-rich entities, generously provided with feedback loops, in the end cope better. Dictators may or may not be better at getting trains to run on time, but in any case, trains are only one among the many actual and potential technologies, and dictations will rarely be able to replicate their achievements across all

of them: it is not in their style to encourage
feedback loops.

An alternative way of seeing the introduction
of new technologies and their public acceptance is
to view them in terms of virtuous and vicious
circles. In the virtuous case, a society
introduces new technologies successfully. Many
things may contribute to this – wise initial
choices, an appropriate concentration of resources,
perhaps disguised protectionism, certainly
effective implementation systems. In any case, the
results are felt in sales, profits and exports, and
the latter in turn are translated into more
research and development and the generation of new
jobs. The buoyancy and self-confidence produced by
the latter, and the obvious material gains in the
society, then strongly predispose the population to
be supportive of further technical change,
completing the circle. By contrast, in the vicious
case, for whatever reason the society is relatively
unsuccessful in introducing new technologies.
Maybe the wrong horse is backed, maybe there is
insufficient concentration, or possibly the
structures and traditions are lacking to follow
through effectively. Whatever the actual reasons,
the results are again felt in sales, profits and
exports, but this time falling ones. The
consequence is that research and development
expenditure becomes squeezed, jobs lost are not
offset by jobs gained, and public attitudes set in
ways inimical to technical change.

These are, of course, again polar models, and
any given society must be expected to have
elements of both, though with one or other tending
to dominate at any given time. Political
controversy enters when one begins to consider at
what point or points in a vicious circle to try and
break out of it, and how to do this, assuming
always that it can in fact be done.

Whichever model one adopts, a given technology
may fail to become established in state, at any
rate initially, for a number of different reasons.
Thus, it may ultimately appear to its sponsors too
risky in an economic sense or too hazardous in
respect of safety to be worth pursuing. Or, on the
other hand, those feeling themselves to be
threatened by the technology, whether economically,
culturally or as regards safety, may mobilise to
halt or at least delay the technology's
introduction: the issue with which above all this
book is concerned. In other circumstances again,

sponsors may come to see new alternatives more or
less as commercially acceptable to them and which
they also feel likely to be more publicly
acceptable. In still other cases, ethical
reservations may be the critical ones. Now the
context within which all these possibilities have
to be considered is that of a highly interconnected
international economy. In this economy, while what
happens at the national level may delay or distort
the introduction and evolution of a given
technology in that particular country, it is most
unlikely to stop the technology altogether, because
different circumstances elsewhere will usually mean
that there is at least one country where the
technology continues to be pressed forward.

> (For while the tired waves, vainly breaking,
> Seem here no painful inch to gain,
> Far back through creeks and inlets making
> Comes, silent, flooding in, the main.)

What this means is that, in the end, adjustments
will often have to be made at the national level
which are more painful than they would have been
had they begun to be made earlier, when they could
have been carried through more smoothly.

The role of the media in shaping public
attitudes to new technologies appears to be
especially interesting. The 1970s was apparently a
decade of particular doubt about technology and its
implications, whereas by contrast in the 1980s, in
the shadow of recession, there seems to have been
some shift in the public's outlook. (In
parentheses one might note both that this ought to
make one more than usually cautious about using
1970s data for predictive purposes in the 1980s,
and that public reservations about science and
technology in the 1970s may, in part at least, have
been a reaction to the very high expectations
widely entertained of science and technology in
many countries in the 1950s and 1960s.)

On the other hand, it is true that automation
was a public concern in the 1960s, and it may well
be that some new area of technology, medical
technology perhaps, judging from present trends,
will become a focus for debate in the 1990s. It is
as if the public mood resonated to different themes
at different times, and in this context, pursuing
the analogy, it is natural to wonder to what extent
the mass media act as neutral conductors and to

426

what extent as transformers and amplifiers of nascent opinion. Ask but not unfortunately answer, at least with true conviction, since the role of the media continues to remain more a matter of opinion than of fact. Another analogy from the world of physics also suggests itself here. Between what one might call the 'true reality' of new technology's impact and the appreciation of this impact at the time by members of the public there is, as it were, a 'refracting layer', made up inter alia of pronouncements by interest groups and political parties, perhaps the publication of some research results, and above all, media analysis and reportage. The opinions of individuals, it must be remembered as well, are also in many cases much influenced by what they take to be the opinions of others whom they respect or with whom they identify. Pressing the analogy, there is a critical distinction to be drawn between the actual impact of new technology and its 'virtual' image – a straight stick held in water appears to be bent. More colloquially, one might say that things may not really be as they seem.

It is a complication that ordinarily the actual impact of technology cannot really be known in detail except retrospectively, and even then this impact may be so mixed up with the effects of other technologies, and also of other social developments more generally, that it cannot be properly dissected out. In any case, from a policy point of view, by then it will not usually matter since policy-makers will long since have had to cope as best they can with the various effects as they have revealed themselves. One could sum all this up by saying that what people, both policy-makers and the public at large, react to at any time can differ quite substantially from what is actually happening. Awkward though this is, it does not vitiate social science research. On the contrary, it simply makes it more incumbent on social scientists to elucidate as fully as they can the complex processes of social and technical change.

Turning now to the array of countries in this book, one is struck by the difference between what one might call the 'thrusters', the 'coasters', and the 'stragglers'. The first group are pace-setters and unequivocally in the van; the second group, broadly content with their role, have one or two areas of technology where they are up with the world leaders, but are otherwise mostly geared to respond to initiatives occurring elsewhere; and the third group, not willing to be coasters or aspiring to be thrusters, find themselves in difficulties and increasingly forced to be selective. There are several factors which between them determine in which group a given country will find itself. Size is obviously one of them, size, that is, above all of population, but also of resources and land area. The historical role of industry and technology is a second factor, though this is perhaps best seen as the expression of a third set of factors relating to the character of the economy, culture and political practices. One would also expect government policies to play their part although, in isolation at least, perhaps a rather smaller part than most governments would generally like to think. And something had also better be put down for luck.

Technology, it has come increasingly to be seen, constitutes power in both a physical and political sense, and the most far-reaching significance attaches to these distinct, yet related, perceptions. It is also material, and salutary, that the historical record contains dramatic and under- and over-estimates of the impact of particular technologies. Because technology was brought forth by a special class, there was for too long a widespread tendency to regard it as a sort of curiosity, and only latterly has it, and the experts and professionals who accompany it, really been treated with the political notice they deserve. But this has now happened, and the organisation of states and their industries to produce, via massive R & D efforts, an unending stream of new technologies has at last come to be seen as one of the major benchmarks of history.

Inevitably, government emerges as a powerful influence on public attitudes towards new technologies, and not simply because of its policies of support and intervention where technologies themselves are concerned. Indeed, government's deeper role in society, its shaping of the agenda for debate, its policies of more diffuse impact, and the general values which it defends and transmits, may cumulatively be far more significant than its specific measures in a given case. This, on the other hand, is by its very nature virtually impossible to prove, and attention is naturally drawn to particular policy measures, especially when their costs and effects seem capable of some quantification. Taking Japan and the United States

428

as the world's two most successful technological
powers over the last three decades, it is clear
simply from the chapters on them in this book that
the role of government in them is very different.
It follows from this alone that there cannot be any
simple, straightforward prescription as to what
government in any particular case 'ought' to do as
regards the promotion of technologies, and it is
going to be of consuming intellectual and practical
interest over the coming decades to see which of
these two models, the Japanese or the American,
retains its technological drive most effectively,
unless, of course, both do.

There has also been a major difference between
Japan and the United States as regards the ethos in
which their technological development occurred.
Japan ended the war as a defeated, and indeed a
devastated, country. As in 1868 so in 1945, if
Japan were ever to recover then there was
absolutely no alternative but to apply every effort
to catching up with the West, by simple imitation,
by clever adaptation, and by intelligent
anticipation. Resources must be husbanded
carefully to this end and nothing be permitted to
stand in its way. The trauma of defeat, and
perhaps also the character of the people, made
possible the single-minded determination the
formula required. In short, Japan began at the
bottom but determined it would get to the top, and
saw in technology the means of doing so.

The American case was quite different. The US
ended the war unequivocally the most
technologically successful nation on earth and its
political leaders resolved, more by reflex than by
conscious decision, that it must and would remain
so. For the most part this has been achieved,
though Sputnik in 1957 caused an almost hysterical
reappraisal, and again in the second half of the
1970s the Americans experienced a mood of pessimism
about their ability to sustain their technological
dominance, especially in the face of what was by
then perceived to be a remarkable Japanese
challenge - the resonance achieved by Vogel's book
Japan as Number One (1979) epitomised this new
concern. But otherwise, the United States, having
begun at the top, has remained there.

The Japanese example no doubt illustrates many
things, but perhaps above all the importance of
having a clear and overriding goal to which an
overwhelming majority of the public adheres more or
less firmly. In Japan's case the goal, as noted, has

429

been to catch up with the West, and it is of course
true that various features of Japanese society and
culture have facilitated the maintenance and
priority of this aim. It follows that there are
wide grounds for argument as regards the
possibilities of other states learning from the
Japanese example. What can be said with confidence
is that (a) it is in the nature of liberal
democracy for interests to be in endless
competition in the processes of allocation and
priority setting; (b) states vary considerably in
the extent to which they encourage or dampen these
competitive processes through their institutional
structures; (c) some states have, or believe
themselves to have, critical priorities deriving
from their security situation and its historical
unfolding. It is when one considers the Japanese
case against each of these three criteria
separately that one begins to appreciate Japan's
great good fortune: constitutionally limited in
its defence commitment; a power structure stable
even in detail over more than a third of a century;
and, at least until recently, the unusually simple
national objective already mentioned against which
to judge policy alternatives.

Education has played a major role in shaping
Japanese attitudes to technology. Japanese
technology has also in an important sense flowed
out of the Japanese way of life: when all have an
abacus the switch to a pocket calculator, one might
say, amounted to evolution not revolution.
Furthermore, acceptance of new technologies remains
to a large extent a continuation of daily life, and
even the language barrier, which some thought might
handicap Japan in the computer field especially,
seems not to have mattered greatly.

The Australian and Canadian cases, more
perhaps than any others considered in this book,
illustrate the limited possibilities for any
essentially national response by any except the
strongest technological powers. In the Australian
instance the decisive factor has indubitably been
geography, or more correctly, geographical
separation in a situation where the society has
identified itself very closely with the norms and
values of the Western world. These two factors,
separation and identity, certainly operate in
general to face Australians with technological
developments already no longer new elsewhere, and
they seem also to militate against the emergence of
a strong indigenous capability. For Canada by

contrast, the crushing influence has been geographical contiguity with what has been the technologically most dynamic power on earth, a power furthermore with a population ten times as large. In such a circumstance it was only to be expected that Canadian attitudes and responses would be mainly derivative, except in respect of those few technologies where protected choice (nuclear power) or circumstance (communication technology) were exceptionally favourable towards a national position emerging. It is also apparent that in both the Australian and Canadian cases, earlier dependence on the 'mother country' tended to predispose towards a responsive mode in respect of later technical change.

With Britain the most self-consciously in crisis of the twelve states covered in this book, as reflected in its contribution to this collection, it is inevitable that for her public acceptance of new technologies should be linked above all to economic performance and the country's position in the notional international league table of growth and success. While debate will no doubt long continue as to the primary, secondary and tertiary causes of Britain's century-old economic decline, and even about what exactly has been cause and what effect in this decline, what is undeniable is that the country has failed repeatedly both to protect its industrial position and to reap full advantage from its scientific strength. It seems that Britain's long political stability and, perhaps, consistent martial successes, have between them masked unresolved political issues of profound significance, frustrated recognition of inherent weakness which turn out to be severe by international comparison, and caused false signals to be made to the nation's most aspiring and enterprising citizens. When, as it appears, so much has for so long been wrong, it is unreasonable to expect a reversal of decline to be either quick or painless. But fundamental to any reversal, it is clear, must be a very different attitude in concrete practice towards the potential of technologies, a different attitude, what is more, reflected throughout the society, in its government, its education and industrial systems, its unions and its general public. Whether or not Britain has as yet turned the corner in this regard is obviously controversial: almost all there would be political agreement about is that there is a crisis and that it is a peculiarly British affair

431

rather than being some malaise of advanced
industrial society more generally.

What perhaps above all the Italian and Spanish
chapters seem to bring out are attitudes to science
and technology less sceptical than elsewhere, and
it may also be that an observation made by the
Italian authors actually holds for both countries,
namely that having missed out on the exaggerated
claims made for science and technology in the 1950s
and 1960s, the public had less to react against
when in the 1970s disappointment ensued. The
Italian public, however, appears by international
standards to be exceptionally well disposed towards
nuclear energy, an outlook definitely not
paralleled in Spain. In both countries, as regards
technological development much, it is evident, is
expected of government.

Among the many interesting features of the
Dutch and Belgian cases, as outlined in this book,
is that in the Netherlands there appear postwar to
have been four more or less distinct periods during
each of which public attitudes towards science and
technology changed, whereas in the Belgian case
stress is rather to be laid on the distinct regions
of Wallonia and Flanders, regions which differ in
their unemployment rates and, at least partly as a
result, in their attitudes towards computerisation
also. If it is the case that the Dutch are more
exercised than most about military technology while
simultaneously less excited about information
technology, the Belgian analysis suggests that this
may be as much as anything to do with the degree of
sensationalism which becomes attracted to the issue
concerned.

Swedish circumstances are especially
arresting. Here is a country in which scientific
rationalism is usually regarded as having
established a special place for itself, yet despite
this, on the evidence in this book alone, the
society and state manifestly had considerable
difficulties in the 1970s in coming to terms with
certain technologies at least. Perhaps the central
lesson to be drawn from Sweden's experience is that
there is nowhere any room for complacency in
respect of technological development and its
appreciation by the public.

And so one comes to the French and West German
instances. One cannot hope to do justice here even
to the material in the two chapters dealing with
them in this book, and it may therefore be best to
confine comment to a single point. This must then

432

be the decisive effect which the shape and
character of the governmental system in each case
has had on both technological development and its
public acceptance. The one a highly centralised
state, the other a federal one, much has followed
both directly and indirectly from this, and the
extent of governmental involvement in industry and
technology being quite markedly different in the
two cases, (though reality here does not always
correspond to appearances), again it becomes
embarrassingly difficult to say by analogy what in
other countries government 'ought' best to be
trying to do.

A Model

If one were to summarise what can be learned about
new technology and its public acceptance, from the
situation in general and this book in particular,
and to do this in the form of a model purporting to
represent some ideal state of affairs, what would
be the key elements one would have to include?

1) One central element would certainly be the
inculcation in as high a proportion of society as
possible of a genuine facility with technology, so
that technology becomes regarded as natural and
potentially benevolent and not as remote and
potentially threatening. This task is the
responsibility above all of the education system.
It is not a matter of replacing the values of the
humanities with those of technology. In that sense
the former values face no challenge and should
continue to be cherished. Nor is it a simple issue
of replacing science education with technology,
though in any given society the balance between the
two may from time to time require adjustment. Good
technology education should rather be seen as being
about practical aptitudes and about attitudes
conducive to socially beneficial innovation and
change, and it is correctly seen as an integral and
not an alternative facility in any well-balanced
and complete educational process. No question of
indoctrination arises here: it is a matter of
making the best possible provision for the public
to make choices, and then trusting it to make them
wisely.
2) There needs to be society-wide confidence that
the effects of technology are not being allowed to
happen casually, and that injustices, whether in

economic or safety consequences, are not being
tolerated by an indifferent, incompetent or unjust
state. As regards the safety of technology, if the
public's confidence is to be retained, regulation
must be kept clearly separate from sponsorship,
regulators must be given adequate teeth, and the
public must be given appropriate opportunities to
satisfy itself as to the conduct of events. As
regards the economic consequences of technology,
there must be well-supervised financial and other
support schemes and compensation for those
unavoidably disadvantaged by new technology, the
one wherever possible financed by the profits of
the other. Retraining and employment facilities
also have their place here - and initial education
should have provided as far as possible for later
flexibility. Categorically, the ugliness and
despair which accompanied the first industrial
revolution should and need have no place, even in
diluted form, in the last decades of the twentieth
century. And there is certainly no excuse for
entering the twenty-first century with approaches
and methods which were already unsatisfactory in
the nineteenth. The genius, however, is to ensure
all this while not depriving technical change and
those mainly responsible for it of the freedom and
restless dynamic which so characterises
contemporary technical and industrial success in
the Western world. Fortunately, there now seems to
be good international evidence, confirmed in a
small way in our book, that technical change and
broad social fairness can, and must in a Western
liberal democracy at least, be married effectively.
Similarly, a genuine sense that personal
improvement is possible through effort needs to be
established as the corollary, and not the
antithesis, of a reasonable equality of
opportunity. For again, there appears to be clear
evidence that people in general want to become more
involved in technological choices bearing upon
their future.
3)   The vision and effective communication of a
clear sense of technological purpose at the
national level, and clear technological goals and
subgoals at the level of organisations, if not yet
absolute prerequisites, are so markedly simplifying
as to offer enormous advantages. There is here a
contrast which is almost a paradox: the pursuit of
new technical ideas must be conducted with high
variety and imagination, yet at the same time it
must be disciplined by the designation and

assimilation of relatively precise targets which nevertheless avoid excessive centralisation. Organisational esprit de corps and a national pride which stops short of becoming even disguised chauvinism can be invaluable supports in this context, and their pursuit is, or at least should be, now an integral part of managing new technologies worldwide, though it may be that these things are more easily fostered in some socio-political cultures than in others. Above all, in all democratic societies of whatever type, the support of electorates must be won and carefully sustained for the view that the steady and responsible advance of technology is the key to national, communal and individual betterment.

4) Leading on from this are more specific things one might look for from wise government, including at least:

(i) official emphasis, going well beyond exhortation and involving in fact a clear vision of the future, on responsible innovation as economically virtuous and as conferring high social status;

(ii) the further evolution of government structures and strategies both in such a way as to make quite unequivocal the priority and genuine commitment accorded to innovation and adaptation, and also congruent with the national culture and the international economic context;

(iii) because of the importance of individual initiative and drive, career and reward systems oriented, so far as possible, to stimulate industrial enterprise and initiative as against simple conformity to low-achieving security;

(iv) the operation of the financial and management systems in such a way as to encourage and reward innovation, balancing this against more conservative goals.

5) The exact sources of new technologies and the circumstances of innovation require to be kept under careful review. Scientific research not closely related in time and relevance to technological development, while essential for advanced industrial countries, must be justified in its own right, and cannot fairly be defended in relation to technological development. On the other hand, technology has itself made the world a smaller place, and as a result parochialism seems increasingly likely to attract penalties, possibly

even very severe ones. Excellent information arrangements, and wisely judged licence and co-operative agreements are both essential aspects of any technological undertaking, and the function of laboratory research and development has to be seen as being as much the adaptation of existing results, of various origins, as the discovery of new ones.

6) It seems to be all too easy for the proponents of new technology to become carried away with their own initial achievements so that they can sometimes become dangerously ambitious and overreach themselves. Here is yet another partial paradox: entrepreneurial and technical flair, to remain successful, need the protection of a responsible financial prudence, and when this is not forthcoming it should hardly be surprising that job losses and bankruptcies can and do produce a bitter scepticism about new technologies. Granted that successful technological innovation is never easy, the real problem is nevertheless one of leadership: a collapse of public confidence on this score is not inherent in the technology itself.

7) When an overall national or organisation strategy has been established, a kind of critical-path analysis can usefully be applied to the introduction of new technologies at the tactical level. The questions to ask are how fast are things going relative to some standard (an original plan or the known performance of competitor companies or countries), and what exactly are the blockages preventing equivalent or even better performances? Myth and reality can differ substantially here. Thus, an inventory of commonly cited obstacles would have to include among other factors such items as producer, management, union or consumer 'resistance' (whether or not the result of misperception), shortages of essential skilled manpower or of knowhow, delays in obtaining hardware or software, problems of finance, and the general depressing effect of worldwide economic recession. But as to which of these in any given national or organisational situation is actually critical, only a careful analysis could hope to reveal, and remedial action should be directed at overcoming the difficulty once identified. Such evidence as this book contains tends to confirm that as between factors of the above kind, union and consumer resistance, where wisely and openly handled, will not normally be paramount.

8) The medium and longer-term employment effects

436

of new technologies are not clear, and certainly not from this book which is concerned with attitudes, but major international and regional disparities seem certain at least to persist and perhaps to intensify. Given this uncertainty, it would seem only prudent for government and society to develop and enhance measures which allow people to maintain their dignity and sense of participation in society even when they have little or no employment.

Above all, government cannot escape its mediating role between the public on the one hand and those directly responsible for technical change on the other. The public will not necessarily blame government for those aspects of technical change which it does not like, any more than it will automatically credit government with those features of technical change which it is glad to welcome. But government would be ill advised to represent itself as less than a close monitor of technical development, willing to intervene with as good a blend of pragmatism and consistency as it can muster. No one philosophy and no single individual holds the key to events. The future, like the past, can be expected to be a compound of many elements in uncertain proportions; and the understanding which the social sciences can offer, though hopefully continuing to improve, will no doubt always fall far short of what decision-makers would wish for. Not even in the most centrally organised of states can government do everything, and in the liberal democracies few would even wish it to try. But it can reasonably be looked to set an example, to establish a tone, and to demonstrate a constant and a just concern. To govern is, perhaps above all, to pilot, and it is apparent that public acceptance of new technologies is closely related to the public perception of the course that is set. The right of governing remains, as Fox averred, a trust.

Postscript: Chernobyl

It is no satisfaction to the editors that their
observations above in respect of nuclear power
should have proved as prescient as they did with
the Chernobyl accident of April 1986.   There is
already at the time of writing (May 1986) no doubt
that this is a benchmark event.   Nothing we could
have imagined could so have underlined the
political importance of the public's attitude to
technology, in East as well as West Europe, and,
indeed, the world around.

CONTRIBUTORS

1. Introduction

Roger Williams is professor jointly in the Departments of Government and Science and Technology Policy in the University of Manchester and has published widely on policies related to technical change.

Stephen Mills is a former Director of the Vienna Centre and Research Fellow at the Technical Change Centre. He is now Deputy Secretary General of the International Social Science Council.

2. UK

Sir Bruce Williams has been Director of the Technical Change Centre in London since its foundation in 1981. He was Professor of Economics at the University of Manchester from 1959 to 1967 and then Vice-Chancellor and Principal of the University of Sydney from 1967 to 1981. He has written extensively on economics, education, and technical change and economic growth.

3. USA

Dorothy Nelkin is a professor at Cornell University. Her research focuses on controversial areas of science and technology as a means to understanding the social and political factors underlying science policy. She is a fellow of the Hastings Center and the AAAS, a Director of the AAAS, and was a Guggenheim Fellow in 1984. She is the author of numerous books, including <u>Science As Intellectual Property</u>, <u>The Creation Controversy</u>, and <u>Workers at Risk</u>.

4. Japan

Professor Shigeki Nisihira is a sociologist and statistician now teaching at Sophia University Tokyo, and Japan's leading analyst of public opinion surveys. He is a former Divisional Director at the Institute of Statistical Mathematics, Tokyo.
Professor Ronald Dore is a researcher at the Technical Change Centre. He has spent most of his career studying Japanese social structure and has written several books about Japan.

5. **West Germany**

Thomas Petermann holds a DPhil in political science. After teaching and research at the University of Freiburg and the Wissenschaftszentrum Berlin, he is now a member of the scientific staff of the parliamentary commission on technology assessment (Deutscher Bundestag, Bonn), and has particular interests in social science research on technology.
Georg Thurn also has a DPhil in political science and is specifically interested in science policy. After various research and teaching positions in German and American universities and as a staff member of the Wissenschaftsrat (German Science Council) he is at present Head of the Research Policy and Planning Office at Wissenschaftszentrum Berlin.

6. France

Eric Barchechath is an economist and consultant specialising in marketing and communication, director at the Centre d'Etudes des Systemes et des Technologies Avancées (CESTA) and contributing author to several publications dealing with the social impact of new technologies.
Corinne Hermant is a social economist and research coordinator at the Centre National de la Recherche Scientifique (CNRS), and has published on telematics, information technology and the diffusion of innovation. She is presently a director at CESTA.
Jean-Paul Moatti is an economist and research coordinator at the Institut National de la Santé et de la Recherche Médicale (INSERM). He has published in the field of technology assessment and on the social implications of energy systems and is

presently in charge of the Bureau of Social and Economic Studies at INSERM.

Pierre Rolle is a chief of research at CNRS and the author of <u>Introduction a la sociologie du travail</u> and of <u>I pavadossi del lavoro.</u>

7. Australia

Peter Stubbs is reader in economics at the University of Manchester with particular interests in technical change, industrial economics and transport economics. He has spent several years in Australia and published extensively on Australian industrial issues.

8. Italy

Gabriele Calvi is Professor of Social Psychology at the University of Pavia and President of EURISKO, Institute for Social and Marketing Research, Milan.
Dr Piero Fazio is an economist and is responsible for the Economic Department of CENSIS (Centro Studi Investimenti Sociali). He is also President of NOVA (Ricerche e Progetti per l'Innovazione), Rome.
Dr Andrea Colombino, previously a researcher in nuclear reactor physics is at present responsible for Internal Information at ENEA, Italian Commission for Nuclear and Alternative Energy Sources, Rome.
Dr Giuseppe Zampaglione is an economist. Previously an official of the UN International Atomic Energy Agency, he is now a member of the staff of the President of ENEA,

9. Canada

Michael Gurstein received his BA from the University of Saskatchewan and his PhD in Sociology from University of Cambridge. He is currently a principal of Socioscope Inc., a research and consulting firm specialising in the applications and implications of new technology systems.
Arthur Cordell, after graduating from McGill University with a BA in economics and psychology, went on to study economics at Cornell University, receiving an MA in 1963 and a PhD in 1965. As a science advisor with the Science Council of Canada since 1968 he has authored and directed a number of studies in a variety of areas including: research and development in the multinational firm, the role and function of government laboratories, the social

and economic impacts of microelectronics and computers.

10. Sweden

Jan **Forslin**, PhD, is a psychologist and Senior Research Fellow at the Swedish Council for Management and Work Life Issues in Stockholm. He has researched extensively in industry on social and organisational adaptation to new technology and has participated in several international projects in this field.

11. **Belgium**

Alain Eraly is a business engineer. He has a PhD in social science and is presently undertaking research at the Sociology of Work Center at Brussels University.

12. Netherlands

Jan **Berting** is a professor in theoretical sociology, Erasmus University, Faculty of Social Sciences, Rotterdam. His research interest is societal transformation and technological change, especially in relationship to social inequality. He is a member of the Board of the European Coordination Center for Research and Documentation in the Social Sciences in Vienna.

13. Spain

Rafael **Lopez-Pintor** has a PhD from the University of North Carolina at Chapel Hill, and is Professor of Sociology at the Universidad Autonoma in Madrid. He is also Director of a consulting firm, OYCOS.
Luis Ramallo has a PhD from Harvard, is sociology professor at the University ESADE in Barcelona, and is the Secretary General of the International Social Science Council of UNESCO.

14. Nuclear Power

Wolfgang Rudig read political science, sociology and economics at the University of Bonn, the Free University Berlin, and the London School of Economics. After completing his doctoral studies at the Department of Science and Technology Policy, University of Manchester, he is now a Research Fellow at the Department of Sociology, University of Edinburgh. His main research interests lie in energy and environmental politics, green parties, and the history of nuclear weapons technologies and strategies.

Acknowledgements
The editors would like particularly to thank Jenny Bryan-Brown at the TCC and Rebecca Smellie and Catherine Smith at Manchester.